Power Systems

More information about this series at http://www.springer.com/series/4622

Sumedha Rajakaruna · Farhad Shahnia
Arindam Ghosh
Editors

Plug In Electric Vehicles in Smart Grids

Charging Strategies

Springer

Editors
Sumedha Rajakaruna
Electrical and Computer Engineering
Curtin University
Perth, WA
Australia

Arindam Ghosh
Electrical and Computer Engineering
Curtin University
Perth, WA
Australia

Farhad Shahnia
Electrical and Computer Engineering
Curtin University
Perth, WA
Australia

ISSN 1612-1287 ISSN 1860-4676 (electronic)
Power Systems
ISBN 978-981-10-1373-7 ISBN 978-981-287-317-0 (eBook)
DOI 10.1007/978-981-287-317-0

Springer Singapore Heidelberg New York Dordrecht London

Softcover reprint of the hardcover 1st edition 2015

Printed on acid-free paper

Springer Science+Business Media Singapore Pte Ltd. is part of Springer Science+Business Media (www.springer.com)

Preface

Plug in Electric Vehicles (PEVs) use energy storages usually in the form of battery banks that are designed to be recharged using utility grid power. One category of PEVs are Electric Vehicles (EVs) without an internal-combustion (IC) engine where the energy stored in the battery bank is the only source of power to drive the vehicle. These are also referred as Battery Electric Vehicles (BEVs). The second category of PEVs, which is more commercialized than the EVs, is the Plug in Hybrid Electric Vehicles (PHEVs) where the role of energy storage is to supplement the power produced by the IC engine. These two types of PEVs are predicted to dominate the automobile market by 2030. Widespread adoption of PEVs allows the world to reduce carbon emissions in transportation needs significantly. Therefore, it is vital to the success of a collective global effort in meeting the climate energy targets and to reduce the dependence on increasingly scarce fossil fuels. However, there are a host of challenges thrust upon utility grid operators on how best to meet, control and coordinate the power demand arising due to charging of PEVs. This book covers the recent research advancements in the area of charging strategies that can be employed to accommodate the anticipated high deployment of PEVs in smart grids. Recent literature has focused on various potential issues of uncoordinated charging of PEVs and methods of overcoming such challenges. These innovative approaches include hierarchical coordinated control, model predictive control, optimal control strategies to minimize load variance, smart PEV load management based on load forecasting, integrating renewable energy sources such as photovoltaic arrays to supplement grid power, using wireless communication networks to coordinate the charging load of a smart grid and using market price of electricity and customers payment to coordinate the charging load. Hence, this book includes many new strategies proposed recently by researchers around the world to address the issues related to coordination of charging load of PEVs in a future smart grid. The book is aimed at engineers, system planners, researchers and graduate students who are searching for the latest developments in research related to charging strategies of PEVs in smart grids.

Contents

Reviewers

K. Ramalingam, Indian Institute of Technology, India
Charalampos Marmaras, Cardiff University, UK
N. Mithulananthan, University of Queensland, Australia
Pol Olivella-Rosell, Universitat Politècnica de Catalunya-BarcelonaTech., Spain
Suryanarayana Doolla, Indian Institute of Technology Bombay, India
Okan Arslan, Bilkent University, Turkey
Mahmud Fotuhi-Firuzabad, Sharif University of Technology, Iran
Daniela Proto, Università degli Studi di Napoli Federico II, Italy
Linni Jian, South University of Science and Technology of China, China
David Ciechanowicz, TUM CREATE, Singapore
Erotokritos Xydas, Cardiff University, UK
Preetham Goli, University of Houston, USA
Abdul Motin Howlader, University of the Ryukyus, Japan
Arye Nehorai, Washington University, USA
Qiang Yang, Zhejiang University, China
Martin Maier, Institut National de la Recherche Scientifique, Canada
Pavan Balram, Chalmers University of Technology, Sweden

Chapter 1
Charging Coordination Paradigms of Electric Vehicles

Alexander Schuller

Abstract The Smart Grid enables bidirectional communication between distributed actors and resources in the power system. In particular Plug-In Electric Vehicles (PEVs) are a new type of load that has a considerable (time) flexibility in its demand. In order to integrate and harvest this flexibility within a DSM (Demand Side Management) program, the charging of PEVs needs to be coordinated. The coordination must occur with respect to a given objective. In addition, the coordination of demand requirements can be performed within different communication and control architectures. The main architectural concepts that can be distinguished are decentralised and centralised control architectures. These categories refer to the level on which the charging decision is made, given an objective and constraints that need to be met given a certain user scenario. This work reviews in detail the recent work with respect to different charging coordination paradigms and distinguishes between the following main objectives: grid integration and technical implications, explicit integration and direct utilization and balancing of renewable energy sources and finally, economic driven decisions. The discussion performed in this chapter shows that in the category with a predominantly technical focus, V2G (Vehicle-to-Grid) and grid load (regional and system-wide) impacts are the main research areas. Work looking into the integration ability of renewable energy sources enabled by PEV demand flexibility is in particular focused on the reduction of imbalances stemming from fluctuating generators, e.g. wind power, on a system and also on regional scales under consideration of grid constraints. Short term storage applications are also discussed, but the coordination of PEV demand flexibility by dynamic price incentives is not covered very extensively. Work from the economic domain focuses on the assessment of regulation market participation and day-ahead wholesale market oriented charging. These approaches in turn do not intensively investigate the effect of cost minimizing charging strategies with respect to the utilization of fluctuating renewable energy sources. By consistently discussing recent work from various areas looking into the versatile facets of charging coordination paradigms of

A. Schuller (✉)
FZI Research Center for Information Technology, Karlsruhe, Germany
e-mail: schuller@fzi.de

S. Rajakaruna et al. (eds.), *Plug In Electric Vehicles in Smart Grids*,
Power Systems, DOI 10.1007/978-981-287-317-0_1

electric vehicles, this work provides an anchor for further investigations that help to harvest the demand side flexibility of Electric Vehicles.

Keywords Electric vehicles · Charging coordination · Grid integration · Renewable energy · Economic incentives

1.1 Background and Motivation: Demand Response

Demand Response (DR) is a concept first introduced to power systems in the 1970s following the 1973 energy crisis. At the time the significant oil price shocks lead to an increased awareness about energy consumption and energy efficiency. The U.S. pioneered in the advancement of this concept by imposing strict programs for energy conservation and demand-side-management measures on its integrated utilities at the time [1]. The programs had their focus on increasing overall energy efficiency, hereby reducing overall demand for energy, and on reducing peak load by enabling large industrial customers to reduce or shift a significant part of their load in order to stabilize the power system. The general load reduction would also contribute to a decrease of needed installed capacity to secure supply at all times. But as demand still varies over the course of every day in a system, and also varies in dependence of weather and season, a considerable number of reserve and peaking generators, often with comparably higher variable costs are needed to allow for the system to function properly. Demand Response is a crucial concept to increase the efficiency of the power system and can be defined as:

> all intentional electricity consumption pattern modifications by end-use customers, that are intended to alter the timing, level of instantaneous demand, or total electricity consumption, [2]

Demand-Side-Management (DSM) is part of the more general concept of Demand Response and is mostly referred to with respect to the explicit measures of utilities that were implemented for larger customers to contribute to technical system stability in a centrally controlled power infrastructure. The term is still employed for these measures, but is also used synonymously for artifacts that in the following will be described as parts of Demand Response. The Smart Grid concept has the potential to enable Demand Response at low general implementation and transactions costs. This in turn will help to integrate a higher share of intermittent generators, increase the system stability and finally tackle one of the most important flaws of power markets: the low or non-existing elasticity of the demand side. EVs can be seen as a large resource for Demand Response as they bring with them a high flexibility for a considerable part of their overall demand.

1.2 Categorising PEV Charging Coordination

In order to harvest the demand flexibility of PEVs within a DR program, their charging process needs to be coordinated. The coordination must occur with respect to a given objective. In addition the coordination of demand requirements can be performed within different communication and control architectures. The main architectural concepts are decentralised and centralised control architectures. These categories refer to the level on which the charging decision is made, given an objective and constraints that need to be met given a certain scenario. Figure 1.1 provides a basic classification of the two charging coordination architectures under inclusion of the mixed hierarchical architecture. Following the predominant centralised control paradigms of the traditional power system, centralised charging control architectures build on scheduling procedures that also consider the requirements of the individual charging jobs. These paradigms often rely on planned schedules that are communicated to a central scheduling instance, or assume that a direct load control (DLC) scheme is in place which can be employed to organize the overall charging process such that in particular technical constraints are met. The Distribution System Operator (DSO) and Transmission System Operator (TSO) are often assumed to be responsible for this form of coordination as technical objectives need to be met for safe and reliable power grid operations [3].

A centralised approach has advantages with respect to reliability of charging control and can be integrated easily into existing power system control paradigms. But centralised control architectures require a high degree of information in order to allow for accurate planning by the central instance. Furthermore central control

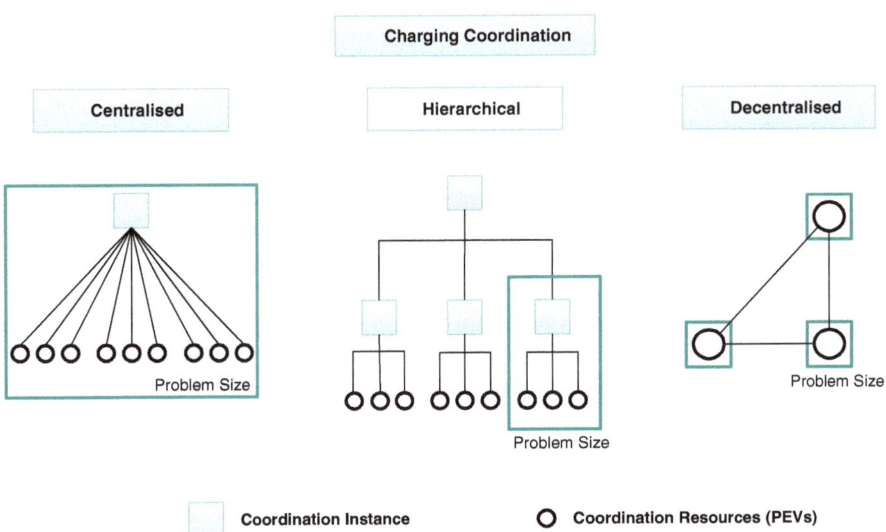

Fig. 1.1 PEV charging coordination paradigms, adapted from [4]

architectures rely on increasingly complex optimisation procedures that do not scale very well in the number of participating units [5], as with every new vehicle additional constraints are added to the optimisation problem. There are many possibilities to reduce the complexity for central coordination procedures or use faster computing algorithms, including heuristics (e.g. genetic algorithms or simulated annealing [6]) or the division of problems into subsets which can be solved easier. Nevertheless this control paradigm might not be very well received by PEV-owners as they do not retain control about the charging process of their vehicle. In order to address the technical complexity and the increasingly more decentralised structure of the power system, hierarchical charging coordination approaches must also be considered.

Hierarchical coordination procedures can be a hybrid form that incorporate aspects of both, centralised and decentralised control paradigms. They can incorporate centralised control and scheduling mechanisms, but in contrast to their system wide counterpart, only address solutions for defined areas or parts of the overall system. This divides the general optimisation problem into a set of interconnected but local, and in the best case optimal, solutions. In a more compact setting this traditional approach is thus still applicable from a technological perspective. The drawbacks with respect to the charging decision being delegated away from the PEV-owner are still prevalent. In this context the role of a so called aggregator, an institution that aggregates the load and thus also the load flexibility of numerous PEVs in order to participate in the power market, ameliorate distribution congestion [7], enhance grid stability through the provision of ancillary services or support the integration of fluctuating renewable energy sources [8], has been extensively proposed as a hierarchical coordination instance.

Charging control in the hierarchical scenario can either follow a schedule based or a price based coordination approach. In the price based scenario, a price is determined either by the aggregator and communicated to his customer PEVs, or it can be determined in a special auction in which the particular PEVs participate [9]. Price based mechanisms can incorporate the system state, and in particular the regional technological constraints if they are designed accordingly. Following the concept of spot pricing (cf. [10]), prices that reflect local capacity constraints and resource availability enable an efficient resource allocation. Prices can vary by location, a concept following the nodal pricing paradigm, or by time, and finally in both dimensions.

Decentralised charging coordination builds predominantly on price based mechanisms. Decentralised charging decisions enable vehicle owners or users to decide when and according to which objective to organize the charging process. The coordination mechanism must therefore incorporate the decisions made by the individual PEVs in order to allow for an effective and reliable operation of the system while guaranteeing supply for the vehicles. In this category prices can either be determined uniformly for all market participants, or discrimination with respect to location and demand time takes place. Decentralised coordination requires more exchange of information, but the number of necessary parameters that need to be communicated is lower, as the decision problem size is confined to one unit, e.g., one PEV.

The different charging coordination architectures cannot always be distinguished sharply. Decentral charging decisions based on centrally communicated uniform prices are one example for a mixed form of charging coordination. Hierarchical and decentralised architectures are inherently combined if price signals are calculated on a regional level by an aggregator, while vehicles still make the decision on how to determine their individual charging schedule. The presented classification is thus giving an overview of the general possibilities on how to organize charging coordination. Considering a more abstracted perspective, this classification can be employed for any resource allocation, including other flexible loads. The next sections provide an overview of the most relevant literature with respect to charging coordination of PEVs looking into primarily technical and economic objectives and the integration ability of fluctuating renewable energy sources.

1.3 Technical Objectives of Charging Coordination

One of the main areas covered in literature of PEV related research is looking into technical questions in particular with a focus on the power grid integration of PEVs. Most of the work mentioned in this section also considers economic constraints, but primarily pursues technical objectives under economic restrictions. Scholars investigating the respective questions in the context of the Smart Grid stem from different professions, and provide insights on similar questions from various perspectives. Traditional power systems engineering, as well as electrical engineering and increasingly researchers from computer science and economics investigated some of the following aspects with a technical focus.

One main branch of research is looking into the assessment of PEV charging load on the power grid on different voltage levels, with a particular focus on distribution grids. Topics in this domain include the investigation of transformer loads following different charging strategies in given standardized distribution grid structures, mainly with households as an inflexible base load. In this context Optimal Power Flow (OPF) methods play an important role. Other main objectives in this context are peak reduction and load shifting in order to minimize distribution system losses and distribution equipment stress. In addition voltage problems and reactive power provision or compensation in distribution grid settings are investigated.

Analyses with respect to the impact of PEVs on distribution system load performed by Lopes et al. [11] and Mets et al. [12] show that controlled charging schemes can help to integrate a higher number of PEVs in the same distribution system (52 % penetration rate in the coordinated as compared to 10 % in the uncontrolled case). In addition peak loads can be significantly reduced by 40 % by coordinated charging. As system peaks are reduced so are losses in the distribution system by around 25 % in the analyzed scenarios [13]. Analyses with respect to optimal charging rates in a residential context show that charging coordination can improve voltage levels and balance phase load in order to reduce transformer equipment wear and integrate higher numbers of PEVs, thus deferring costly line and distribution system upgrades, [14, 15]. Most coordinated charging approaches follow the centralised or hierarchical control architecture with rather high information requirements regarding the individual PEV-user [16]. Further investigations are looking into the interaction between distribution and transmission systems and thus show that local load situations can be quite different from overall system status and require different integration strategies [17, 18].

Besides the regional impact assessment there is also work looking into the system wide impacts of PEVs. In particular the impact of considerable PEV penetration rates on existing power systems and the corresponding unit commitment models in the U.S. are at the center of attention, [19, 20]. These analyses are either looking into operational aspects like additional $C0_2$-emissions and costs in the European [21], or U.S. systems caused by the integration of PEVs. Other analyses are estimating the reductions in primary energy consumption enabled by PEVs and the effects on overall system load [22].

Another technical branch of research is focused on the storage and energy feedback aspect of PEVs, known as vehicle-to-grid, (V2G). This notion introduced by Kempton and Letendre [23] has received a high level of attention. In particular the question if a profitable participation of PEV fleets, coordinated by an aggregator, in a direct control scheme has been addressed in different settings. The necessary communication architecture has been assessed in Quinn et al. [24], the main application domain for V2G is the provision of ancillary services, since regulation and spinning reserve products appear as the economically most stable options under consideration of high battery investment and degradation costs, [25–27]. In addition energy arbitrage under nodal and wholesale prices in the U.S. and Germany have been investigated [28].

These analyses show in particular that it can be profitable for PEVs to provide certain regulation and spinning reserve products, as both the U.S. and the European markets include capacity and energy payments for regulation market participants. Sortomme and El-Sharkawi [29], Dallinger et al. [30] also show that the most profitable option to participate in regulation markets is the provision of negative regulation, which means that charging occurs at times when the grid has surplus energy that needs to be withdrawn.

This operation strategy incurs no additional battery costs and can be profitable in particular because of the capacity payments that are paid for being available to the power grid at the contracted times. Positive regulation can also be slightly profitable, but needs to consider additional investments in grid and communication infrastructure. In this context is has also been shown that frequency regulation support can be performed by the vehicles [11].

Table 1.1 provides an overview of different approaches in the technical domain. As there are vast amounts of at least partly relevant literature this table provides a general overview of the main areas covered in in PEV research with a primarily technological perspective. The table provides an overview of the main research objective addressed, the coordination approach and the scope covered by the model. The categories covered in the scope are the consideration of technological grid

Table 1.1 Characteristic related literature with a predominantly technical focus

Technical focus	Main objective (RQ)	Coordination approach	Model scope					Main finding
Authors			Grid constraints	Ancillary services	RES utilization	PEV trip modeling	Dynamic prices	
Acha et al. (2010) [13]	Energy and emission cost minimisation through PEV grid integration	Central	Yes	No	(Yes)	(Yes)	(Yes)	Losses are reduced, 30 % PEV penetration is no problem, emission reductions are not high due to UK gen eration mix
Galus et al. (2010) [27]	Evaluate aggregator management model for provision of ancillary services	Hierarchical	(Yes)	Yes	No	(Yes)	No	State based short term system operation planning must consider individual trip patterns and resulting energy demand
Galus et al. (2011) [7]	Integrated assessment of PEV charge scheduling	Hierarchical	Yes	No	No	Yes	No	Integrated planning model and scheduling approach reduces operative peak load from PEVs
Gonzales et al. (2012) [18]	Assessment of PEV charging on transmission and distribution level	Central/ decentral	Yes	No	No	(Yes)	Yes	Local distribution system conditions must be considered for charging coordination
Lopes et al. (2010) [11]	Distribution grid integration of PEVs under different charging strategies	Central	Yes	(Yes)	(Yes)	No	No	Charging control allows for higher PEV numbers, voltage support, loss and peak reduction

(continued)

Table 1.1 (continued)

Technical focus	Main objective (RQ)	Coordination approach	Model scope					Main finding
Authors			Grid constraints	Ancillary services	RES utilization	PEV trip modeling	Dynamic prices	
Saber et al. (2010) [31]	Unit commitment with PEVs using an PSO approach	Central	No	Yes	No	No	Yes	PSO solver can help to successfully integrate large numbers of deterministically avail-able PEVs
Sandels et al. (2010) [32]	Assessment of EV aggregation archi-tectures for AS provision	Hierarchical	No	Yes	No	(Yes)	No	Control electronics should be located in PEVs, negative regula-tion is the most cost efficient operation mode
Sortomme et al. (2011) [29]	Assessment of unidirectional V2G for PEVs parked at work	Hierarchical	No	Yes	Yes	(Yes)	Yes	Unidirectional controlled charging is beneficial for all stakeholders, no additional storage costs
Sundstrom and Bind-ing (2012) [16]	Charging cost minimization under consider-ation of grid con-straints and market bid formulation	Hierarchical	Yes	No	No	Yes	Yes	Distributed optimisation by different roles can support voltage levels, while keeping cost low. High information requirements

constraints (e.g. voltage, power ratings, power flows), consideration of ancillary services, the ability to integrate or support RES utilization. In addition the categories trip modeling of PEVs and the application of dynamic prices are taken into account. Finally a short synopsis of the findings is given.

1.4 Renewable Energy System Integration

Making PEVs more sustainable with respect to green-house-gas (GHG) emissions, reducing fossil fuel dependency and assisting the power grid in the integration of fluctuating renewable generation are some of the core advantages of charging coordination with a focus on higher utilization shares of renewable energy [33]. The literature in this field is often intertwined with economic and technical objectives. Most analyses are focusing on the coupling of PEV demand flexibility with intermittent renewable generation. Starting from an overall power system perspective, assessments of PEV charging load impacts in systems with a high share of wind-power generation have been conducted, e.g. in [34, 35], where the impact of renewable sources (predominantly wind-power) on the merit order of the conventional power plants or the integration ability of additional wind power capacity is assessed. An analysis in the impact for the German case in 2030 was performed by Dallinger and Wietschel [36]. They show that coordinated PEV charging, based on a variable pricing scheme and assuming responsive PEV-owners can contribute to balance intermittent generation.

Besides a cost assessment in different scenarios, the capability of PEVs to reduce system-imbalance e.g. in the UK and Danish system have been analyzed in Druitt and Frueh [37] and Goeransson et al. [38]. Druitt and Frueh [37] show that with a wind-power share covering 30 % of the UK electricity demand, one million PEVs can supply about half of the balancing power required. With higher PEV adoption rates of up to 10 million vehicles, about 70–85 % of the balancing requirements can be met only by the vehicle fleet. Goeransson et al. [38] in turn show that emissions in the Danish system can be reduced by PHEVs by a coordinated charging pattern by 4.7 % when vehicles have an overall demand share of 20 %. Emission reductions in this case are due to more efficient thermal generation, avoiding additional startups and part load operation. Additional analyses by Ekman [39] show that coordinated charging and V2G capabilities of 500,000 and 2.5 million vehicles in Denmark are capable to reduce AS and system reserve requirements if wind-power generation covers 50 % of the Danish demand. In this case the authors also find that PEVs cannot provide the necessary demand side flexibility alone, but still need additional controllable generation for back up or other demand side flexibility options in order to reduce the excess wind energy provided.

Other work with a focus on the V2G domain from Kempton and Tomic [40] shows that PEVs can help to provide short term storage in the case of the US. power system for up to 2 h, but are not capable to serve as a medium term energy storage which allow for a compensation of daily and weekly generation shortages in wind-power production (assuming an installed capacity of 700 GW wind-power and 38 % of the U.S. vehicle fleet being PHEVs that serve as an operating reserve).

Another U.S. case analysis performed by Valentine et al. [41] for the NYISO area shows that coordinated charging according to wind power availability improves system balance, but might slightly improve costs. This study shows that coupling of PEV load and wind-power infeed should not be performed in a mandatory way but that they should be treated as independent resources in pool markets with unit commitment models. Markel et al. [42] show that centralised charging coordination with respect to a renewable energy availability signal from the utility can reduce ramp rates for conventional generation by 5 % when a 5 % PEV adoption rate and 15 % RES share of demand is assumed. In addition they show that the communication requirements for centralised fleet control can securely be covered by existing mobile communication infrastructure.

Relevant work with a focus on the operative decisions of single actors in a regional setting has been performed e.g. by Finn et al. [43], Vandael et al. [44] and Galus and Andersson [45]. Finn et al. [43] show that in the Irish case DSM measures including PEVs can increase the absolute share of utilised wind-power for charging. Vasirani et al. [46] propose a coalition formation approach to directly map the demand of PEVs and the production of wind-generators in a VPP. Galus and Andersson [45] show that in the region of Zurich PEVs coordinated by an aggregator can help to balance the production forecast error of a 500 MW wind-farm. Vandael et al. [44] present a hierarchical approach for the reduction of local renewable energy balancing requirements in a distribution network setting. Their analysis shows that while the charging intentions of the individual PEVs are still met, imbalances can be reduced by up to 44 % as compared to the uncoordinated case.

Table 1.2 provides a short comparative overview of some of the main related analyses. Charging coordination for renewable energy integration has been investigated in different settings, most of the reviewed papers were either focused on balancing fluctuating renewable production, while considering technological and economic constraints. As balancing of renewable energy production must be performed on a short term basis most approaches assume centralised or at least hierarchical control architectures. Balancing occurs for time intervals of 15 min, therefore the provision of ancillary services is only partly considered, in particular primary regulation is thus not considered. Besides the assessment of PEV demand flexibility employment for RES integration, most approaches also evaluate the changes in demand patterns based on the prevalent market model, or on simple tariffs with respect to the economic impact of the demand shift. Most papers assume that PEVs are price responsive and have an automated charging control unit which acts on behalf and according to the preferences of the PEV-user. Nevertheless most studies only make basic assumptions about the trip behavior of the vehicles and rather focus on active inclusion of PEVs into the power grid. In this respect future

Table 1.2 Characteristic related literature with a focus on integration of renewable energy sources

Technical focus / Authors	Main objective (RQ)	Coordination approach	Model scope					Main finding
			Grid constraints	Ancillary services	RES utilization	PEV trip modeling	Dynamic prices	
Druitt and Frueh (2012) [37]	System wide wind-imbalance reduction through PEVs in UK	Central	No	Yes	Yes	(Yes)	Yes	PEVs can reduce balancing requirements for wind-balancing by 50 % while charging costs are lowered
Finn et al. (2012) [43]	Evaluation of DSM-signals for maximum wind power usage while keeping costs low	Decentral: scheduling	No	No	Yes	(Yes)	Yes	Cost based minimisation strategy promises most benefits in the analysed case, wind share maximisation leads to slightly higher costs
Galus and Andersson (2011) [45]	Balancing of wind generation with large PHEV fleet under consideration of grid topologies	Hierarchical	Yes	(Yes)	(Yes)	Yes	(Yes)	The MPC model enables the aggregator to balance the wind in-feed error for a 500 MW wind farm
Göransson et al. (2010) [38]	Comparison of different charging strategies in the system of Denmark with respect to total emissions	Central	(Yes)	No	Yes	(Yes)	No	Emission reductions in wind-thermal systems are only possible if PEVs are actively managed

(continued)

Table 1.2 (continued)

Technical focus Authors	Main objective (RQ)	Coordination approach	Model scope					Main finding
			Grid constraints	Ancillary services	RES utilization	PEV trip modeling	Dynamic prices	
Kempton and Tomic (2005) [47]	Assessment of storage options and grid support by PEVs	(Central)	No	Yes	Yes	No	No	PEVs are well suited for regulation, can be used for peak power, not for medium term storage to compensate for RES shortages
Markel et al. (2009) [42]	Assessment of charging strategies for direct RES utilization	Central	No	No	Yes	Yes	Yes	Coordinated charging can reduce peaks, increase RES use, communication infrastructure is sufficient for central control
Mets et al. (2012) [48]	Distributed charging for wind energy utilization	Hierarchical	No	No	Yes	(Yes)	Yes	Wind energy utilisation can be doubled by the distributed decision mechanism with a hierarchical coordinator
Valentine et al. (2012) [41]	Wind energy balancing and energy price impact of PEVs in NY-ISO	Central	No	No	Yes	(Yes)	Yes	PEVs can support wind integration and contribute to lower prices. Must take wind policy is not cost optimal
Vandael et al. (2011) [44]	Imbalance reduction with PEVs, PV case study	Decentral: economic	(Yes)	No	Yes	No	Yes	The distributed mechanism could reduce balancing costs by 14–44 %

work could enhance the existing analyses if real-life driving profiles are employed for the assessment of PEV charging demand flexibility with respect to the renewable energy integration potential.

1.5 Economic Objectives

The last main group of relevant related work is concerned with the economic evaluation of charging coordination in different market settings and the assessment of allocation mechanisms from an economic perspective. The papers discussed in the following are thus primarily focused on operative economic objectives with some considering technical and renewable energy integration aspects.

Employing the demand flexibility of PEVs for the provision of AS was discussed above, one of the main economic assessments for the general viability of the V2G concept was performed in [47]. Based on data from 2003 an economic evaluation of the provision of regulation and spinning reserve products in the CAISO market area shows that PEVs, in particular those with a high power connection can generate quite high profits mainly due to capacity payments they receive.

This analysis is quite static and does not consider the dynamics of driving behavior. Work by Andersson et al. [26] and Dallinger et al. [30] (both assuming a hierarchical aggregator approach) shows that when the daily variations of prices and mobility patterns are considered, V2G activity is profitable only for certain regulation products. In particular down or negative regulation (in the European context negative secondary and tertiary reserve) can profitably be implemented by PEVs.

These analyses show that the capacity payment is a crucial part of the revenue that can be generated by the individual EVs. As mentioned above these approaches consider full availability and control of the participating vehicles. In addition PEVs are modeled as price takers, not influencing the price determination of regulation products. Following the analysis of Druitt and Frueh [37], Quinn et al. [24] and the sources mentioned above, one can see that the complete capacity requirements for regulation (and thus balancing) can be supplied by less than 10 % of the respective vehicle fleets, assuming all of them would be electric, technically capable and willing to participate. V2G can thus be a profitable option for the first movers and can even be performed without too high battery degradation costs, (cf. [28, 49]), but will eventually not be a viable option for all PEV-owners over time.

Following this observation, the interaction of PEVs adjusting their demand (mostly without V2G operations) in accordance with economic signals emitted from the power market is one of the main topics covered in literature. In particular the optimal operation of charging in the U.S. setting within the frame of unit commitment (UC) based pool market models was investigated by Sioshansi [50], Caramanis and Foster [51], Foster and Caramanis [52]. Sioshansi [50] compares two operation strategies, one that includes the demand requirements of 1 % of the vehicle fleet of the ERCOT service area as PHEVs (75,000 vehicles) in the ISOs unit commitment model and a tariff based charging strategy for TOU and RTP schemes. The results show that the charging costs in the centralised overall cost minimization UC scheme are lower than in the tariff based scenario. In addition the analysis of the ERCOT case shows that RTP schemes are efficient in communicating the marginal costs of power production to the demand side, but cannot capture the non-convexities of generator startup costs in a system with high shares of coal generation, leading to higher overall charging costs than in the other cases. The work of Caramanis and Foster [51] shows that a load aggregator for vehicles can develop efficient charging control strategies for his PEV fleet, which allows for successful hedging in the day-ahead market but still permits to consider intra-day charging flexibility in the real-time market. This analysis shows that charging costs can be reduced by at least 20 % as compared to uncontrolled charging, and that PEVs can successfully reschedule their demand on a short term basis, under consideration of new information about prices, grid constraints and in particular their own demand requirements. When aggregators consider shorter optimisation horizons and grid capacity constraints in their optimisation calculus, results from Foster and Caramanis [52] show that charging costs can be reduced, and the demand flexibility of the vehicles can also be employed in hour-ahead energy and regulation products. This shorter charging decision dispatch allows to choose the most appropriate commitment of the available PEV demand resources, and shows that accounting for uncertainty in the power system state and the resulting prices needs further investigation in particular in the European (or German) market scenario.

Following the hierarchical and decentralised charging decisions based on day-ahead and spot prices, the following approaches should be mentioned. Rotering and Ilic [53] are considering PHEVs in the Californian day-ahead market and present optimal smart charging strategies based on dynamic programming, that help to reduce daily energy costs by more than 50 %. In addition they analyze a firm commitment in the regulation market which allows the vehicles to generate additional profits that outweigh the driving energy costs. For another case in which PEV owners perform arbitrage accommodation based on the respective LMPs, Peterson et al. [28], find that when battery degradation costs are considered in V2G operation strategies, the annual profit per PEV would range between 12 and 118 USD for historical price data from NYISO, PJM and 1SO-New England areas from 2003 to 2008. This work performs a benchmark analysis and compares the values from a perfect foresight scenario with a naive forecasting technique building on a moving average of 2 weeks for the respective hours. When uncertainty is accounted for in this manner, the annual profits decrease to values of 6–72 USD. Energy arbitrage is thus only slightly profitable, but could be an option if additional infrastructure for grid interaction would be available to the vehicles, since the analysis builds on the assumption that EVs are not available in the time between 8:00 a.m. and 4:59 p.m., a time that is most likely to incorporate the daily peak prices.

Further Work from Verzijlbergh et al. [54] compares different charging strategies that are likely to be implemented by different actors and have been described in the related work mentioned before. In particular charging strategies from the perspective of an aggregator, the DSO and a wind-farm operator are considered in the setting of the Dutch power system. The aggregator performs a wholesale cost minimization to satisfy the demand of his customers at a minimum cost level, the DSO in turn distributes load in order to minimise the distribution system losses, and the wind-farm operator employs the charging flexibility to reduce the imbalance between planned and actual production of the wind-generators. In all cases a hierarchical or centralised control paradigm is implemented. The results based on the Dutch case show, that in particular the imbalance reduction strategy highly deviates from the load patterns of the traditional cost minimal and loss minimisation approaches. The imbalance strategy leads to highly accentuated peaks in the system that could in turn, if interaction of fleets with differing objectives takes place, increase the overall system balancing costs or create additional stress on distribution system components. Besides the technical comparison a basic cost assessment with respect to wholesale prices shows, that the loss oriented strategy incurs the highest costs. Considering interactions in the respective settings is thus an important aspect for the assessment of charging strategies.

Flath et al. [55] investigate how decentralised, cost minimising charging strategies can be improved by the concept of area prices. The study analyses how different charging strategies perform with respect to average costs and local distribution grid load. Besides the cost minimizing optimal strategy, heuristics that require less price and trip information based on specified price thresholds and a charging strategy incorporating an "as late as possible" charging scheme are also assessed. Results show that uniform pricing based on wholesale prices leads to new peaks in the total load of the vehicle fleet, which could lead to overload of distribution assets if the vehicles are regionally clustered.

When an additional local price component reflecting the current load of the local transformer is added, the load peaks can be reduced by more than 80 % while average costs for charging only increase by 15 %. This approach thus demonstrates that PEVs can be potentially coordinated very well by a dynamic pricing scheme if they are price responsive, rational actors.

Work from Vandael et al. [44], Fan [56] and Gerding et al. [9] emphasises the decentralised charging decision approach and also evaluates the mechanisms incorporated with respect to their economic or game theoretical properties. Important properties of a mechanism are its incentive compatibility, economic and in particular pareto efficiency, budget-balance, individual rationality and strategic robustness, [57].

These concepts from the algorithmic-mechanism design domain are important in order to apply distributed decision processes in the critical infrastructure of the Smart Grid. If charging decisions are made in a decentralised manner, mechanisms need to be designed to set incentives for the PEVs to participate (rather than not), thus making it individually rational to participate. Incentive compatibility reflects the fact that the information e.g. w.r.t. the demand of the individual vehicle is communicated truthful to the mechanism, making this property one of the most important ones if strategic decision behavior of PEVs is considered. Most approaches sketched in the previous section do not assume untruthful behavior of PEVs in order to address other explicit questions from the technical domain.

Table 1.3 Characteristic related literature with an economic focus

Technical focus	Main objective (RQ)	Coordination approach	Model scope					Main finding
Authors			Grid constraints	Ancillary services	RES utilization	PEV trip modeling	Dynamic prices	
Andersson et al. (2010) [26]	Economic assessment of PEVs for regulation in Sweden and Germany	Hierarchical	No	Yes	No	(NO)	Yes	PEVs can be profitable for regulation service provision in Germany, mainly due to capacity payments
Caramanis and Foster (2009) [51]	Cost minimisation for aggregator under consideration of day-ahead, real time market and grid capacity constraints	Hierarchical	Yes	(Yes)	(No)	No	Yes	Rolling horizon optimisation enables cost minimal charging for PEVs under realistic US market conditions
Flath et al. (2013) [55]	Evaluation of price based charging strategies under different conditions	Decentral: economic	Yes	No	No	Yes	Yes	Including locational price components reflecting grid capacity reduces system load and individual cost
Gerding et al. (2011) [9]	Evaluation of a distributed online mechanism for PEV charging	Decentral: economic	Yes	No	No	(Yes)	Yes	Online mechanism enables efficient distributed coordination with little information requirements
Goebel (2013) [58]	Assessment of economic value of	Hierarchical	No	No	No	Yes	Yes	Charging coordination allows for up to

(continued)

Table 1.3 (continued)

Technical focus / Authors	Main objective (RQ)	Coordination approach	Model scope					Main finding
			Grid constraints	Ancillary services	RES utilization	PEV trip modeling	Dynamic prices	
	charging coordination for PHEVs							45 % reduction of charging costs
Peterson et al. (2010) [28]	Economics of PEV energy arbitrage for different LMPs	Decentral: economic	(Yes)	No	No	Yes	Yes	Energy arbitrage at given LMP scenarios can be profitable, but overall returns are low given battery degradation
Rotering and Ilic (2010) [53]	Individual cost minimal charging considering ancillary service provision	Decentral: economic	No	Yes	No	(Yes)	Yes	Smart charging is beneficial, but ancillary service provision allows for additional profit (no price impact)
Sioshansi (2012) [50]	Comparison of tariff and UC based charging of PEVs	Central	No	(Yes)	No	Yes	Yes	RTP can lead to higher cost and emissions, DLC performs better in the ERCOT setting

Fan [56] is investigating a distributed (PEV) demand response approach, based on the idea of congestion pricing of communication networks. In particular a discriminatory pricing approach is presented which enables every EV to act according to its individual willingness to pay for the charging rate in a particular time slot. This pricing mechanism is shown to be capable to reduce local load peaks while maintaining computational tractability.

Table 1.3 presents a selective overview of relevant related work with a primary focus on economic assessment or objectives of charging coordination. A considerable part of PEV charging coordination literature is concerned with the economic possibilities for the provision of ancillary services by PEVs. Most of these V2G approaches employ centralised or at least hierarchical control architectures in order to allow for a reliable provision of the contracted AS-products. Some of them consider uncertainty aspects, or short term dispatch but the main body of literature is considering day-ahead or longer optimisation horizons. Further analyses focusing only on the coordinated withdrawal of power from the grid is increasingly build around decentralised, price based decision and optimisation mechanisms. These approaches rely on the individual to decide whether or not charging in a particular time frame is aligned with his budget constraints and economic preferences. Technical attributes are mostly considered as constraints in most models, but an explicit economic evaluation with respect to the real time utilisation of renewable energy by PEVs has not been performed so far.

1.6 Conclusion

The literature reviewed in the previous sections showed that PEV charging coordination can be categorised in particular with respect to its objectives and its control architecture. In the category with a predominantly technical focus V2G and grid load (regional and system-wide) impacts are the main research area. Work looking into the integration ability of renewable energy sources enabled by PEV demand flexibility is in particular focused on reducing imbalances stemming from fluctuating generators, e.g. wind power, on a system and also on regional scales under consideration of grid constraints. Short term storage applications are also discussed, but the coordination of PEV demand flexibility by dynamic price incentives is not covered very extensively. Work from the economic domain focuses on the assessment of regulation market participation and day-ahead wholesale market oriented charging. These approaches in turn do not intensively investigate the effect of cost minimising charging strategies with respect to the utilization of fluctuating renewable energy sources. Further work should thus focus on decentralised price based coordination of PEV demand for real time integration of renewable energy sources into the power system.

References

1. Sioshansi F, Vojdani A (2001) What could possibly be better than real time pricing? Demand response. Electr J 14(5):39–50
2. Albadi M, Elsaadany E (2008) A summary of demand response in electricity markets. Electr Power Syst Res 78(11):1989–1996
3. Gonzalez Vaya M, Andersson G (2012) Centralized and decentralized approaches to smart charging of plug-in vehicles. In: 2012 IEEE power and energy society general meeting, pp 1–8
4. Malone TW (1987) Modeling coordination in organizations and markets. Manag Sci 33 (10):1317–1332
5. Li T, Shahidehpour M (2005) Price-based unit commitment: a case of lagrangian relaxation versus mixed integer programming. IEEE Trans Power Syst 20(4):2015–2025
6. Padhy N (2004) Unit commitment—a bibliographical survey. IEEE Trans Power Syst 19 (2):1196–1205
7. Galus MD, Waraich RA, Andersson G (2011) Predictive, distributed, hierarchical charging control of PHEVs in the distribution system of a large urban area incorporating a multi agent transportation simulation. In: Proceedings of the 17th power systems computation conference, Stockholm
8. Caramanis M, Foster JM (2009b) Management of electric vehicle charging to mitigate renewable generation intermittency and distribution network congestion. In: Proceedings of the 48 h IEEE conference on decision and control (CDC) held jointly with 2009 28th Chinese control conference, Number iii, pp 4717–4722
9. Gerding EH, Robu V, Rogers A, Stein S (2011) Online mechanism design for electric vehicle charging. In: Tumer YS (ed) Proceedings of the 10th international conference on autonomous agents and multiagent systems, Number Aamas, Taipei, Taiwan, pp 2–6
10. Schweppe FC, Caramanis MC, Tabors RD, Bohn RE (1988) Spot pricing of electricity. Kluwer Academic Publishers, Boston
11. Lopes JAP, Soares FJ, Almeida PMR (2010) Integration of electric vehicles in the electric power system. In: Proceedings of the IEEE
12. Mets K, Verschueren T, Haerick W, Develder C, Turck FD (2010) Optimizing smart energy control strategies for plug-in hybrid electric vehicle charging. In: 2nd IEEE/IFIP international workshop on the management of the smart grid
13. Acha S, Green TC, Shah N (2010) Effects of optimised plug-in hybrid vehicle charging strategies on electric distribution network losses. In: IEEE PES T&D 2010, pp 1–6
14. Richardson P, Flynn D, Keane A (2010) Impact assessment of varying penetrations of electric vehicles on low voltage distribution systems. In: Proceedings of the 2010 IEEE PES general meeting, Minneapolis
15. Huang S, Safiullah H, Xiao J, Hodge B-MS, Hoffman R, Soller J, Jones D, Dininger D, Tyner WE, Liu A et al (2012) The effects of electric vehicles on residential households in the city of Indianapolis. Energy Policy 49:442–455
16. Sundstrom O, Binding C (2012) Flexible charging optimization for electric vehicles considering distribution grid constraints. IEEE Trans Smart Grid 3(1):26–37
17. Salah F, Ilg JP, Flath CM, Basse H, van Dinther C (2013) Impact of electric vehicles on high-voltage grids: a swiss case study. In: Working Paper of KIT
18. Gonzalez Vaya M, Galus MD, Waraich RA, Andersson G (2012) On the interdependence of intelligent charging approaches for plug-in electric vehicles in transmission and distribution networks. In: 2012 3rd IEEE PES international conference and exhibition on innovative smart grid technologies (ISGT Europe)
19. Sioshansi R, Fagiani R, Marano V (2010) Cost and emissions impacts of plug-in hybrid vehicles on the ohio power. Energy Policy 38(11):6703–6712
20. Sioshansi R, Denholm P (2010) The value of plug-in hybrid electric vehicles as grid resources. Energy J 31:1–22

21. Kiviluoma J, Meibom P (2011) Methodology for modelling plug-in electric vehicles in the power system and cost estimates for a system with either smart or dumb electric vehicles. Energy 36(3):1758–1767. doi: 10.1016/j.energy.2010.12.053
22. Kintner-Meyer M., Schneider K, Pratt R (2007) Impacts assessment of plug-in hybrid vehicles on electric utilities and regional us power grids part 1: technical analysis. Pacific Northwest National Laboratory
23. Kempton W, Letendre S (1997) Electric vehicles as a new power source for electric utilities. Transp Res Part D: Trans Environ 2(3):157–175
24. Quinn C, Zimmerle D, Bradley TH (2010) The effect of communication architecture on the availability, reliability, and economics of plug-in hybrid electric vehicle-to-grid ancillary services. J Power Sources 195(5):1500–1509
25. Tomic J, Kempton W (2007) Using fleets of electric-drive vehicles for grid support. J Power Sources 168(2):459–468
26. Andersson SL, Elofsson AK, Galus MD, Göransson L, Karlsson S, Johnsson F, Andersson G (2010) Plug-in hybrid electric vehicles as regulating power providers: case studies of Sweden and Germany. Energy Policy 38(6):2751–2762. doi:10.1016/j.enpol.2010.01.006
27. Galus MD, Zima M, Andersson G (2010) On integration of plug-in hybrid electric vehicles into existing power system structures. Energy Policy (2008)
28. Peterson SB, Apt J, Whitacre J (2010) Lithium-ion battery cell degradation resulting from realistic vehicle and vehicle-to-grid utilization. J Power Sources 195(8):2385–2392
29. Sortomme E, El-Sharkawi MA (2011) Optimal charging strategies for unidirectional vehicle-to-grid. IEEE Trans Smart Grid 2(1):119–126
30. Dallinger D, Krampe D, Wietschel M (2011) Vehicle-to-grid regulation reserves based on a dynamic simulation of mobility behavior. IEEE Trans Smart Grid 2(2):302–313
31. Saber AY, Venayagamoorthy GK (2010) Intelligent unit commitment with vehicle-to-grid—a cost-emission optimization. J Power Sources 195(3):898–911. doi:10.1016/j.jpowsour.2009.08.035
32. Sandels C, Franke U, Ingvar N, Nordstrom L, Hamren R (2010) Vehicle to grid—reference architectures for the control markets in sweden and germany. IEEE
33. Richardson DB (2013) Electric vehicles and the electric grid: a review of modeling approaches, impacts, and renewable energy integration. Renew Sustain Energy Rev 19:247–254
34. Pehnt M, Helms H, Lambrecht U, Dallinger D, Wietschel M, Heinrichs H, Kohrs R, Link J, Trommer S, Pollok T, Behrens P (2011) Elektroautos in einer von erneuerbaren Energien geprägten Energiewirtschaft. Zeitschrift für Energiewirtschaft 35(3):221–234
35. Short W, Denholm P (2006) Preliminary assessment of plug-in hybrid electric vehicles on wind energy markets. Tech report, NREL
36. Dallinger D, Wietschel M (2012) Grid integration of intermittent renewable energy sources using price-responsive plug-in electric vehicles. Renew Sustain Energy Rev 16(5):3370–3382
37. Druitt J, Frueh W-G (2012) Simulation of demand management and grid balancing with electric vehicles. J Power Sources 216:104–116
38. Goeransson L, Karlsson S, Johnsson F (2010) Integration of plug-in hybrid electric vehicles in a regional wind-thermal power system. Energy Policy 38(10):5482–5492
39. Ekman CK (2011) On the synergy between large electric vehicle fleet and high wind penetration–an analysis of the danish case. Renew Energy 36(2):546–553
40. Kempton W, Tomic J (2005) Vehicle-to-grid power implementation: from stabilizing the grid to supporting large-scale renewable energy. J Power Sources 144(1):280–294
41. Valentine K, Temple WG, Zhang KM (2012) Electric vehicle charging and wind power integration: coupled or decoupled electricity market resources? In: Proceedings of the IEEE PES IEEE 2012 PES general meeting
42. Markel T, Kuss M, Denholm P (2009) Communication and control of electric drive vehicles supporting renewables. In: 2009 IEEE vehicle power and propulsion conference, Number August, pp 27–34. IEEE

43. Finn P, Fitzpatrick C, Connolly D (2012) Demand side management of electric car charging: benefits for consumer and grid. Energy 42(1):358–363 (8th World energy system conference 2010)
44. Vandael S, Boucké N, Holvoet T, De Craeamer K, Deconinck G (2011) Decentralized coordination of plug-in hybrid vehicles for imbalance reduction in a smart grid. In: Proceedings of 10th international conference on autonomous agents and multiagent systems innovative applications track (AAMAS 2011), Number section 3, pp 803–810
45. Galus MD, Andersson G (2011) Balancing renewable energy source with vehicle to grid services from a large fleet of plug-in hybrid electric vehicles controlled in a metropolitan area distribution network. In: Proceedings of the Cigré 2011 Bologna Symposium, Bologna, pp 1–18
46. Vasirani M, Kota R, Cavalcante R, Ossowski S, Jennings N (2011) Using coalitions of wind generators and electric vehicles for effective energy market participation. In: The tenth international conference on autonomous agents and multiagent systems (AAMAS-2011)
47. Kempton W, Tomi´c J (2005) Vehicle-to-grid power fundamentals: calculating capacity and net revenue. J Power Sources 144(1):268–279
48. Mets K, Turck F, De, Develder C (2012) Distributed smart charging of electric vehicles for balancing wind energy. http://users.atlantis.ugent.be/cdvelder/papers/2012/mets2012sgc.pdf
49. Schuller A, Dietz B, Flath CM, Weinhardt C (2014) Charging strategies for battery electric vehicles: economic benchmark and V2G potential. IEEE Trans Power Syst 1–9. doi:10.1109/TPWRS.2014.2301024
50. Sioshansi R (2012) OR Forum—modeling the impacts of electricity tariffs on plug-in hybrid electric vehicle charging, costs, and emissions. Oper Res 60(3):506–516
51. Caramanis M. Foster J (2009) Management of electric vehicle charging to mitigate renewable generation intermittency and distribution network congestion
52. Foster JM, Caramanis MC (2013) Optimal power market participation of plug-in electric vehicles pooled by distribution feeder. IEEE Trans Power Syst 28(3):2065–2076. doi:10.1109/TPWRS.2012.2232682
53. Rotering N, Ilic M (2010) Optimal charge control of plug-in hybrid electric vehicles in deregulated electricity markets. IEEE Trans Power Syst 1–9
54. Verzijlbergh R, Lukszo Z, Ilic M (2012) Comparing different EV charging strategies in liberalized power systems. In: 2012 9th International conference on the european energy market (EEM), pp 1–8
55. Flath CM, Ilg J, Gottwalt S, Schmeck H, Weinhardt C (2014) Improving electric vehicle charging coordination through area pricing. Transp Sci 48(4):619–634. doi:10.1287/trsc.2013.0467
56. Fan Z (2012) A distributed demand response algorithm and its application to PHEV charging in smart grids. IEEE Trans Smart Grid 3(3):1280–1290
57. Steinle J (2008) Algorithmic mechanism design—Eine Einführung. Algorithmic Mechanism Design
58. Goebel C (2013) On the business value of ICT-controlled plug-in electric vehicle charging in California. Energy Policy 53:1–10. doi:10.1016/j.enpol.2012.06.053

Chapter 2
Control and Management of PV Integrated Charging Facilities for PEVs

Preetham Goli and Wajiha Shireen

Abstract The ongoing research in the field of plug-in electric vehicles (PEVs) and the growing global awareness for a pollution free environment, will lead to an increase in the number of PEVs in the near future. The proliferation of these PEVs will add stress to the already overloaded power grid creating new challenges for the distribution network. To mitigate this issue several researchers have proposed the idea of charging PEVs using renewables coupled with smart charging strategies. This chapter reviews the current literature on the state of the art infrastructure proposed for PEV charging facilities integrated with photovoltaic system. The proposed control algorithms, various smart charging techniques and different power electronic topologies for photovoltaic charging facilities (PCFs) are reviewed. Studies assessing the ability of photovoltaic charging stations to minimize the loading on distribution transformers are assessed. Finally, a simple and unique energy management algorithm for a PV based workplace charging facility based on dc link voltage sensing is presented. The power needed to charge the plug-in electric vehicles comes from grid-connected photovoltaic (PV) generation or the utility or both. The efficacy of the proposed algorithm is validated through simulation and experimental results.

Keywords Plug-in electric vehicles (PEVs) · Photovoltaic charging facility (PCF) · Distribution network

P. Goli (✉)
Department of Electrical and Computer Engineering, University of Houston, Houston, USA
e-mail: pgoli2@central.uh.edu; preethamgoli@gmail.com

W. Shireen
Department of Electrical and Computer Engineering, Department of Engineering Technology, University of Houston, Houston, USA
e-mail: wshireen@central.uh.edu

S. Rajakaruna et al. (eds.), *Plug In Electric Vehicles in Smart Grids*,
Power Systems, DOI 10.1007/978-981-287-317-0_2

2.1 Introduction

With the growing global awareness for a pollution free environment, rising energy costs, PEVs are being introduced by many automotive makers. It is known that if 25 % of the 176 million fleets of light vehicles in U.S were converted to PEVs, it will rival the entire U.S power generation capacity [1]. The proliferation in PEVs requires charging stations to fulfil their battery requirements. Though PEVs are being marketed with the goal of minimizing the pollution from automobiles, the energy requirements for charging the batteries is still met by power generated by fossil fuel sources. Hence many researchers have proposed the idea of charging PEVs using renewable energy sources like wind and photovoltaic. Many pilot projects are also underway to charge PEVs from solar photovoltaic system [2–5]. Charging stations based on wind power is still in the nascent stages though few ventures have been announced [6]. Due to the social and economic benefits, research on charging stations featuring photovoltaic system has attracted researchers around the world.

Using solar power to charge batteries is not a new idea. It is a reliable source for charging light duty electric vehicles such as golf carts, scooters and airport utility vehicles [7]. Large-scale deployment of photovoltaic chargers in a parking lot is analysed in [8]. A 2.1 kW photovoltaic charging station integrated with the utility at Santa Monica is described in [9]. An experimental control strategy for electric vehicle charging system composed of photovoltaic array, emulated power grid and programmable dc electronic load representing lithium ion battery emulator is presented in [10]. PV parking lot charging and other business models to charge PEVs with solar energy are discussed in [11]. Economics of PV powered workplace charging station has been studied in [12, 13]. The analysis shows the feasibility of a PV based workplace parking garage with benefits to the vehicle owner as compared to home charging, such that the garage owner will get the payback of installations and maintenance cost and profit within the lifetime of the PV panels. According to [13] integrating a solar collector into a parking lot would result in a much more rapid payback-period, encouraging widespread installation of solar capacity. Reference [14] describes how smart control strategies can help PEVs and PV to integrate with the present electricity systems. Co-benefits of large scale deployment of PEVs and PV systems has been studied in [15]. The study concludes that PV provides a potential source of midday generation capacity for PEVs, while PEVs provide a dispatchable load for low value or otherwise unusable PV generation during periods of low demand (particularly in the spring).

As per the National household travel survey vehicles are parked for at least 5 h in workplaces [16]. Hence these places are favorable for developing charging station infrastructure but this would lead to serious overloading issues at the distribution level. Since upgrading of transformers is an expensive option for the utilities, this issue needs close attention as the PEV penetration increases. Several papers have been published to address the overloading of distribution transformers while charging the PEVs [17–19]. Nevertheless, not much study has been reported to be

Table 2.1 PV characteristics [23]

PV type	Module price ($/WP)	Efficiency (%)	Peak energy (WP)	Total cost of PV ($)
Crystalline silicon	2.14	22	264	565
Polycrystalline silicon	1.74	15.5	186	324
Thin film	0.93	12	144	134

tightly related to the case of reducing the loading on distribution transformers using a photovoltaic system. Though few papers exist in the literature, they are mostly confined to residential distribution networks [20, 21]. There is plenty of parking area in the U.S—a reasonable fraction of which is suitable for PV installation. This chapter reviews the current literature on the state of the art infrastructure proposed for PEV charging facilities integrated with photovoltaic system. The proposed control algorithms, various smart charging techniques and the economic benefits of photovoltaic charging facilities (PCFs) are reviewed. Various power electronic topologies, control algorithms and charging strategies will be discussed. It will be shown that a network of PCFs will accelerate the deployment of PEVs through economic and environmental benefits to the utilities and vehicle owners. The impact of grid connected photovoltaic system on the utility distribution networks is analyzed. The suitability of using PV power for charging PEVs is accessed in this chapter.

Determining the size and type of PV panel is an important consideration for a solar carport. Few papers [22, 23] have recommended the use of monocrystalline silicon as the most cost-effective solar cell type for PV charging facilities. Table 2.1 shows the PV characteristics of various modules, the peak energy produced and the total cost of the PV module. The PV panel can be sized by taking the best and worst months into consideration. As described in [24], the initial cost of the PV panel would be $20,000 when it is designed based on the worst month of the year and $10,000 when it is designed based on the best month of the year. However, for the first case, surplus energy can be injected into the grid, to balance the final cost.

2.2 Impact of PEV Charging on the Distribution System

Large-scale penetration of PEVs can have a detrimental and destabilizing effect on the electric power grid. With the variation in demand, the production of power can vary significantly. Variation in charging time of PEVs can result in distinct differences in fuels and generating technologies [25]. Figure 2.1 illustrates the impact of charging one million PEVs on the Virginia—Carolinas electric grid in 2018 on the various generation technologies. As shown in Fig. 2.1, at low specific power and late in the evening, coal was the major fuel used, while charging more heavily

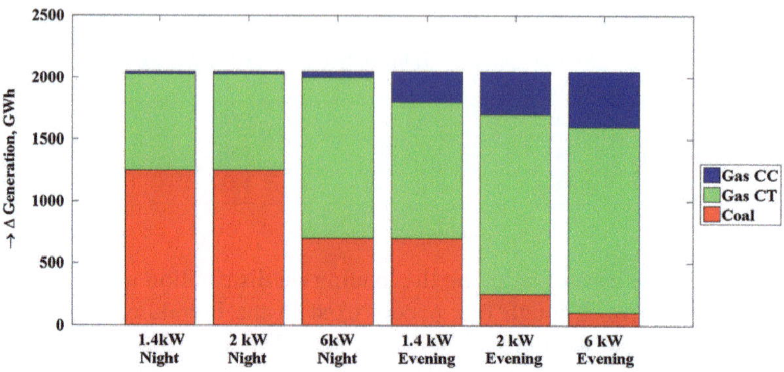

Fig. 2.1 Generation shares by plant type for PEV charging level and timing [25]

during peak times led to more use of combustion turbines and combined cycle plants. Since the initial deployment of PEVs is assumed to be clustered to a particular neighborhood, many authors have focused their research on the study of distribution transformer impacts. Depending upon the time, place of vehicle charging, various charging methods and the charging power levels there could be several ramifications on the distribution network. Various analytical techniques and different simulation tools were employed by several authors to estimate the transformers loss-of-life, average lifetime and harmonic losses. The percentage of transformers loss-of-life and average lifetime are important factors to be considered while studying the charging behavior of PEVs on future distribution system. High penetration of PEVs in the future will increase the loss-of-life factor of distribution transformers [26–29].

The impact of controlled and uncontrolled charging of PEVs on the average lifetime of a transformer is described in [24, 26]. As per [24], the average lifetime of a transformer is reduced by 4–20 % under uncontrolled charging for a PEV penetration of 10 %. At 50 % penetration of PEVs, the average lifetime is reduced by 200–300 % with uncontrolled charging. On the other hand controlled charging increases the lifetime by 100–200 % with respect to uncontrolled charging for 50 % penetration of PEVs. Plug-in electric vehicle charging rates can have a significant impact on the lifetime of a transformer [24, 25]. Table 2.2 summarizes the sensitivity of transformer lifetime to different charging rates (3.6 and 7.7 kW) under controlled and uncontrolled charging for various levels of PEV penetration. As expected, transformer life degradation is exacerbated when the charging rate is increased from 3.6 to 7.7 kW.

The percentage of transformers loss-of-life can be minimized through distributed charging and controlled off-peak charging which requires coordination among utilities, customers and charging stations. Simulation results in [17] show that distributing the load profile of the battery charging helps to minimize the distribution transformer loss-of-life. Power management of the PEV battery charge profile can help manage the loss-of-life of the distribution transformer. Controlled

Table 2.2 Impact of charging PEVs [27]

Charger rating	3.6 kW				7.7 kW			
Charge type	Uncontrolled (years)		Controlled (years)		Uncontrolled (years)		Controlled (years)	
PEV penetration (%)	Minimum lifetime	Average lifetime	Minimum lifetime	Average lifetime	Minimum lifetime	Average lifetime	Minimum lifetime	Average lifetime
10	29.2	61.1	32.7	56.3	27	49.8	34.1	59.6
25	26.8	37.4	25.7	55.5	16.8	36.5	33.3	52.9
50	12.2	38.9	10.7	45.2	11.2	24.3	29.9	48.4
75	16	31.4	11.7	49.1	5.8	14.5	42.9	62.2
100	8.3	18.4	12.8	37.8	2.3	10.5	26.8	38.8

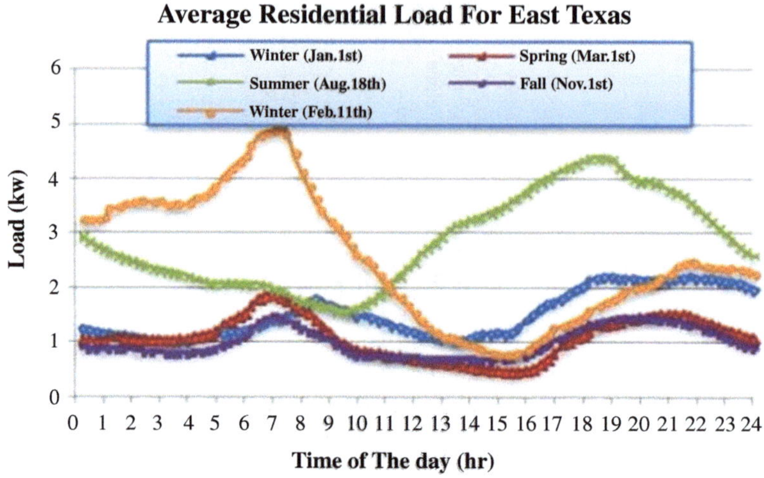

Fig. 2.2 15 min interval data of average residential individual customer in East Texas [26]

off-peak charging can shift PEVs charging load to an off-peak time. Usually charging PEVs at night time is proposed as the best way to mitigate the loss-of-life issues of distribution transformers. However, PEVs can also introduce a new peak or near peak in early off-peak time. Generally, the impact of extra load on transformers in summer is much greater than that in winter. However, some winter mornings with peak load may be an exception. Charging from midnight through early morning in those days may exert strong impact on transformers. Figure 2.2 describes this effect by taking the average residential load for East Texas into consideration. As shown in the figure for a particular day in winter, February 11th, the load consumed in the early morning is higher than that in summer days. Therefore, it is not always appropriate to charge electric vehicles at 1 am in those days. The required control strategy should depend on the actual load profile in a particular area for a particular time period.

2.3 Mitigating the Impact of PEV Charging on the Distribution System

To mitigate the issue of transformer loss-of-life due to PEV charging, integration of renewables like rooftop PV systems into the existing power grid has been proposed. In [30], a case study for the year 2030 was built based on demand increase, forecasted PEV and DG units. The results showed that PEV battery charging would prove onerous for the constraints studied. DG penetration would be able to provide support for PEV battery charging but PEV battery charging management would be necessary to minimize the impact in order to reach high levels of PEV penetration.

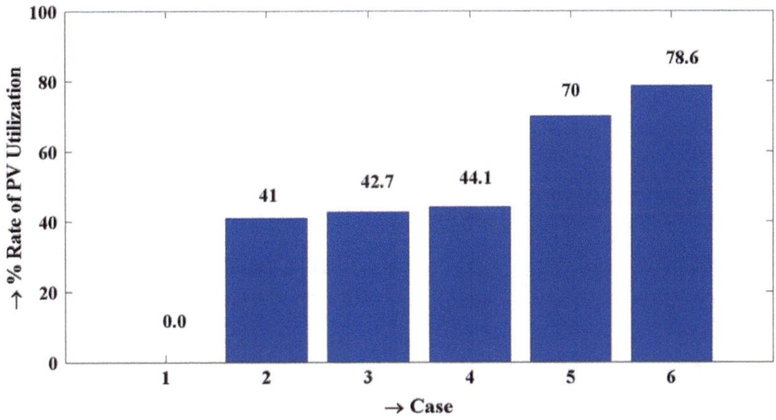

Fig. 2.3 The rate of PV utilization [32]

The possibility of smoothing out the load variance in a household microgrid by regulating the charging patterns of family PEVs is investigated in [31]. A case study is presented, which demonstrates that, by regulating the charging profiles of the PEVs, the variance of load power can be dramatically reduced. Integration of residential PV system with PEVs is studied in [32]. A residential PV system was simulated for various charging schemes of PEVs and the results are shown in Fig. 2.3. Several cases with different combinations of PV, PEV, V2H (Vehicle to Home i.e. discharge of PEV) and various charging schemes were analyzed. Case 1 describes a residential facility without PV and PEV. Case 2 describes a residential facility integrated with rooftop PV system without PEV. Therefore, these 2 cases analyze the effects of using PV while using gasoline vehicles instead of PEVs. Case 3 represents a residential facility with PV and PEV without V2H capabilities. Cases 4–6 have all the facilities (i.e. PV, PEV and V2H capabilities) but their charge-discharge schemes are different. As shown in Fig. 2.3 the local consumption rate of PV output increased by 1.7 % when gasoline vehicles are replaced with PEVs. On the other hand the rate of PV utilization increased by 8.6 % when the charging scheme changed from fixed (fixed target of SOC) to variable (variable target value of SOC).

The integration of PV rooftop in PCFs can relieve the burden on the distribution networks, by reducing the effective load seen by the distribution grid peak charging, as well as supplying power to the grid when excess power is generated by PV rooftops. A PV parking lot for PEVs is proposed in [33], in which the PEVs can be charged from the PV source as well as the distribution grid. Mathematical models are developed to estimate the electric power capacity for PV parking lot. An evaluation of impacts resulting from expected scenarios are performed through stochastic sequential simulations of the distribution system with load and PV generation in [34]. Figure 2.4 shows the LOL (loss-of-life) experienced at a particular distribution transformer, for change in stochastic load and PV generation

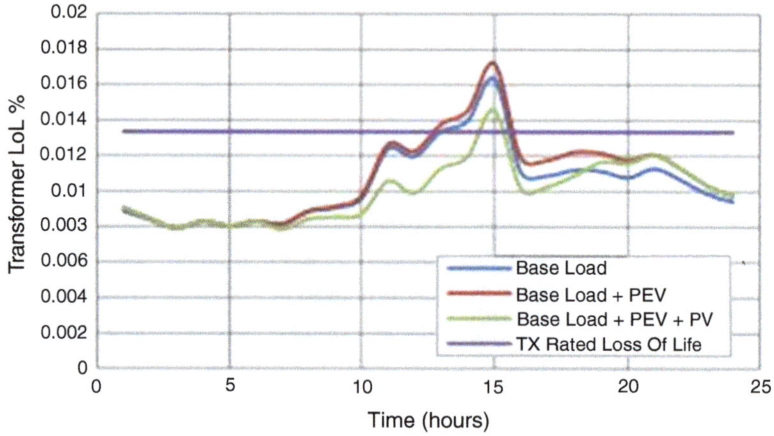

Fig. 2.4 Stochastic distribution transformer loss of life [34]

units. It is evident from the figure that rooftop PV coupled with PEVs can reduce LOL of the distribution transformer. These studies have shown that PV generation coupled with PEV charging can delay and reduce the temperature rise of distribution transformers.

2.4 Proposed Architectures for PV Based PEV Charging Facilities

The charging units for PEVs can be either on-board or off-board. In case of an off-board charger, the charger is an external unit while in the case of an on-board charger it is a component of the vehicle. On-board chargers are supplied with ac power and they consist of an AC/DC rectifier, DC/DC boost converter for power factor correction and a DC/DC converter to charge the battery as shown in Fig. 2.5. Currently AC charging is being employed to charge PEVs by means of on-board chargers. The major drawback of this technology is that it does not support fast charging as it is required to increase the power capability of the on-board charger thereby increasing the cost and weight of the PEV. Hence to support fast charging

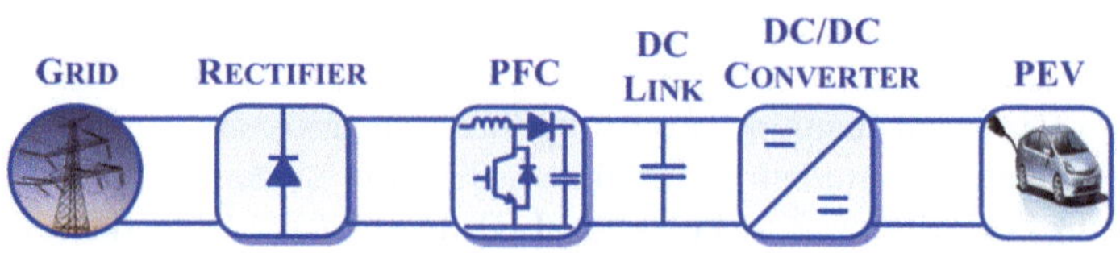

Fig. 2.5 Conventional PEV charger

of PEVs off-board chargers are proposed which directly supply dc power to the PEV charging inlet. It is to be noted that in case of an off-board charger the entire power conversion (AC/DC) takes place in an external unit and therefore it is feasible to increase the ratings of the power converters in order to support fast charging.

AC system is being used since years for power distribution and there are well developed infrastructure-standards and technologies. DC system on the other hand has many advantages, starting with the fact that overall efficiency of the system could be higher and it facilitates the integration of renewable energy sources with fewer power converters. Since PV arrays generate dc power, a charging facility featuring PV power facilitates the charging of PEVs from a dc bus which is more effective, economical and efficient since it does not involve more power conversion stages unlike AC charging. Various methods have been proposed for integrating PEV chargers within a photovoltaic system. Several power electronic topologies for a PCF have been proposed in the literature based on the type and the number of converters which are classified as:

A. Centralized architecture
B. Distributed architecture
C. Single stage conversion with Z-converter

A. Centralized architecture

Detailed block diagram representing the centralized architecture is shown in Fig. 2.6. It consists of a central DC/DC boost converter which performs the function of maximum power point tracking. The DC/DC chargers are integrated with the PV charging facility at the dc link. Multiple PEVs can be charged by increasing the corresponding ratings of PV panels and the associated power converters. Each parking spot must have a dedicated DC/DC buck converter which is connected to the dc link. This configuration is suitable for charging stations in the range of several kilowatts. It is applicable for charging vehicles like golf carts, campus utility vehicles etc. which commute for very short distances with low battery capacities. Battery switch station powered by PV is a good candidate for adopting centralized architecture. But this kind of configuration does not support fast charging since

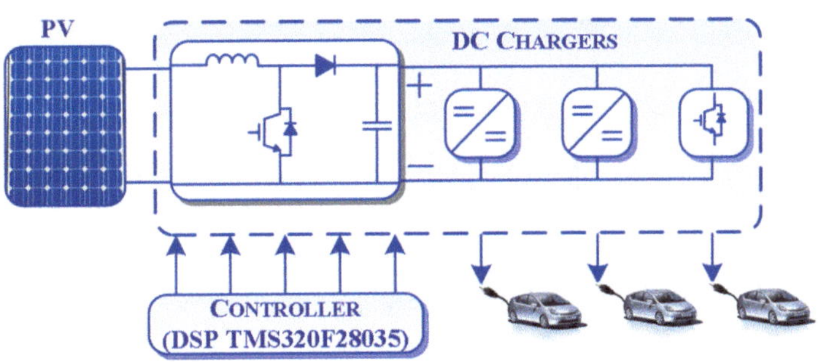

Fig. 2.6 Centralized architecture

installation of a very high power DC/DC converter is very expensive and it is vulnerable to single fault shutdown.

B. Distributed Architecture

Presence of DC/DC converters with high power ratings is an important criterion for fast charging of PEVs. This can be achieved economically through distributed architecture as shown in Fig. 2.7. In this case several strings of PV panels are connected in series. Each parking spot has a dedicated PV panel to support the charging of PEV and each string of PV panels is interfaced with their own DC/DC converter and shares a common dc bus, which connects to an AC utility grid through a bi-directional DC/AC converter. The DC/DC battery chargers are connected to the dc bus. Each parking spot requires an individual DC/DC converter to charge the PEVs. The proposed architecture is suitable for installation at places such as workplace, universities, shopping malls etc. where the demand of PEVs and their duration of stay in the parking lots are highly probabilistic in nature. It is more reliable since the PEVs can be charged from the grid during the periods of low insolation or cloudy weather. Also, it is important to note that the extra energy generated by PV can be injected into grid, which can be used to balance the PV costs.

A PCF requires constant power from the PV or the grid to meet the high demand of PEVs. The reliability of a PCF can be improved by including an energy storage unit such as a battery bank, ultra capacitor, fuel cell etc. For instance in [35] the power generated by roof top photovoltaic system is stored in VRLA (valve-regulated lead-acid) batteries and fuel cells in a PEV docking station. The PEVs arriving

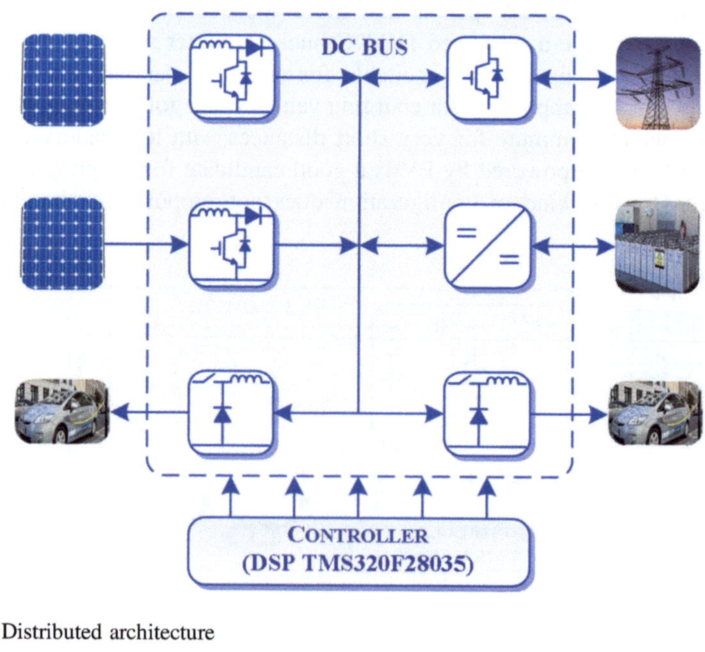

Fig. 2.7 Distributed architecture

at the docking station can be charged from two separate tracks i.e. using the energy from the VRLA batteries or the fuel cells. The use of storage capacity in PCFs has the following advantages [36]:

- Efficient use of renewable energy sources
- Maximization of renewable energy sources contribution
- Better demand and production match, better auxiliary service supply and improved overall reliability

The core idea of including an ESU (energy storage unit) is that the power demand by PEVs can be either supplied by the PV or the utility or through a local energy storage unit. Energy derived from the ESU can charge the PEVs during certain contingencies such as islanding condition without the availability of PV power. It facilitates the charging of PEVs using minimum energy from the grid. The charging station appears as a dc microgrid with local generation from the PV system, PEVs' as loads and battery bank representing the storage system.

C. Single stage conversion with Z-converter
The double stage conversion described in the above architectures is replaced by a single stage using a Z-converter [22] as shown in Fig. 2.8. It does not require an additional DC/DC converter for each charging spot and a single DC/DC converter is employed to provide galvanic isolation. The Z-converter has double modulation capability, and can shape the grid current while simultaneously regulating PEV battery charging. The unit can be employed for both power absorption and injection, with simultaneously controlled battery charging. This ensures close to unity power factor for all operating modes and power flow paths; achieving this with a single conversion stage can be considered a unique advantage of the Z-converter. Furthermore, this topology possesses inherent buck-boost capability, allowing increased voltage range on the PV or grid. Despite the single conversion stage, reliability, rather than efficiency or cost, is the strong point of the Z-topology. Also the single phase power processed by the Z-converter consists of 120 Hz double line frequency ripple. This ripple can be mitigated by placing an additional decoupling capacitor across the PV source which introduces possible deviation from perfectly constant power extraction at the PV panels.

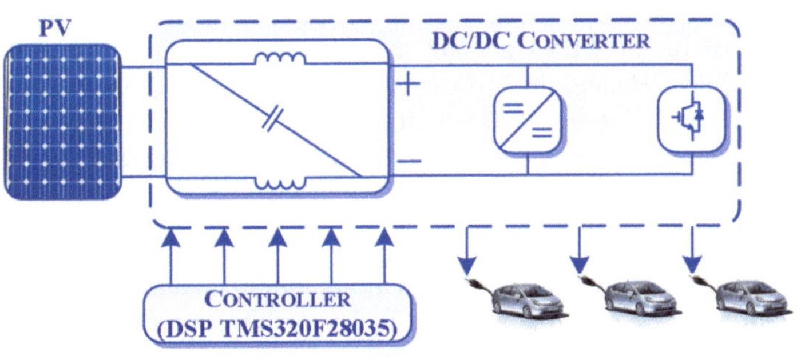

Fig. 2.8 Single stage conversion with Z-converter

2.5 Control Strategies

Workplace based photovoltaic charging facilities and residential PV charging are the two available options for charging PEVs using solar power. Depending on the solar irradiation, PEVs can be charged either from the photovoltaic or the distribution grid or both. The solar charging station should distribute the power available at the PV panels to the PEVs effectively and safely. Typically PEVs arrive at the charging facility with different State-of-Charge (SOC). More than often, the amount of PV power available to charge multiple PEVs is limited. Furthermore, the PV source is stochastic in nature, its power characteristic is nonlinear and the PEV batteries to be charged should be within certain voltage and current limits. Therefore, this process necessitates intelligent control of the power conditioning unit to manage the direction of power flow in PV integrated charging stations. Several algorithms have been proposed in the literature which differ significantly based on the type and location of the PCF. The algorithms also differ based on the various control parameters such as PV power, load demand, state-of-charge etc. Accordingly they can be classified as follows:

A. Residential Photovoltaic Charging
B. Battery Switch Stations
C. Workplace Photovoltaic Charging

A. Residential Photovoltaic Charging
Few authors [37–40] have proposed an architecture for a grid-connected residential photovoltaic system that can be used to charge PEVs as well as to supply the existing household loads. The control algorithms depend on the power generated by the PV and the SOC of the PEV battery. Raul et al. [39] proposed a residential load coordination mechanism to charge PEVs. Depending on the load demand of the distribution transformers, the PEVs can be charged using renewable energy (PV/Wind) or the power from the grid. Each household is installed with a rooftop PV system and a small scale wind turbine. A residential microgrid composed of rooftop panels and a biodiesel generator to charge PEVs and supply AC/DC household loads is described in [41]. In order to share the load among the sources, master-slave control method is employed. The operation of the residential microgrid depends on the PV power, load demand, SOC of the battery storage and tariff set by the utility. Most of the PEVs are not available for charging during daytime at residential facilities. Hence, this process demands for an additional component in the form of an energy storage unit which might not be economically attractive for an individual home owner. Residential charging is advantageous for households with more than one PEV.

B. Battery Switch Stations:
A PV based battery control strategy for charging multiple batteries in a solar battery charging station (SBCS) is proposed in [42]. The architecture of the SBCS is similar to the one shown in Fig. 2.6 but the DC/DC chargers are replaced by bi-directional switches. The proposed control strategy first charges each individual battery until

they reach the same voltage level and then charges the multiple batteries in parallel simultaneously according to the battery charging period and the available solar energy. This control strategy eliminates the use of multiple DC/DC converters per battery connection, making the SBCS less complicated and economical. Though being economical, the proposed architecture does not consider the scenarios when the PV panel is not generating any power or generating power in excess. Hence it cannot be considered for charging PEVs. A PV-based battery switch station (BSS) is proposed in [43]. The energy exchange strategy depends on the battery swapping demand of the PEVs and power generated by the PV. An algorithm is proposed to charge PEV batteries using the maximum energy from PV.

C. Workplace Photovoltaic Charging
In few cases, authors have proposed the idea of inserting a DC/DC battery charger at the dc link of the grid-connected PV system. By measuring the power generated by the PV and the power demand of the PEV, the control algorithm ensures the charging of the PEV battery from the appropriate source as described in [44]. Based on the imbalance between the PV power and the load demand, various possible scenarios are described. In case of [45], the power flow in a PV parking lot is managed through a set of computer controlled relays. PV panels of different ratings are interfaced with PEV chargers and the power grid through computer controlled relays. Depending on the irradiation levels, the relays direct the entire PV power to the PEVs or the grid or both. Hamilton et al. [46, 47] proposed an extension to this method for a modular dc PV charging station. Several PV panels are interfaced with the dc bus through a set of DC/DC converters. The DC/DC converter intelligently controls the power flow to the PEVs based on a certain preset limits of the dc bus voltage. Based on the preset limits the energy conversion unit facilitates three way energy flow among the power grid, PV modules and PEVs.

The concept of dc bus signalling has been proposed by several authors to schedule power to dc loads in a microgrid [48–50]. Few of them have extended this concept to charge PEVs in a microgrid environment [48, 49]. A smart charging station architecture integrated with PV power is proposed in [51, 52]. The smart charging station can operate in standalone mode and grid-connected mode. The switching between various modes is facilitated by the variation in dc link voltage levels induced due to the change in solar insolation. During the period of low solar insolation and peak load on distribution transformer, the controller shifts the charging of PEVs to non-peak period. The proposed control algorithm is simple as it involves only a single parameter i.e. dc link voltage to manage the direction of power flow in the charging station. It facilitates the charging of PEVs using minimum energy from the grid without any adverse impacts on the distribution transformer. The following sections explain the concept of dc link voltage sensing and its application for control and management of PV powered charging facilities.

2.5.1 Concept of DC Link Voltage Sensing

The primary requirement for a microgrid operation is to maintain the common dc bus voltage within an acceptable range. The terminals within a microgrid can be generally categorized into four types: generation, load, energy storage unit (ESU), and grid connection using voltage-source converters (VSCs). These four types of terminals can be further divided into two groups in terms of their contribution to system control and operation which are the power terminal and the slack terminal.

A power terminal is the one which outputs or absorbs power to/from the microgrid on its own and usually does not take the system's need into account. Typical examples would be variable dc loads (PEVs) and nondispatchable (variable) generation, such as wind turbines and photovoltaic based generations, when operating purely according to environmental conditions. Conversely, a slack terminal is the one which is responsible for balancing the power surplus/deficit caused by power terminals and maintaining stable system operation. Typical examples include a grid-connected VSC terminal (G-VSC) and ESU when they are actively supporting the dc microgrid system.

As previously described, different measures shall be taken by each terminal according to system operating conditions, thus a fast and reliable scheme for acknowledging system operation status is essential. Apart from using as communication means, dc link voltage is a good indicator of the system's operational status. The simplified equivalent circuit of the dc bus including the ESU and PEV is shown in Fig. 2.9, where P_{DC} and P_{AC} refer to the total power on the dc side (PV panel and DC/DC converter) and the ac side (inverter and the grid) of the dc bus respectively.

From Fig. 2.9, the instantaneous power relationship in a grid-connected PV system is given by

$$p_{dc}(t) = p_{ESU}(t) + p_c(t) + p_{PEV}(t) + p_{ac}(t) \tag{2.1}$$

where p_{dc} is the output power of the DC/DC converter on the dc side, p_{ESU} is the power delivered to (or by) the ESU, p_c is the power to the dc link capacitor, p_{PEV} is

Fig. 2.9 DC power flow

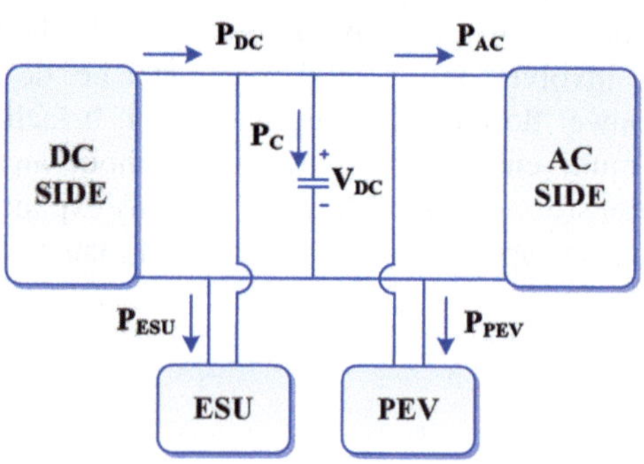

the power consumed by the plug-in electric vehicles, and p_{ac} is the power extracted by the inverter on the ac side. The instantaneous ac power (output of the inverter) can be written as

$$p_{grid}(t) = (V_m \sin \omega t)(I_m \sin \omega t) \tag{2.2}$$

$$= \frac{V_m I_m}{2} - \frac{V_m I_m}{2} \cos \omega t \tag{2.3}$$

where p_{grid} is the power injected into the grid, V_m is the amplitude of the phase voltage and I_m is the amplitude of the grid current. The ac power includes a dc term and a second-order ripple in the dc voltage. The average input power to the ac side can be written as

$$P_{AC} = V_{DC} I_{AC} \tag{2.4}$$

where I_{AC} is the average input current to the ac side (i.e. on the dc side of the inverter). Equating the average power on the input of ac side to the dc term on the output of ac side

$$\frac{V_m I_m}{2} = \eta V_{DC} I_{AC} \tag{2.5}$$

where η is the efficiency of the inverter. If V_{dc} and $V_{dc(ref)}$ are the actual and reference values of dc link voltage, respectively, the change in energy ΔE_{dc} stored in the dc link capacitor C_{dc} can be written as

$$\Delta E_{dc} = \frac{C_{dc}}{2} \left(V_{dc(ref)}^2 - V_{dc}^2 \right) \tag{2.6}$$

To inject the PV power to the grid while maintaining a constant V_{dc}, the following energy balance should be satisfied:

$$\Delta E_{dc} = T(p_{dc} - p_{esu} - p_{PEV} - \frac{V_m I_m}{2\eta}) \tag{2.7}$$

where T is the time period of ac supply.

Combining (2.6) and (2.7)

$$V_{dc}^2 = V_{dc(ref)}^2 - \frac{2T}{C_{dc}}(p_{dc} - p_{PEV}) + \frac{2T}{C_{dc}} p_{dc} + \frac{2T}{C_{dc}\eta} V_m I_m \tag{2.8}$$

$$V_{dc} = \sqrt{V_{dc(ref)}^2 - \frac{2T}{C_{dc}}(p_{dc} - p_{PEV}) + \frac{2T}{C_{dc}} p_{ESU} + \frac{2T}{C_{dc}\eta} V_m I_m} \tag{2.9}$$

$$V_{dc} = \sqrt{V_{dc(ref)}^2 - \frac{2T}{C_{dc}}(\eta_{boost}p_{PV} - p_{PEV}) + \frac{2T}{C_{dc}}p_{ESU} + \frac{2T}{C_{dc}\eta}V_m I_m} \tag{2.10}$$

where η_{boost} is the efficiency of the DC/DC converter on the dc side.

From (2.10), it is clear that the fluctuations in PV power due to the change in solar irradiance causes variations in the dc link voltage. For a workplace based charging facility PEVs can be assumed to stay in the parking lot from morning till evening. Hence the variation in PEV load is considered.

Also from (2.7), the charging power of the dc capacitor can be written as

$$p_c = p_{dc} - p_{esu} - p_{PEV} - \frac{V_m I_m}{2\eta} \tag{2.11}$$

$$\frac{1}{2}cv_{dc}^2 = p_{dc} - p_{esu} - p_{PEV} - \frac{V_m I_m}{2\eta} \tag{2.12}$$

$$cv_{dc}\frac{dv_{dc}}{dt} = p_{dc} - p_{esu} - p_{PEV} - \frac{V_m I_m}{2\eta} \tag{2.13}$$

From 2.13, it can be inferred that a constant dc voltage indicates a balanced power flow among all the terminals, and a rising or dropping dc voltage indicates power surplus or deficit, respectively. Since the dc voltage can be used as an effective indicator of power-flow status, the control scheme of the proposed charging facility can be designed according to dc link voltage variation. Assuming the PEV demand to be constant over a period of time, the variation in dc link voltage occurs only due to the fluctuation in solar insolation. The operational voltage range can be divided into several levels. Based on the voltage level the charging facility has several modes of operation.

Figure 2.10 shows the variation in the dc link voltage and the power from the PV array with step changes in irradiation. A PV panel of rating 5.5 kW was modeled in Matlab taking the battery capacity of a single PEV into consideration. The reference dc link voltages have been chosen taking into consideration the change in sun conditions from early morning to late evening. As shown in Fig. 2.10 the PV array starts delivering power when the dc link voltage is greater than 50 V. At 250 V the PV system delivers 4,500 W which is the power requirement of standard PEV battery. Between 300 and 350 V the power delivered by the PV array is greater than 5,000 W, exceeding the power requirement of the PEV. This excess power can be sent to the grid. By taking the dc link voltage and the corresponding power delivered by the PV array into consideration, three reference voltage levels have been chosen as $V_{DC\text{-}1}$ = 50 V, $V_{DC\text{-}2}$ = 250 V and $V_{DC\text{-}3}$ = 350 V. The modes of operation of the charging station are classified depending on the change in the dc link voltage. As the dc link voltage is the only criteria for switching between various modes the overall complexity of the system is reduced.

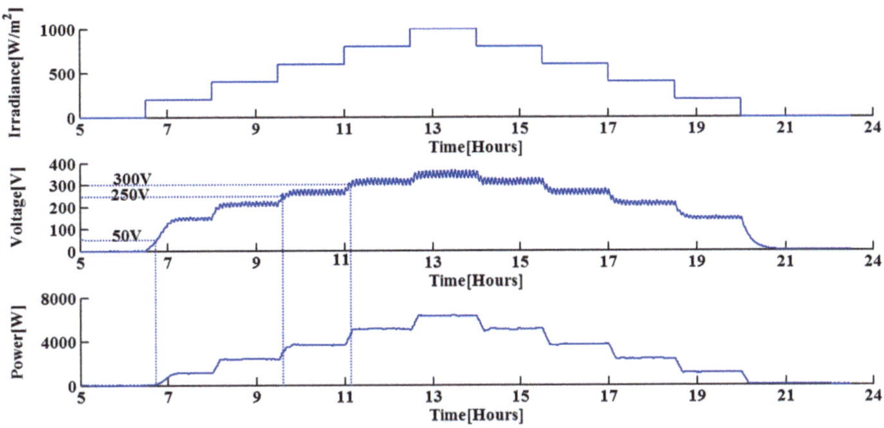

Fig. 2.10 Change in the dc link voltage and power generated by the PV with the change in sun condition

2.6 Power Management Algorithm for PV Charging Facility

The detailed circuit configuration for the proposed workplace based charging facility is shown in Fig. 2.11. The architecture consists of several strings of PV panels interfaced to their own DC/DC converters which share a common dc bus. The DC/DC boost converter performs the function of maximum power point tracking (MPPT) to facilitate the operation of PV panel at the maximum power point. The energy storage unit (ESU) is connected to the dc bus via a bi-directional DC/DC buck-boost converter. The ESU will support the charging of PEVs when there is no power available either from the grid or the PV. The battery pack in the ESU can be charged either from the grid during off peak hours or from the PV after all the PEVs have been charged in the charging facility. DC/DC buck converter connected to the dc bus controls the charging of the PEV. The control description shown for the charging facility in Fig. 2.11 is based on the requirements for two PEVs. Multiple PEVs can be charged by having separate buck converters installed for each charging point. The charging facility is connected to the power distribution network through a DC/AC bi-directional grid tied converter.

The control unit monitors and controls the power flow between the source and PEV. As shown in Fig. 2.11 the control unit generates the switching signals to control the various power converters in the charging facility based on the voltage and current values sensed by the voltage and current sensing units. V_{PV}, voltage across the PV array and I_{PV}, the current flowing from the PV array are used to implement MPPT by means of incremental conductance algorithm. V_{DC} is the magnitude of the voltage at the dc bus, V_{B1} and V_{B2} are the detected battery voltages of the PEVs which give a measure of the state-of-charge (SOC) while

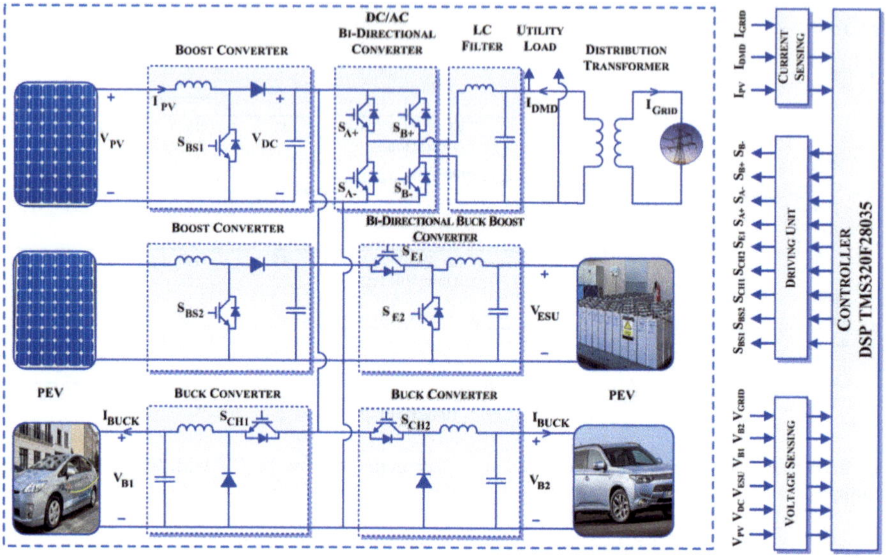

Fig. 2.11 Detailed circuit configuration of the proposed architecture

V_{ESU} gives the measure of the SOC of the ESU. I_{DMD} represents the loading condition of the distribution transformer, I_{grid} is the current fed into the grid by the DC/AC converter and V_{Grid} is the grid side voltage.

2.6.1 Modes of Operation

The operation of the charging station can be categorized into four modes: Mode-1 (grid-connected rectification), Mode-2 (PV charging and grid-connected rectification), Mode-3 (PV charging) and Mode-4 (grid-connected inversion). A set of variables I_{DMD}, $I_{DMD\text{-}max}$, $V_{DC\text{-}1}$, $V_{DC\text{-}2}$, $V_{DC\text{-}3}$, V_B and V_{BH} are used to describe the modes of operation. I_{DMD} represents the distribution transformer load and $I_{DMD\text{-}max}$ represents the peak load condition of the transformer. V_{DC} is the voltage at the dc bus. $V_{DC\text{-}1}$, $V_{DC\text{-}2}$ and $V_{DC\text{-}3}$ are the three chosen reference voltage levels of the dc bus. V_B and V_{ESU} are the detected battery voltages of the PEV and the ESU. V_{BH} is the battery voltage corresponding to the threshold value of the state-of-charge (T_{SOC}). The charging of PEV should be terminated once the battery voltage V_B is equal to V_{BH}. Figure 2.12 shows the direction of power flow during various modes of operation of the charging station.

The four modes of operation are described as follows:

Mode-1: $V_{DC} < V_{DC\text{-}1}$: *Grid-connected rectification*

Case-1: $V_{DC} < V_{DC\text{-}1}$ *and* $I_{DMD} < I_{DMD\text{-}max}$

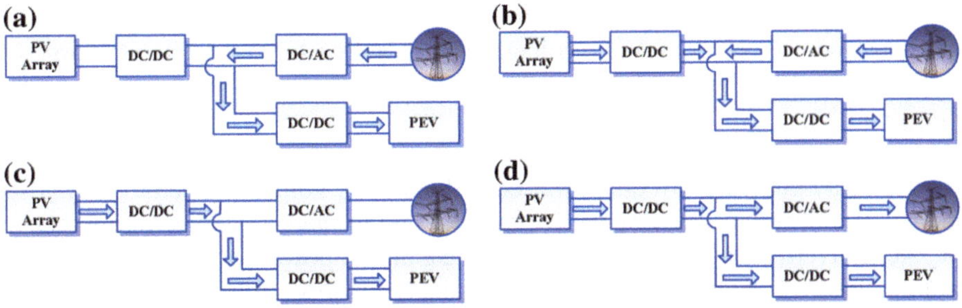

Fig. 2.12 Direction of power flow during the operation modes

In this mode the photovoltaic system does not generate any power either due to low radiation or bad weather conditions. The DC/DC boost converter is isolated and the power required to charge the PEV is provided by the grid. Anytime during this mode if the dc link voltage exceeds $V_{DC\text{-}1}$, the control shifts to Mode-2. The DC/DC buck converter regulates the output voltage to charge the PEV. As the grid is at off peak, it continues to supply power till the vehicle is completely charged. The controller terminates the charging of PEV by disabling the DC/DC buck converter when V_B exceeds V_{BH} and the grid supplies power to charge the battery pack in the ESU.

Case-2: $V_{DC} < V_{DC\text{-}1}$ *and* $I_{DMD} \geq I_{DMD\text{-}max}$

This mode is similar to Case-1 but with an increase in local demand on the distribution transformer. In order to reduce the stress on the grid, the charging of PEV is terminated temporarily by de-activating the grid-connected bi-directional DC/AC converter. As the distribution transformer is relieved from the additional burden of charging the PEV, it can continue supplying power to the local loads during the peak time. During this period the PEV can be charged by the ESU if the stored energy is sufficient to cater the needs of PEV charging. Once the grid is back to off peak condition (i.e. $I_{DMD} < I_{DMD\text{-}max}$) the charging of the PEV is restored and the controller monitors its charging.

Mode-2: $V_{DC\text{-}1} \leq V_{DC} < V_{DC\text{-}2}$: *PV charging and grid-connected rectification*

In this mode the power generated by the photovoltaic system is less than the power required to charge the PEV. Therefore all the power generated by the PV is transferred to the PEV and the deficit is supplied by the grid. The dc link voltage varies with the change in irradiation. This instantaneous change in the dc link voltage is sensed by the controller to generate an equal voltage at the output of the DC/AC bi-directional converter through the process of rectification. If at any point I_{DMD} exceeds $I_{DMD\text{-}max}$ the bi-directional DC/AC converter is isolated from the grid. The PV system continues charging the PEV whereas the grid caters the peak load demand.

Mode-3: $V_{DC\text{-}2} \leq V_{DC} < V_{DC\text{-}3}$: *PV charging mode*

In this mode the PV system generates all the power required to charge the PEV. As the grid does not supply any power it is isolated by the bi-directional DC/AC converter. The controller ensures that the PEV is not over charged by terminating its charging once V_B exceeds V_{BH} (voltage corresponding to 95 % state of charge of the PEV battery). This mode occurs as long as the dc link voltage is in between $V_{DC\text{-}2}$ and $V_{DC\text{-}3}$.

Mode-4: $V_{DC\text{-}link} \geq V_{DC\text{-}3}$: *PV charging mode and Grid inversion mode*

The PV array generates excess power once the dc link voltage exceeds $V_{DC\text{-}3}$. This additional power generated by the PV array is sent to the grid via the bi-directional DC/AC converter. Once the PEVs are charged, all the power from the PV source is sent to the grid. The mode then resembles normal operation of PV generation systems.

2.6.2 Control Description

2.6.2.1 DC/DC Boost Converter

The control method for DC/DC boost converter is summarized in Fig. 2.13. A single phase boost stage is used to boost the PV voltage and track the MPP of the panel. To track the MPP, input voltage (V_{PV}) and input current (I_{PV}) are sensed. The two values are then used by the MPPT algorithm. The MPPT is realized using an outer voltage loop that regulates the input voltage i.e. panel voltage by modulating the current reference for the inner current loop of the boost stage.

Two 2-pole 2-zero controllers, $G_V(S)$ and $G_I(S)$ are used to close the inner DC-DC boost current loop and the outer input voltage loop. MPPT algorithm provides reference input voltage, V_{MPPT} to the boost stage to enable panel operation at maximum power point. The sensed input voltage is compared with the voltage command (V_{MPPT}), generated by MPPT controller, in the voltage control loop. The voltage controller output, $I_{boostsw\_Ref}$ is then compared with the output current ($I_{boostsw}$) feedback in the current controller. The current loop controller output determines the PWM duty cycle so as to regulate the input voltage indirectly.

2.6.2.2 DC/AC Inverter

The control method for grid-connected DC/AC converter is shown in Fig. 2.14. This stage uses two nested control loops—an outer voltage loop and an inner current loop. V_{dc_Ref} is the reference voltage for the DC link, V_{DC} is the detected

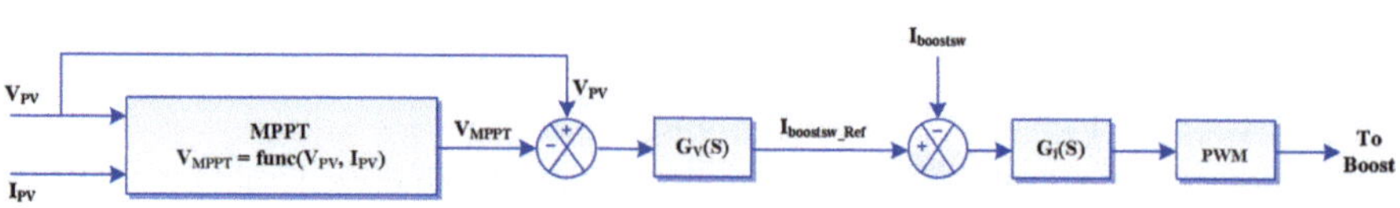

Fig. 2.13 Control diagram of DC/DC boost converter

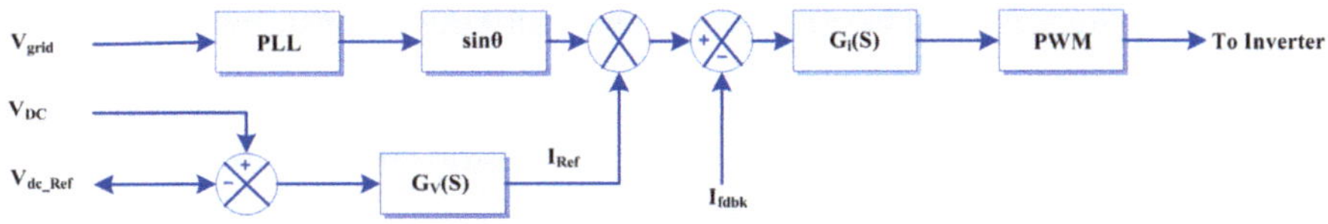

Fig. 2.14 Control diagram of DC/AC inverter

DC link voltage, V_{grid} is the voltage at the secondary of the distribution transformer, θ is the grid phase angle, I_{REF} is the reference current for the DC/AC converter generated by the voltage loop and I_{fdbk} is the current fed into the grid by the DC/AC converter.

Two PID controllers, $G_V(S)$ and $G_i(S)$ are used to close the outer voltage loop and the inner current loop. The voltage loop generates the reference command (I_{Ref}) for the current loop as increasing the current command will load the stage and hence cause a drop in the DC link voltage the sign for reference and the feedback are reversed. The current command is then multiplied by the AC angle to get the instantaneous current reference. Since the inverter is grid connected the grid angle is provided by the PLL. The instantaneous current reference is then used by the current compensator along with the feedback current (I_{fdbk}) to provide duty cycle for the full bridge inverter.

2.6.2.3 DC/DC Buck Converter

The control method for DC/DC buck converter for PEV charging is based on V_B, V_{BH}, I_{DMD} and $I_{DMD\text{-}max}$ as shown in Fig. 2.15. V_B is the detected battery voltage, V_{BH} is the battery voltage corresponding to 95 % SOC. I_{DMD} is the load on the distribution transformer and $I_{DMD\text{-}max}$ represents the peak load condition. The control mode is determined by the detected battery voltage of the PEV and the loading condition of the distribution transformer. The charging of the PEV is turned

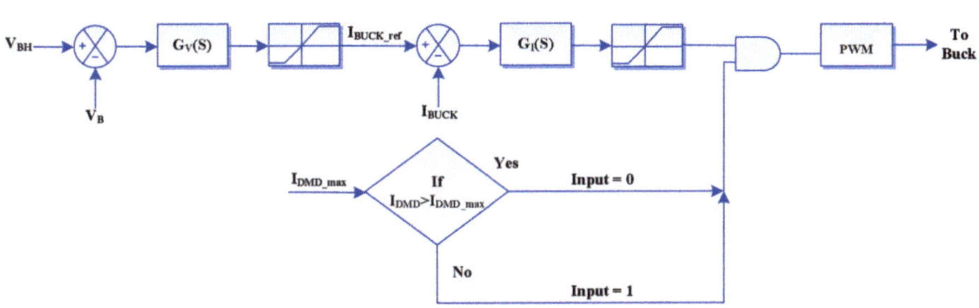

Fig. 2.15 Control diagram of DC/DC buck converter

off once the battery voltage reaches V_{BH} or the distribution transformer reaches the peak load condition.

2.6.3 Simulation Studies

In order to validate the proposed control algorithm simulations were done in Matlab Simulink using the simpowersystems toolbox. The reference dc bus voltages i.e. $V_{DC\text{-}1}$, $V_{DC\text{-}2}$ and $V_{DC\text{-}3}$ are set at 50, 250 and 350 V. The reference dc link voltage levels are selected based on a training mode wherein the PEV load is kept constant and the solar irradiation is allowed to vary in steps. The values of $I_{DMD\text{-}max}$ and T_{soc} are set at 80 A (peak to peak) and 95 %. Toyota Prius plug-in hybrid has been chosen as the PEV which has a total battery capacity equal to 4.5 kWh and nominal voltage equal to 48 V. The rms value of AC grid voltage is 240 V. A PV panel of rating 5.5 kW has been modelled taking the battery capacity of the PEV into consideration. The reference dc bus voltages have been chosen taking into consideration the change in sun conditions from early morning to late evening (Fig. 2.10). As the dc bus voltage varies, the source from which the PEV is charged also varies accordingly. Simulation results describing the transitions between various modes are shown below.

Figure 2.16 shows the transition of the grid from off peak to on peak when the charging station is operating in mode 1. The loading condition is accessed by measuring the current (I_{DMD}) on the secondary side of the distribution transformer. Initially the grid is at off peak and hence the AC grid delivers the power required to charge the PEV and other local loads. As shown in Fig. 2.16, from 1.5 to 2.0 s the

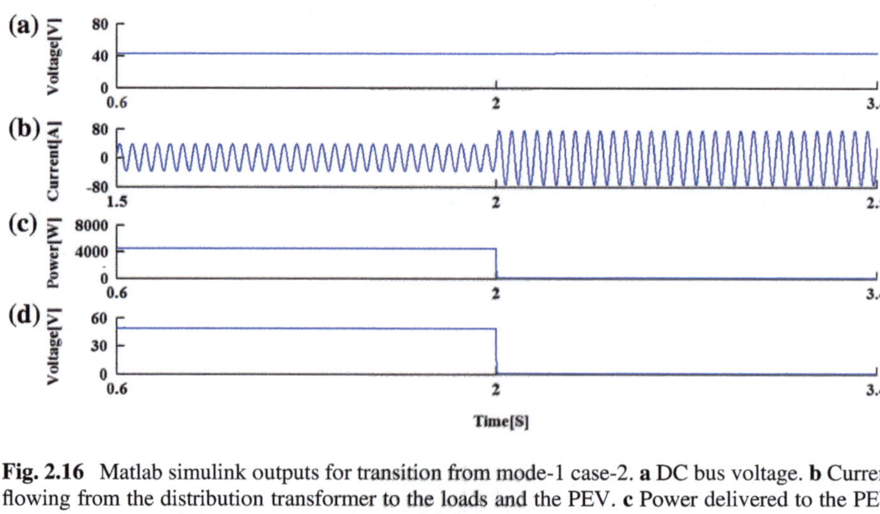

Fig. 2.16 Matlab simulink outputs for transition from mode-1 case-2. **a** DC bus voltage. **b** Current flowing from the distribution transformer to the loads and the PEV. **c** Power delivered to the PEV (charging power). **d** Output voltage of the DC/DC buck converter

current flowing in the secondary side of the distribution transformer is less than 80 A. With the increase in utility load at 2.0 s, I_{DMD} exceeds 80 A ($I_{DMD\text{-}max}$). The charging of the PEV is terminated when the current flowing from the distribution transformer, I_{DMD} exceeds $I_{DMD\text{-}max}$. This is done to reduce the stress being imposed on the AC grid during the peak time. Hence the power consumed by the PEV reduces to zero during the peak time as shown in the figure.

The simulation results for the transition from mode 2 to mode 3 are shown in Fig. 2.17. During the initial stages the dc bus voltage is less than 250 V and grid continues to supply the deficit power to charge the PEV. Once the dc bus voltage exceeds 250 V, the PV system alone caters the charging of PEV. The power flowing from the PV and the Power Grid is shown in Fig. 2.17. As shown in the figure, the deficit power of 1,000 W to charge the PEV is supplied by the grid in mode-2 and it does not supply any power in mode-3 as the PV alone caters to the demand of the PEV.

The transition from mode 3 to mode 4 is shown in Fig. 2.18. With the dc bus voltage exceeding 350 V there is an increase in power flowing from the PV in mode 4. The PV system feeds this excess power to the grid in addition to charging the PEV. The sinusoidal output of the DC/AC bi-directional converter shows that it acts as an inverter in this case. In order to maintain the energy balance the dc link voltage is kept constant at 360 V. Finally Fig. 2.19 shows the termination of the vehicle charging when SOC = T_{soc}.

The simulation results validate the modes of operation and the control algorithm described in this section. As described in Sect. 1.5.1, the modes of operation change due to the change in the dc bus voltage which in turn changes due to the change in the irradiation levels according to the time of the day.

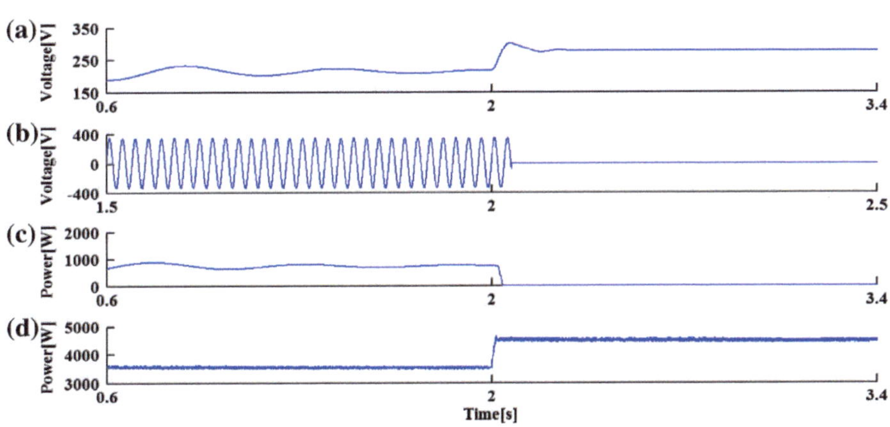

Fig. 2.17 Matlab simulink outputs for transition from mode 2 to mode 3. **a** DC bus voltage. **b** Voltage of the grid. **c** Power delivered by the grid. **d** Power delivered by the PV array

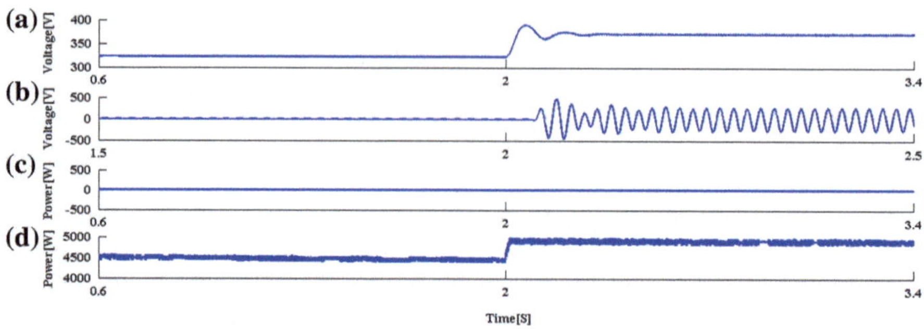

Fig. 2.18 Matlab simulink outputs for transition from mode 3 to mode 4. **a** DC bus voltage. **b** Output voltage of the DC/AC bi-direction converted (inverter). **c** Power delivered by the grid. **d** Power delivered by the array

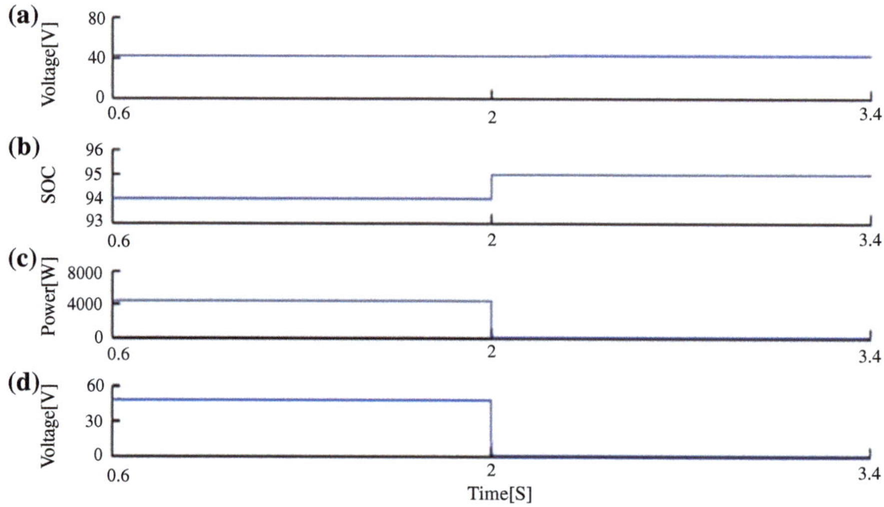

Fig. 2.19 Matlab simulink outputs for transition in state-of-the-charge during mode-1. **a** DC bus voltage. **b** State-of-charge of the PHEV battery. **c** Charging power delivered to the PHEV. **d** Output voltage of the DC/DC buck converter

2.6.4 Experimental Verification

To verify the practical feasibility and effectiveness of the proposed control strategies experimental tests have been carried out in the laboratory. C2000 microcontroller, TMS320F28035 by TI (Texas Instruments) is used to generate all the required control signals.

Figure 2.20 shows the experimental setup of the system. The components include the Solar Explorer Kit by TI (Texas Instruments), power pole board in buck configuration by Hirel, an isolation transformer and a battery. The DC/DC boost

Fig. 2.20 Experimental setup

converter for the PV stage and the inverter is a part of the solar explorer kit; and the DC/DC buck converter for PEV charging is the Power-Pole board from Hirel. A synchronous buck boost stage which is integrated on the board (solar explorer kit) is used to emulate the PV panel. In the place of a PEV a 9 V 1,200 mAh battery is used. By changing the value of irradiation different modes of operation are emulated. Since this is a scaled down version the dc link reference voltage levels are chosen as $V_{DC\text{-}1} = 15$ V, $V_{DC\text{-}2} = 20$ V and $V_{DC\text{-}3} = 25$ V. The value of $I_{DMD\text{-}max}$ is chosen as 1.5 A. Depending on the reference voltage levels the different modes of operation are classified as follows:

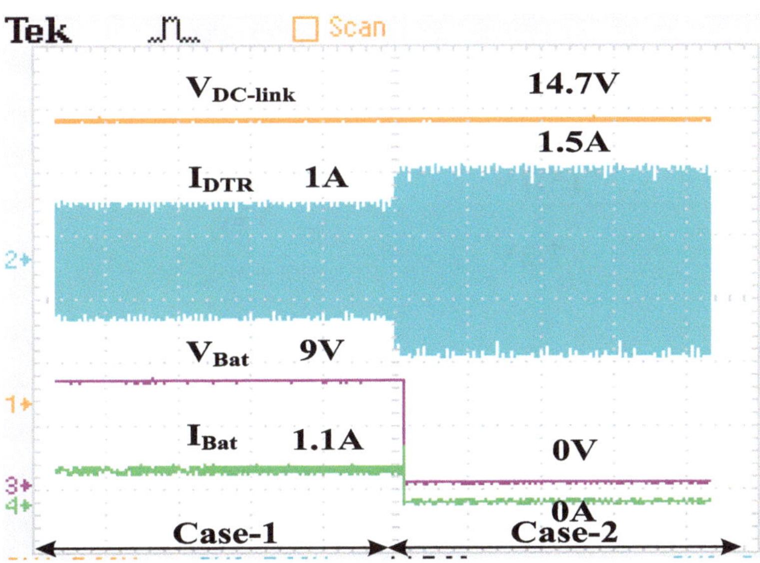

Fig. 2.21 Experimental outputs describing the loading of distribution transformer in Mode-1

$$V_{dc-Link} < 15\,V - \text{Mode-1}$$
$$15\,V < V_{dc-Link} < 20\,V - \text{Mode-2}$$
$$20\,V < V_{dc-Link} < 25\,V - \text{Mode-3}$$
$$V_{dc-Link} > 25\,V - \text{Mode-4}$$

Experimental tests have been carried out in terms of steady-state performance and transient-performance between different modes and the results are provided below. Figure 2.21 through 2.25 explain the experimental results for the various modes of operation.

Experimental results for Mode-1 are shown in Fig. 2.21. With the increase in the loading of distribution transformer, I_{DMD} increases from 1 to 1.5 A as shown in Fig. 2.21 and accordingly the PEV is turned off so that the grid can cater to other loads without overloading the distribution transformer (assuming that $I_{DMD\text{-}max} = 1.5$ A). The turning-off of the PEV is illustrated by the fact that V_B and I_B go to zero with the increase in distribution transformer loading. This is done by generating a duty cycle of zero for the buck converter switch.

Experimental results for the transition between Mode-2 and Mode-3 are shown in Fig. 2.22. In the initial state, the dc link voltage is around 15.7 V and current flows from both the PV as well as the grid to charge the PEV. Once the dc link voltage increases to 22.4 V (Mode-3) no power is drawn from the grid.

Transition from Mode-3 to Mode-4 is shown in Fig. 2.23. With the change in dc link voltage from 22.4 to 29.9 V the bi-directional converter goes from off-state to

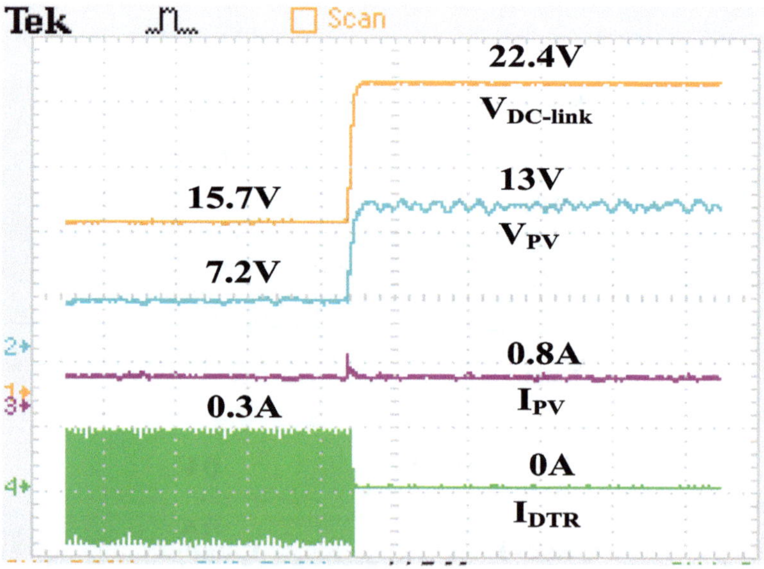

Fig. 2.22 Experimental outputs for transition from Mode-2 to Mode-3

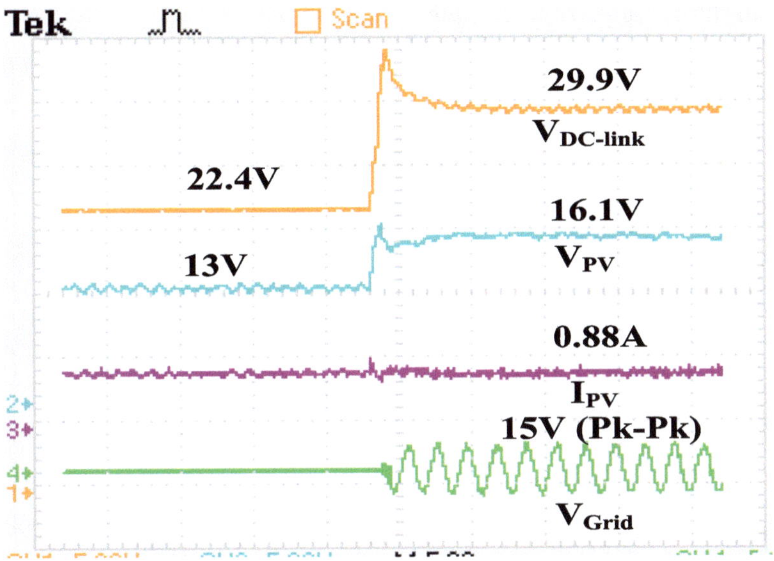

Fig. 2.23 Experimental outputs for transition from Mode-3 to Mode-4

on-state. Mode-4 resembles the normal operation of a grid connected PV system. In this case the battery has been completely charged and hence the entire power generated by the PV is delivered to the grid. Figure 2.24 shows the steady state experimental results of Mode-4. The dc link voltage is 29.9 V and the output

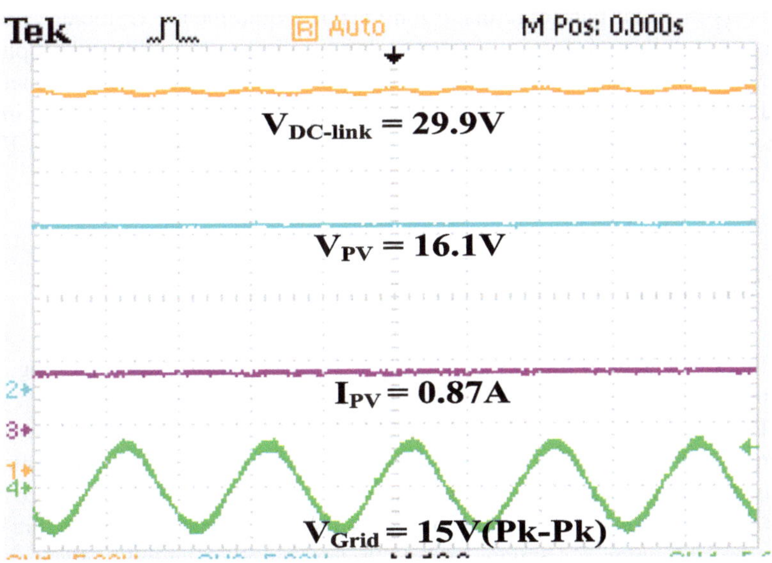

Fig. 2.24 Experimental outputs Mode-4

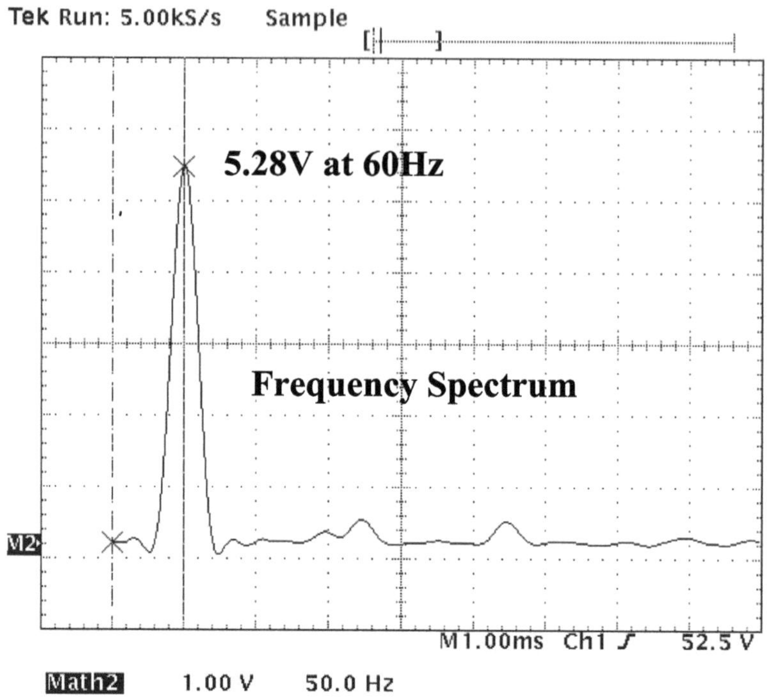

Fig. 2.25 FFT of inverter output voltage

voltage of the inverter is a sine wave. A high switching frequency along with LCL filter meets the total harmonic distortion (THD) requirements. Unipolar switching strategy was followed for inverter switching and the switching takes place at 20 kHz. The inverter switching at 20 kHz together with the LCL filter generates a filtered single phase AC output. The total harmonic distortion (THD) of the inverter output voltage is calculated to be 5.4 %. Figure 2.25 shows the Fast Fourier Transform (FFT) of the inverter output voltage.

2.7 Conclusion

To mitigate the loading on distribution transformers due to PEV charging, smart charging strategies coupled with renewable energy resources are the need of the hour. This chapter discussed the current state of the infrastructure for PV powered charging facilities for PEVs. Several power electronic topologies are presented and compared. Control strategies are reviewed for residential and workplace based photovoltaic charging. The chapter proposed a charging station architecture based on distributed topology. A unique control strategy based on dc link voltage sensing, which decides the direction of power flow is presented and the various modes of operation have been described. The practical feasibility and effectiveness of the proposed control strategy has been validated by simulation and experimental results. The proposed control method based on the change in dc link voltage level due to the change in irradiation of the sun, is simple and unique. The energy management algorithm facilitates charging of the PEVs using minimum energy from the utility with a kind of demand management to improve the energy efficiency. Smart charging techniques like the one proposed in this chapter will help avoid major expense to upgrade distribution transformers and other substation equipment with the increase in PEV loads on the distribution system.

References

1. Kempton W, Tomic J (2005) Vehicle-to-grid power implementation: from stabilizing the grid to supporting large-scale renewable energy. J Power Sources 144(1):280–294
2. EPRI (2012) Tennessee valley authority smart modal area recharge terminal (SMART) station project. http://www.epri.com/abstracts/Pages/ProductAbstract.aspx?ProductId=000000000001026583. Accessed Nov 2012
3. Gorton M (2011) Solar-powered electric vehicle charging stations sprout up nationally. http://www.renewableenergyworld.com/rea/news/article/2011/11/solar-powered-electric-vehicle-charging-stations-sprout-up-nationally. Accessed Nov 2011
4. Motavalli J (2013) Here now: the world's first portable, self-contained solar EV charging station. http://www.plugincars.com/here-now-worlds-first-portable-self-contained-solar-ev-charging-station-128015.html. Accessed Aug 2013
5. Ayre J Solar-integrated EV fast charging station (eco station) gets CODA energy storage system. http://cleantechnica.com/2013/09/02/solar-integrated-ev-fast
6. Giges NS (2012) Wind-powered charging stations coming soon. https://www.asme.org/engineering-topics/articles/renewable-energy/wind-powered-charging-stations-coming-soon. Accessed Oct 2012
7. Liu K, Makaran J (2009) Design of a solar powered battery charger. In: Electrical power and energy conference (EPEC), 22–23 Oct 2009
8. Neumann H, Schar D, Baumgartner F (2012) The potential of photovoltaic carports to cover the energy demand of road passenger transport. Prog Photovolt Res Appl 20:639–649
9. Ingersoll JG, Perkins CA (1996) The 2.1 kW photovoltaic electric vehicle charging station in the city of Santa Monica, California. In: The twenty fifth IEEE photovoltaic specialists conference, 13–17 May 1996
10. Locment F, Sechilariu M, Forgez C (2010) Electric vehicle charging system with PV grid-connected configuration. In: IEEE vehicle power and propulsion conference (VPPC), 1–3 Sept 2010
11. Letendre S (2009) Solar electricity as a fuel for light vehicles. In: Proceedings of the 2009 American solar energy society annual conference, Boulder, CO
12. Tulpule PJ, Marano V, Yurkovich S, Rizzoni G (2013) Economic and environmental impacts of a PV powered workplace parking garage charging station. J Appl Energy 108:323–332
13. Birnie III DP (2009) Solar-to-vehicle (S2 V) systems for powering commuters of the future. J Power Sources 186(2):539–542
14. Zhang Q, Tezuka T, Ishihara KN, Mclellan BC (2012) Integration of PV power into future low-carbon smart electricity systems with EV and HP in Kansai Area, Japan. J Renew Energy 44:99–108
15. Denholm P, Kuss M, Margolis RM (2012) Co-benefits of large scale plug-in hybrid electric vehicle and solar PV deployment. J Power Sources 236:350–356

16. National Houshold Travel Survey (NHTS) (2009). Summary of travel trends-2009. http://nhts.ornl.gov/2009/pub/stt.pdf. Accessed Oct 2009
17. Rutherford MJ, Yousefzadeh V (2011) The impact of electric vehicle battery charging on distribution transformers. In: Twenty-sixth annual IEEE applied power electronics conference and exposition (APEC), 2011
18. Green RC, Wang L, Alam M (2011) The impact of plug-in hybrid electric vehicles on distribution networks: a review and outlook. J Renew Sustain Energy Rev 15(1):544–553
19. Shao SN, Pipattanasomporn M, Rahman S (2009) Challenges of PHEV penetration to the residential distribution network. In: IEEE power and energy society general meeting, 26–30 July 2009
20. Zhao J, Kucuksari S, Mazhari E, Son YJ (2013) Integrated analysis of high-penetration PV and PHEV with energy storage and demand response. J Appl Energy 112:35–51
21. Erol-Kantarci M, Mouftah H (2011) Management of PHEV batteries in the smart grid towards a cyber-physical power infrastructure. In: IEEE workshop on design, modeling and evaluation of cyber physical systems, July 2011
22. Carli G, Williamson SS (2013) Technical considerations on power conversion for electric and plug-in hybrid electric vehicle battery charging in photovoltaic installations. IEEE Trans Power Electron 28(12):5784–5792
23. Giannouli M, Yianoulis P (2012) Study on the incorporation of photovoltaic systems as an auxiliary power source for hybrid and electric vehicles. J Sol Energy 86(1):441–451
24. Li X, Lopes LAC, Williamson SS (2009) On the suitability of plug-in hybrid electric vehicle (PHEV) charging infrastructures based on wind and solar energy. In: IEEE power and energy society general meeting, 26–30 July 2009
25. Hadley, SW (2007) Evaluating the impact of plug-in hybrid electric vehicles on regional electricity supplies. In: Bulk power system dynamics and control—VII. Revitalizing operational reliability, Aug 2007
26. Yan Q, Kezunovic M (2012) Impact analysis of electric vehicle charging on distribution system. In: North American power symposium (NAPS), Sept 2012
27. Moghe R, Kreikebaum F, Hernandez JE, Kandula RP, Divan D (2011) Mitigating distribution transformer lifetime degradation caused by grid-enabled vehicle (GEV) charging. In: IEEE energy conversion congress and exposition (ECCE), Sept 2011
28. Masoum AS, Deilami S, Moses PS, Abu-Siada (2010) A impacts of battery charging rates of plug-in electric vehicle on smart grid distribution systems. In: IEEE PES innovative smart grid technologies conference Europe (ISGT Europe), Oct 2010
29. Qian K, Zhou C, Allan M, Yuan Y (2010) Load model for prediction of electric vehicle charging demand. In: International conference on power system technology (POWERCON), Oct 2010
30. Papadopoulos P, Skarvelis-Kazakos S, Grau I, Cipcigan LM, Jenkins N (2010) Predicting electric vehicle impacts on residential distribution networks with distributed generation. In: IEEE vehicle power and propulsion conference (VPPC), Sept 2010
31. Jian L, Xue H, Xu G, Zhu X, Zhao D, Shao Z (2013) Regulated charging of plug-in hybrid electric vehicles for minimizing load variance in household smart microgrid. IEEE Trans Industr Electron 60(8):3218–3226
32. Osawa M, Yoshimi K, Yamashita D, Yokoyama R, Masuda T, Kondou H, Hirota T (2012) Increase the rate of utilization of residential photovoltaic generation by EV charge-discharge control. In: IEEE innovative smart grid technologies—Asia (ISGT Asia), May 2012
33. Chukwu UC, Mahajan SM (2014) V2G parking lot with pv rooftop for capacity enhancement of a distribution system. IEEE Trans Sustain Energy 5(1):119–127
34. Geiles TJ, Islam S (2013) Impact of PEV charging and rooftop PV penetration on distribution transformer life. In: IEEE power and energy society general meeting (PES), 21–25 July 2013
35. Robalino DM, Kumar G, Uzoechi LO, Chukwu UC, Mahajan SM (2009) Design of a docking station for solar charged electric and fuel cell vehicles. In: International conference on clean electrical power, 9–11 June 2009

36. Alamri BR, Alamri AR (2009) Technical review of energy storage technologies when integrated with intermittent renewable energy. In: International conference on sustainable power generation and supply, SUPERGEN, 6–7 April 2009
37. Gurkaynak Y, Khaligh A (2009) Control and power management of a grid connected residential photovoltaic system with plug-in hybrid electric vehicle (PHEV) load. In: Twenty-fourth annual IEEE applied power electronics conference and exposition (APEC), 15–19 Feb 2009
38. Nagarajan A, Shireen W (2010) Grid connected residential photovoltaic energy systems with plug-in hybrid electric vehicles (PHEV) as energy storage. In: IEEE power and energy society general meeting, 25–29 July 2010
39. Oviedo RM, Fan Z, Gormus S, Kulkarni P (2014) A residential PHEV load coordination mechanism with renewable sources in smart grids. J Electr Power Energy Syst 55:511–521
40. Gurkaynak Y, Li Z, Khaligh A (2009) A novel grid-tied, solar powered residential home with plug-in hybrid electric vehicle (PHEV) loads. In: IEEE vehicle power and propulsion conference VPPC 09, 7–10 Sept 2009
41. Roggia L, Rech C, Schuch L, Baggio JE, Hey HL, Pinheiro JR (2011) Design of a sustainable residential microgrid system including PHEV and energy storage device. In: Proceedings of the 2011—14th European conference on power electronics and applications (EPE 2011), 30 Aug–1 Sept 2011
42. Sheng S, Hsu CT, Li P, Lehman B (2013) Energy management for solar battery charging station. In: IEEE 14th workshop on control and modeling for power electronics (COMPEL), 23–26 June 2013
43. Chen Z, Liu N, Xiao X, Lu X, Zhang J (2013) Energy exchange model of PV-based battery switch stations based on battery swap service and power distribution. In: IEEE vehicle power and propulsion conference (VPPC), Oct 2013
44. Choe GY, Kim JS, Lee BK, Won CY, Lee TW (2010) A bi-directional battery charger for electric vehicles using photovoltaic PCS systems. In: IEEE vehicle power and propulsion conference (VPPC), 1–3 Sept 2010
45. Vaidya M, Stefanakos EK, Krakow B, Lamb HC, Arbogast T, Smith T (1996) Direct DC-DC electric vehicle charging with a grid connected photovoltaic system. In: Twenty fifth IEEE photovoltaic specialists conference, 13–17 May 1996
46. Gamboa G, Hamilton C, Kerley R, Elmes S, Arias A, Shen J, Batarseh I (2011) Control strategy of a multi-port, grid connected, direct-DC PV charging station for plug-in electric vehicles. In: IEEE energy conversion congress and exposition (ECCE), 12–16 Sept 2011
47. Hamilton C, Gamboa G, Elmes J, Kerley R, Arias A, Pepper M, Shen J, Batarseh I (2010) System architecture of a modular direct-DC PV charging station for plug-in electric vehicles. In: 36th annual conference on IEEE industrial electronics society (IECON), 7–10 Nov 2010
48. Schönberger J, Duke R, Round SD (2006) DC-Bus signaling: a distributed control strategy for a hybrid renewable nanogrid. IEEE Trans Industr Electron 53(5):1453–1460
49. Sun K, Zhang L, Xing Y, Guerrero JM (2011) A distributed control strategy based on dc bus signaling for modular photovoltaic generation systems with battery energy storage. IEEE Trans Power Electron 26(10):3032–3045
50. Jin C, Wang P, Xiao J, Tang Y, Choo FH (2014) Implementation of hierarchical control in DC microgrids. IEEE Trans Industr Electron 61(8):4032–4042
51. Preetham G, Shireen W (2012) Photovoltaic charging station for plug-in hybrid electric vehicles in a smart grid environment. Paper presented at IEEE PES innovative smart grid technologies, 16–20 Jan 2012
52. Goli P, Shireen W (2014) PV integrated smart charging of PHEVs based on DC link voltage sensing. IEEE Trans Smart Grid 5(3):1421–1428

Chapter 3
Hierarchical Coordinated Control Strategies for Plug-in Electric Vehicle Charging

Zechun Hu, Yonghua Song and Zhiwei Xu

Abstract Currently in most of the works in the literature, a group of plug-in electric vehicles (PEVs) is controlled by an "aggregator". The aggregator is responsible for making the charging schedule for each PEV and also participates in power system regulation or electricity market bidding. However, practically, to coordinate the charging of large scale PEVs in power system, the diversities in charging infrastructure, PEV types and local operational constraints in the power system should also be well considered. Therefore, hierarchical control of PEVs is regarded as an effective way to achieve charging cost minimization and system operational security. This book chapter introduces hierarchical control frameworks for PEV charging, which includes coordinated charging strategy for charging station (or virtual charging station), coordinated charging strategy for battery swapping station, hierarchical coordinated charging strategy for multiple charging stations and a three level coordinated charging framework for large scale of PEVs. The detailed mathematical formulations for each level operator in the proposed hierarchical control framework, which jointly optimize system load profile and charging costs, are clearly presented. The inter-relationships between various levels of operators in terms of energy transaction and information exchanged are also specified. Finally, case studies are carried out on three cases. The simulations results demonstrate the effectiveness of the hierarchical charging control framework and optimization methods in reducing peak demand and charging costs.

Keywords Charging station · Battery swapping and charging station · Coordinated charging · Hierarchical control

---

Z. Hu (✉) · Y. Song · Z. Xu
Department of Electrical Engineering, Tsinghua University, Beijing
People's Republic of China
e-mail: zechhu@tsinghua.edu.cn

Y. Song
e-mail: yhsong@tsinghua.edu.cn

Z. Xu
e-mail: xu-zw07@mails.tsinghua.edu.cn

S. Rajakaruna et al. (eds.), *Plug In Electric Vehicles in Smart Grids*,
Power Systems, DOI 10.1007/978-981-287-317-0_3

3.1 Introduction

When large scale of PEVs penetrates into the power system, their charging load can be a substantial burden to the operations of existing power systems if their charging is not properly managed. Voltage drops, excessive power losses and overloading of distribution transformer and lines are likely to occur. In some severe situations, additional investment might even be needed for electrical equipment upgrading. There is a breadth of literature to investigate the impacts of PEV charging on power grids under various charging strategies and operational scenarios. Authors of [1] evaluated the impacts of the integration of PEVs on power grid operations and electricity market. Su et al. [2] comprehensively summarized the electrification of transportation. Roe et al. [3] studied various aspects of how PEVs could impact power grid infrastructure. Authors of [4] evaluated the impact of different levels of PEV penetration on incremental investment and energy losses of distribution network.

However, the PEV charging load has flexibility. For example, the vehicles are normally idle for an average of more than 10 h during the night, while the time needed to get them fully charged is only several hours. Therefore, through coordinating the charging of large scale of PEVs, the PEV charging stress on power systems can be effectively mitigated. In addition, coordinated charging is also viewed as a cost efficient approach to help smoothen the system load profile, reduce renewable generation curtailment and reduce charging costs through energy arbitrage. As a result, the researches on coordinated PEV charging have attracted wide interests and a large number of research papers have been published in recent years. Clement et al. [5] proposed a coordinated charging strategy to minimize power losses in distribution systems. Han et al. [6] proposed an optimal V2G aggregator for frequency regulation services. Sundstrom and Binding [7] presented a centralized PEVs charging coordination framework considering interactions among charging service provider (aggregator), retailer and distribution system operator. Richardson et al. [8] proposed a linear optimization model based on network sensitivities to maximize the total charging power while satisfying network voltage and thermal limits. Authors of [9] proposed three coordinated PEV charging methods to minimize the distribution system power losses, load factor or load variance. Wu et al. [10] formulated a model to minimize energy costs of aggregator in the day-ahead market based on electricity prices and charging needs predictions. Vagropoulos and Bakirtzis [11] sought to centrally find the optimal bidding strategy for PEV aggregators both in energy and ancillary service markets and maximize aggregator's profits by charging coordination. Luo et al. [12] proposed a two-stage optimization model to jointly minimize the peak load and the load fluctuation. Yao et al. [13] presented a hierarchical decomposition method to coordinate the charging/discharging of PEVs. Qi et al. [14] applied Lagrangian relaxation method to optimize the charging schedule of PEVs across multiple charging stations.

As the number of PEV charging stations increases to beyond hundreds, it is apparent that the single level centralized control as reported in most of the literature may not be applicable for large-scale PEV coordinated charging control. Thus, distributed control strategies are proposed. In [15], Wen et al. developed a decentralized charging coordination strategy based on distributed alternating direction method of multipliers algorithm. In [16], Hamid et al. proposed a distributed recharging rate control algorithm, which combines the objectives of regulating frequency and improving the utilization of electric generators. Authors of [17] formulated a decentralized control strategy to provide frequency regulation service for power system operation. In [18], Ma et al. applied the game theory to coordinate the charging of PEVs. Gan et al. [19] proposed a decentralized algorithm to coordinate the PEV charging loads to fill the valleys in load profiles. Sheikhi et al. [20] proposed game theory based optimal decentralized control strategies.

However, the single level centralized coordinated charging and distributed coordinated control strategies proposed in most of the literature have not provided an adequate and complete solution to realize the coordinated charging of large scale PEVs. On one hand, when the number of PEVs grows very large in the future, it will be extremely difficult for one single level centralized coordinated PEV charging control algorithm to get an optimal solution in reasonable computation time. On the other hand, though decentralized strategies mentioned before can efficiently handle coordinated charging control problems of large-scale PEVs, their control results may be not as good as those of the centralized control strategies because of lack of global information.

Since a single PEV's battery capacity is limited, it is not easy or economic for PEVs to participate in the electricity market separately; hence large-scale PEVs tend to be jointly aggregated by one or several aggregators in the electricity market. When the number and the range of aggregated PEVs grow, the charging of PEVs aggregated by a single aggregator may have significant impacts both on the transmission systems and the distribution systems. Yet till now, the comprehensive coordinated PEV charging schemes at all levels (e.g., transmission system, distribution system, and charging stations) have not been well studied. Especially for vertically regulated power market, such as China [21], where the state-of-the-art power grid and the communication infrastructure are applicable to centralized operations, the implementation of hierarchical coordinated control of large scale PEV charging may be more practically possible. On the demand side, some utilities (e.g. Shenzhen, China) specially design TOU tariff incentives for PEVs charging, which make charging coordination more attractive to investors. There is a need to propose the optimal solution that is important to various stakeholders in PEV charging ecosystem, including grid operators, utility companies, aggregators, charging station owners, and PEV owners, etc. The major contribution of this book chapter to the literature is to propose a three-level hierarchical framework for coordinated PEV charging. The energy transaction and information exchange between various levels are clearly presented. Moreover, by applying charging demand aggregation techniques at the station and the distribution level operators, we effectively reduce the computation and communication requirements, which make our three-level hierarchical framework scalable to large scale PEV coordinated charging control.

3.2 Concept and Structure for Hierarchical Coordinated Charging

As mentioned in the introduction part, a single level centralized coordinated charging and distributed coordinated control strategies proposed in most of the literature have not provided an adequate and complete solution to realize the coordinated charging of large scale PEVs. In this book chapter, we endeavor to present a hierarchical control framework that is (1) scalable to consider various kinds of charging infrastructure, PEV types and power system operational constraints and (2) capable of coordinated controlling large scale PEVs in real time. Specifically, the hierarchical coordinated charging framework mainly includes three levels, namely charging station level control, distribution level control and transmission level control.

First, charging stations, battery swapping stations and parking decks with multiple charging points are at the lowest level of our control framework. These places naturally aggregate the PEVs and can serve as idea places to implement coordinated charging. Moreover, in cities where special time of use (TOU) tariffs are designed for PEV charging, charging service providers of these places have incentives to make profit through shifting charging demands to off-peak periods by means of charging coordination. In particular, in this book chapter, we first design coordinated charging strategies for charging stations and battery swapping stations. By dynamically responding to the TOU prices, the station level operator optimizes the charging schedules of PEVs or batteries and effectively realizes better load control and charging cost minimization.

Coordinated charging operations within only individual parking decks or charging stations, which are merely subject to local operational constraints may not be adequate in response to time varying distribution system operational states. In some cases, without coordination across multiple aggregators or charging stations, the occurrence of unexpected overlapped charging hours in multiple parking decks or charging stations could even have more severe negative impacts on power grid operation than uncoordinated charging. Therefore, we further develop hierarchical coordinated charging strategies for multiple aggregators in the distribution systems [22]. In particular, the strategy seeks to (1) coordinate the aggregate charging load of different aggregators or stations to achieve system load controlling and total electricity cost minimization, (2) coordinate the charging of PEVs within each aggregator under the constraints of charging requirements and local transformer capacity limits. In order to facilitate the coordination computation and control in real time, only the information of charging load boundaries of each aggregator is required to be revealed for centralized coordination at the distribution system operator (DSO) in our control framework. This helps different aggregators protect customer charging requirement privacies and reduce the computational burden at the distribution system operator level. At the aggregator level, we design an effective scheduling algorithm to optimally allocate the power to each PEV and thus the efficiency of the overall charging control system is further improved.

As more and more PEVs are integrating into the power systems, the system operator can potentially further exploit the charging flexibilities of large scale PEVs and optimize the overall system operational performances. For example, when grid scale renewable generations are integrated into transmission systems, coordinating the charging flexibility of large scale PEVs with the variability of renewable generations can effectively improve system level renewable energy penetration and reduce energy costs. Therefore, we further present a three-level coordinated charging control framework, where the energy transaction and information exchange between various levels (i.e. transmission, distribution and station) are clearly identified. Specifically, in the day ahead, the day-ahead forecast of the aggregated charging demand of PEVs and the base load profiles are first predicted. A day-ahead aggregate charging load trajectory is then determined for each distribution system operator. In real time, the distribution system operator and station operators dynamically communicate and coordinate with each other with real-time charging requirements and finally achieve three-level charging coordination. Similarly, we apply charging demand aggregation techniques at the station and the distribution level operator and thus the computation and communication requirements are effectively reduced.

In our hierarchical control framework, time is discretized with step size Δ. We assume the power consumption of PEV is constant over Δ. We use subscript $\{i : i \in I\}$ to index DSO, subscript $\{j : j \in J_i\}$ to index aggregator under DSO i (charging station, battery swapping station or parking deck with multiple charging points), subscript $\{k : k \in K_{ij}\}$ to index charging port at aggregator j under DSO i and $\{t : t \in T\}$ to index time step. We further use K_{ij}^t to denote the set of charging ports which are occupied by PEV (battery) at time t. The power consumption of the PEV (battery), which is connected with the k-th charging port of aggregator j under DSO i, at time step t is denoted as $p_{ijk}(t)$ $(\forall k \in K_{ij}^t)$.

In the sequel, this book chapter is organized as follows. Sections 3.3 and 3.4 introduce the charging coordination strategies for charging stations and battery swapping stations, respectively. Section 3.5 presents the hierarchical control framework for multiple aggregators. We further show the three levels coordinated charging framework for large scale of PEVs in Sect. 3.6 and finally conclude this chapter in Sect. 3.7.

3.3 Coordinated Charging Strategy for Charging Station

To begin with, we focus on the coordinated charging strategy for each charging station j, $(j \in J_i)$ which is installed with multiple charging points. In urban areas, most of the PEVs are parked at the parking lots or charging stations instead of private garages for recharging. Charging stations are ideal places for the implementations of PEVs' coordinated charging by monitoring the real-time conditions of the PEVs in stations.

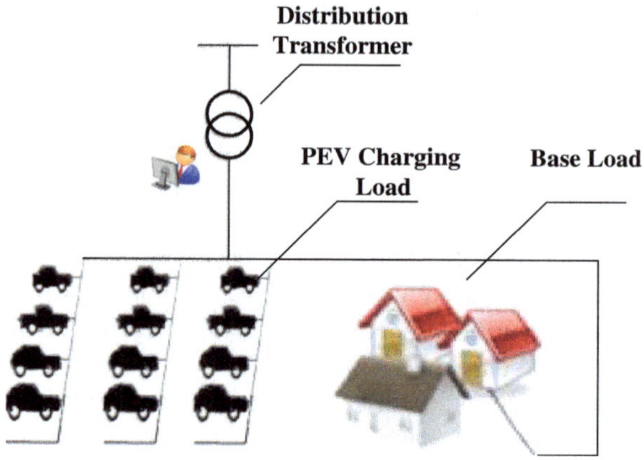

Fig. 3.1 Schematic illustration of charging stations for PEVs

Figure 3.1 is a schematic illustration of charging stations or parking lots with rechargeable piles for PEVs. We call these two kinds of places both "charging station" in general. The charging load and base load are connected to the distribution transformer. For a charging station with independent distribution transformer, the base load is viewed as zero.

Acting as charging service provider, the charging station purchases electricity from the utility at TOU purchase prices. The charging service provider seeks to coordinate the charging of PEVs to off-peak periods to reduce its charging costs.

3.3.1 Control Strategy Overview

Based on the historical load (base load) data of the distribution transformer, the base load could be predicted for each day. $\xi_{ij}(t)$ ($\xi_{ij}(t) \in [0, 1]$) is the ratio of the available capacity for charging to the total capacity of distribution transformer at time step t. For the charging station with its own supplying distribution transformer, $\xi_{ij}(t) = 1, \ \forall t \in T$.

Once an arrival PEV connects to the k-th charging port at time t ($k \in K_{ij}^{t}$), the battery capacity of the PEV B_{ijk} and arrival state of charge (SOC) SOC_{ijk}^{A} are obtained through battery management system (BMS) on board. In order to achieve the coordinated charging of PEVs in the charging station, the customers need to inform the coordinated charging system of the expected parking time duration for charging d_{ijk} and expected SOC when they departs SOC_{ijk}^{D}. At the beginning of each time interval, the coordinated charging control system of the charging station calls the charging optimization program to determine the charging power of the PEVs in station in the following $H_{ij}(t)$ time intervals based on the parking state of the charging station, customers charging needs, available capacity of the distribution transformer for charging and electricity prices. Through this strategy, the operational energy costs of the charging station are minimized; the needs of the customers are satisfied as much as possible and the charging load of the distribution transformer is flattened. Specifically, the charging optimization program involves two stage optimization models.

3.3.2 First Stage Optimization Model

Specifically, the following first stage optimization model is to be minimized at time step t.

$$\min \sum_{l=0}^{H_{ij}(t)-1} \sum_{k \in K_{ij}^t} p_{ijk}(t+l) \times c_{ij}(t+l) \times \Delta + \gamma \times \sum_{k \in E} \alpha_k \tag{3.1}$$

$$\text{s.t.} \sum_{k \in K_{ij}^t} p_{ijk}(t+l) \le A_{ij} \times \xi_{ij}(t+l), \ \forall l \in \{0, 1, \ldots, H_{ij}(t)-1\} \tag{3.2}$$

$$\rho \times \sum_{l=0}^{H_{ijk}(t)-1} p_{ijk}(t+l) \times \Delta + SOC_{ijk}(t) \times B_{ijk} = SOC_{ijk}^D \times B_{ijk}, \forall k \in K_{ij}^t \backslash E \tag{3.3}$$

$$\begin{aligned}(SOC_{ijk}^D - \alpha_k) \times B_{ijk} \le \rho \times \sum_{l=0}^{H_{ijk}(t)-1} p_{ijk}(t+l) \times \Delta + SOC_{ijk}^A \\ \times B_{ijk} \le SOC_{ijk}^D \times B_{ijk}, \forall k \in E\end{aligned} \tag{3.4}$$

$$0 \le p_{ijk}(t+l) \le P_{ijk}^{\max}, \forall k \in K_{ij}^t, \forall l \in \{0, 1, \ldots, H_{ij}(t)-1\} \tag{3.5}$$

where $H_{ij}(t)$ is the length of the planning horizon at time t, which is normally selected as the maximum value of the remaining parking durations of PEVs. $H_{ijk}(t)$ is the length of planning horizon for PEV at charging port k. $c_{ij}(t)$ is the electricity purchase price of charging station at time step t. γ is a penalty coefficient for customer charging demand de-rating. Set E is the index set of the charging ports which are connected with newly arrived PEVs at time t and α_k is charging requirement de-rating factor of PEV at charging port k ($\forall k \in E$). A_{ij} is the capacity of the distribution transformer. ρ is charging efficiency and $SOC_{ijk}(t)$ is the SOC of PEV at charging port k at time t. $P_{ijk}^{\max}$ is the rated power of charging port k.

The first part of the objective function is to minimize the overall energy purchase costs of the charging station over the planning horizon. While the second part is used to minimize the demand de-rating with a relatively large weight to guarantee that the charging need of the arrived customer is satisfied as much as possible.

The constraints (3.2) imply the constraints on capacity of distribution transformer. At any time interval during the planning horizon, the summation of the base load and charging load of PEVs should not exceed the rated capacity of the supplying distribution transformer. The constraints (3.3) suggest the charging requirement constraints of the customers that connected before the last time step, while constraints (3.4) set the charging requirement constraints for the newly connected PEVs. For the newly arrived customer connected to the charging station at the very last time period, the parameter α_k is introduced to guarantee the problem feasibility. Finally, the values of charging power for each PEV are limited in constraints (3.5).

3.3.3 Second Stage Optimization Model

In the first stage optimization model, by responding to the time varying electricity prices dynamically under the constraints of distribution transformer capacity and customers' needs, the charging costs of charging station are minimized. We could further build the second stage optimization model to flatten the charging load without sacrificing the quality of charging service or increasing charging costs that obtained from the first stage optimization.

$$\min \ L_{\max} \tag{3.6}$$

$$\text{s.t.} \ \sum_{l=0}^{H_{ij}(t)-1} \sum_{k \in K_{ij}^t} p_{ijk}(t+l) \times c_{ij}(t+l) \times \Delta \leq C_{\min} \tag{3.7}$$

$$\sum_{k \in K_{ij}^t} p_{ijk}(t+l) + \left(1 - \xi_{ij}(t+l)\right) \times A_{ij} \leq L_{\max}, \forall l \in \left\{0, 1, 2, \ldots, H_{ij}(t) - 1\right\} \tag{3.8}$$

$$\rho \times \sum_{l=0}^{H_{ij}(t)-1} p_{ijk}(t+l) \times \Delta + SOC_{ijk}(t) \times B_{ijk} = SOC_{ijk}^{D} \times B_{ijk}, \forall k \in K_{ij}^{t} \backslash E \tag{3.9}$$

$$\begin{aligned} (SOC_{ijk}^{D} - \sigma_k) \times B_{ijk} &\leq (\rho \times \sum_{l=0}^{H_{ijk}(t)-1} p_{ijk}(t+l) \times \Delta + SOC_{ijk}^{A} \times B_{ijk}) \\ &\leq SOC_{ijk}^{D} \times B_{ijk}, \forall k \in E \end{aligned} \tag{3.10}$$

where L_{max} is the total peak demand over the planning horizon and C_{min} and σ_k is the optimal cost and charging requirement de-rating factor computed at the first stage, respectively.

The objective of this model is to minimize the peak load in the planning horizon. Constraints (3.7) indicate that the costs of the charging station should be maintained while trying to flatten the load. Constraints (3.8) suggest the load of the distribution transformer should not exceed the maximum load. Constraints (3.9) and (3.10) guarantee the customer service level will not be degraded.

The two optimization models formulated above are both linear programming (LP) problems, which can be solved by software package CPLEX efficiently [23].

3.3.4 Flow Chart of Coordinated Charging Control

Based on the models formulated above, the optimal charging power for each PEV can be obtained at the beginning of each time step and the coordinated charging of PEVs is therefore realized. The system updates the state of charging station every Δ amount of time and control orders are generated based on the optimization results. If no PEV enters the charging station during the previous time period, the system will automatically change the state of the charging station based on the results computed previously. Otherwise, the charging strategy will be calculated again at the beginning of the next time interval. If a PEV enters in the middle of the current time interval, the charging states of the other PEVs will not change during this time interval. Based on the descriptions of the control decision process, the detailed coordinated charging control flowchart is illustrated in Fig. 3.2.

3.4 Coordinated Charging Strategy for Battery Swapping and Charging Station

Although PEV sales increase in many countries all over the world, the limited recharging infrastructure is still one of the main obstacles limiting the mass adoption of PEVs. Better Place was the pioneer to build battery swapping networks to refuel PEVs, although it was bankrupt [24]. In China, the electric grid companies put a lot of effort on promoting battery-swapping technologies. Although they are not successful in the private PEV market, battery-swapping technologies win acceptance for public transportation, e.g. public buses and taxies, to a certain extent. Now, there are more than 20 electric bus routes that are refueled by battery swapping stations.

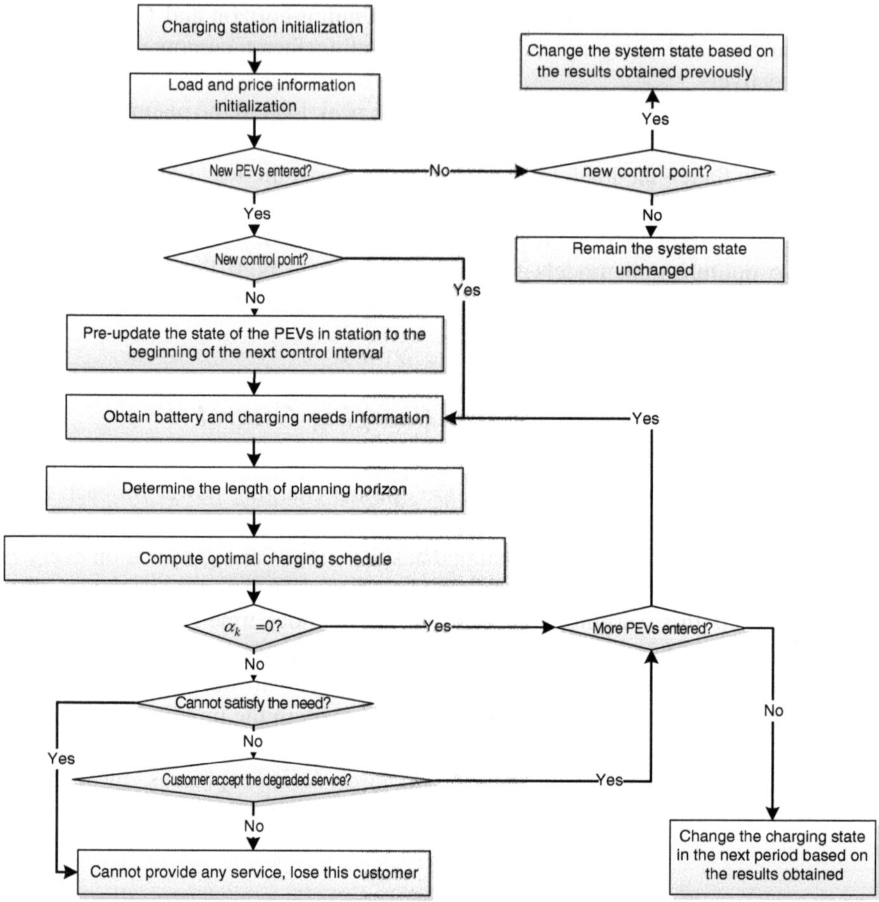

Fig. 3.2 Flowchart of the coordinated charging control for a charging station

The most evident advantage of battery swapping technology is that the used battery of a PEV can be exchanged with a fully charged one within a few minutes even for an electric bus. Furthermore, after the used batteries swapped off the vehicle, their recharging process can be controlled centrally. For the electric bus fleet that is not operating during the night, the batteries within the charging station can be used for coordinated charging or even V2G (vehicle-to-grid). In this section, the coordinated charging strategy for the bus battery swapping and charging station (BBSCS) is briefly introduced.

3.4.1 Operation of Bus Battery Swapping and Charging Station

3.4.1.1 Basic Assumptions

The key assumptions for the operation of a BBSCS are listed as follows:

(1) Electric buses run along a fixed round route, the BBSCS is located close to the starting or ending stop. A BBSCS can be shared by several bus routes.
(2) If the remained energy in the battery of an electric bus is not enough for its next round trip, the battery should be swapped in the BBSCS. The bus driver is responsible for checking the battery SOC.
(3) The swapped batteries with low SOC will be recharged in the BBSCS. The charging power can be controlled.
(4) The battery capacities of electric buses belonging to the same bus route are the same.

3.4.1.2 Battery Swapping Demand Analysis

Considering that the electric bus fleet is operated similar to the conventional bus fleet, the bus departing schedule is arranged based on passenger flow and traffic status. When an electric bus departs, its remaining travel distance available with the energy in the battery must be longer than the length of the bus route. Otherwise, the battery must be replaced with a fully charged one. Thus, daily battery swapping demand can be quantitatively analyzed.

Basic input information for battery demand analysis includes the bus timetable, battery capacity, initial State of Charge (SOC), driving distance, average driving speed, electricity consumption per kilometer and so on. The battery swapping demands only occur during the daily operating hours.

The time of each electric bus k arriving at the BBSCS j after a round trip can be approximately calculated as follows:

$$t_{ijk}^{A} = t_{ijk}^{D} + \frac{L_{ijk}}{V_{ijk}} \times \frac{1}{\Delta} \tag{3.11}$$

where t_{ijk}^{A} and t_{ijk}^{D} are respectively the arrival time and departure time of electric bus k. L_{ijk} is the length of the operating route in km. V_{ijk} is the average driving speed in km/h.

The minimum battery SOC to meet the requirement of a single round trip is about:

$$SOC_{ijk}^{r} = \frac{L_{ijk} \times Q_{ijk}}{B_{ijk}} \times 100\,\% + SOC_{ijk}^{\min} \tag{3.12}$$

where Q_{ijk} and $SOC_{ijk}^{\min}$ are respectively the average energy consumption rate of bus k, in kWh/km, and required minimum state of charge. When the battery SOC of a bus is lower than SOC_{ijk}^{r}, the battery should be replaced with a fully charged one before it departs from the BBSCS for the next trip. So the battery swapping time of each bus within a whole day can be predicted, the battery swapping demands of a bus fleet over time can be calculated by accumulation.

3.4.2 Optimal Charging Strategy Within BBSCS

With the battery swapping demands available, the charging process within the BBSCS can be optimized. Two-stage optimization models are built, which is very similar to those given in Sect. 3.3. The objective of the first stage optimization model is to minimize the total charging cost respecting the battery swapping demands within a whole day.

$$\min \sum_{l=0}^{H_{ij}(t)-1} \sum_{k \in K_{ij}^t} p_{ijk}(t+l) \times c_{ij}(t+l) \times \Delta + \gamma \times \sum_{k \in K_{ij}^t} \alpha_k \tag{3.13}$$

$$\text{s.t.} \sum_{k \in K_{ij}^t} p_{ijk}(t+l) \le A_{ij} \times \xi_{ij}(t+l), \forall l \in \{0, 1, \ldots, H_{ij}(t)-1\} \tag{3.14}$$

$$\begin{aligned} (SOC_{ijk}^D - \alpha_k) \times B_{ijk} \le \Big(\rho \times \sum_{l=0}^{H_{ijk}(t)-1} p_{ijk}(t) \times \Delta + SOC_{ijk}(t) \times B_{ijk}\Big) \\ \le SOC_{ijk}^D \times B_{ijk}, \forall k \in K_{ij}^t \end{aligned} \tag{3.15}$$

$$0 \le p_{ijk}(t+l) \le P_{ijk}^{\max}, \forall k \in K_{ij}^t, \forall l \in \{0, 1, \ldots, H_{ij}(t)-1\} \tag{3.16}$$

The constraints (3.14) and (3.15) are similar to (3.2) and (3.3), respectively. The constraints (3.15) indicate that the charging demand of battery k should be satisfied as much as possible. The charging time horizon of each battery $H_{ijk}(t)$ is decided based on the detailed battery swapping demand analysis.

The second stage optimization model can be similarly built as follows.

$$\min L_{\max} \tag{3.17}$$

$$\text{s.t.} \sum_{l=0}^{H_{ij}(t)-1} \sum_{k \in K_{ij}^t} p_{ijk}(t+l) \times c_{ij}(t+l) \times \Delta \le C_{\min} \tag{3.18}$$

$$\sum_{k\in K_{ij}^t} p_{ijk}(t+l) + \left(1-\xi_{ij}(t+l)\right) \times A_{ij} \leq L_{\max}, \forall l \in \left\{0,1,2,\ \ldots\ ,H_{ij}(t)-1\right\} \tag{3.19}$$

$$\begin{aligned}(SOC_{ijk}^D - \sigma_k) \times B_{ijk} \leq (\rho \times \sum_{l=0}^{H_{ijk}(k)-1} p_{ijk}(t+l) \times \Delta + SOC_{ijk}(t) \times B_{ijk}) \\ \leq SOC_{ijk}^D \times B_{ijk}, \forall k \in K_{ij}^t\end{aligned} \tag{3.20}$$

3.4.3 Dealing with Uncertainties

The above two stage optimization models are built based on the assumption that when and which batteries entering and leaving the BBSCS can be predicted precisely. However, there are many uncertain factors, such as bus driving speed, different electricity consumption rate of each bus, and even unexpected equipment failures. So the optimization models should take the uncertainties into account. One of the solutions to this problem is to solve the two-stage optimization models using the updated data when necessary. Another solution is to keep several fully charged spare batteries, i.e., there will be a few more fully charged batteries than the batteries required during the operating hours of the bus route. This solution can be realized by setting appropriate constraints (3.15) and (3.20).

3.5 Hierarchical Coordinated Charging for Multiple Aggregators

As our discussions earlier, by coordinating the charging load to off-peak periods, cost minimization of individual aggregator can be achieved. From the perspective of the DSO, whereas, a better load profile with low peak-to-average ratio through charging coordination is expected. However, in lack of effective coordination across different aggregators, system load profile might still be undesirable. For example, authors of [25] propose to coordinate the charging of PEVs within one charging station. Through simulations, it is found that if coordination between aggregators is absent, charging rebound effect (another load peak) is likely to occur. Therefore, in this section, we develop a centralized, hierarchical framework to coordinate the charging of PEVs in multiple aggregators [22].

The hierarchical control framework seeks to (1) coordinate the aggregate charging load of different aggregators under a common DSO i with the objective of system load controlling and total electricity cost minimization, (2) coordinate the charging of PEVs within each aggregator considering various local constraints. We also develop techniques to derive the aggregate charging load boundaries of each aggregator to help different aggregators protect their customers charging requirement privacies and reduce the computational burden at the DSO level. Moreover, at the aggregator level, we present an efficient heuristic scheduling algorithm to intelligently allocate the aggregate reference power to each PEV, which further improves the efficiency of the coordinated charging control system.

3.5.1 Hierarchical PEVs Charging System Architecture

Before we proceed to the detailed formulation of our hierarchical control framework, we first present the schematic illustration of the decentralized charging system in an urban area we are focusing here, as depicted in Fig. 3.3. The voltage is stepped down twice by the primary distribution transformer at the substation and local distribution transformers. Each community j $(j \in J_i)$ with multi-family

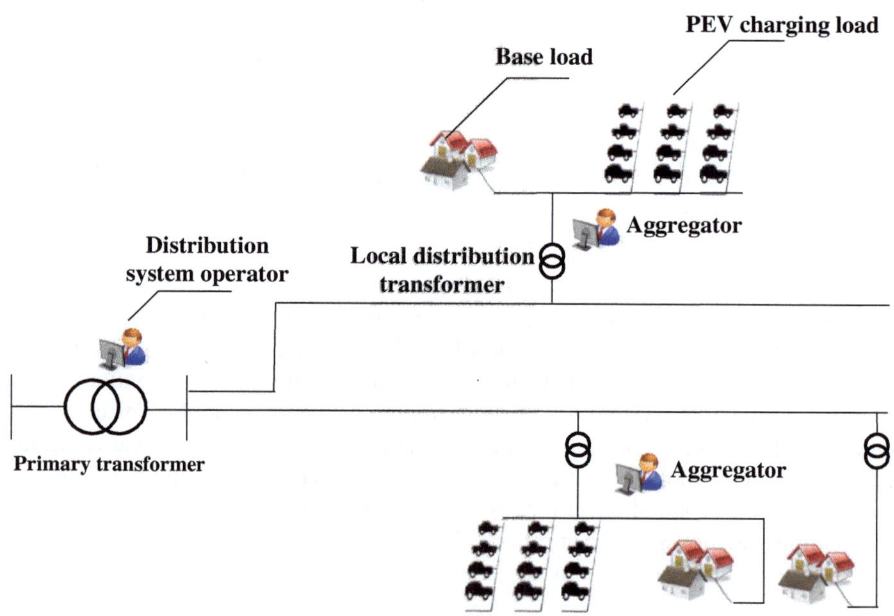

Fig. 3.3 Schematic illustration of decentralized charging system in an urban area

dwellings also has an aggregator (charging station, parking deck, etc.) equipped with multiple charging points. Both the base loads (e.g. loads excluding the charging loads) and PEV charging loads are supplied by the local distribution transformer, which is constrained by a loading upper bound.

This aggregator at community j ($j \in J_i$) operates and controls the switch on and off states as well as the charging power of each charging point within the parking deck and is also required to follow the charging regulation requirements dictated by the DSO under a pre-specified incentive. The aggregators under the same primary distribution transformer are not necessarily operated by the same company; hence, for privacy protection purposes, we assume the aggregator cannot reveal the detailed information of their customer charging requirements. Similarly, we suppose that each aggregator purchases electricity from the utility at time-of-use (TOU) rates and sells it to the PEV charging customers at retail prices to make profits by providing charging services.

Likewise, we assume once a PEV connects to the k-th charging point at aggregator j, its battery capacity value B_{ijk} (in kWh) and initial SOC (SOC_{ijk}^{A}) can be instantly obtained by the aggregator's charging management system through communicating with battery management system (BMS) on PEV's board. The customer is also assumed to inform its expected parking duration d_{ijk} and the desired SOC upon departure SOC_{ijk}^{D}.

Finally, each aggregator predicts local base loads in the day-ahead based on historical data. Moreover, distribution system level base load predictions are also implemented at the DSO level.

Based on the above information, the PEVs coordinated charging strategy dynamically determines charging power of each charging port in multiple aggregators by following three-step coordination procedure which we will elaborate on in the next subsections.

3.5.2 Control Strategy Overview

This control framework for multiple aggregators under the DSO i is similarly designed in a rolling horizon fashion where charging load requirements and base loads along the horizon are considered. Specifically, the optimal control problem is solved at discrete time step. The charging power $p_{ijk}(t)$ at port k is kept constant within each interval Δ and can vary from zero to rated power of the charging port $P_{ijk}^{\max}$. After solving the problem, only the computed charging power for interval $(t, t+1]$ is sent to the charging ports for implementation. At the end of interval $(t, t+1]$, the states of all the accommodated PEVs under the DSO i (e.g. PEVs arrival/departure, SOC etc.) and base load forecasting results will be updated over the planning horizon and the above procedure is repeated again to obtain the optimal charging strategy for all charging ports in the next time interval $(t+1, t+2]$.

The hierarchical coordinated charging strategy implemented at the beginning of each time interval include three steps, namely, (1) charging load boundaries computation at the aggregator level, (2) centralized aggregate charging load optimization at the DSO level and (3) intelligent charging power allocation within the aggregator. Detailed explanations of these three steps are elaborated in the following three subsections, respectively [22].

3.5.3 Charging Load Aggregation Method

We propose a method to represent the aggregate flexibility of PEV charging demands within an aggregator at this step [22]. In fact, the charging load of each PEV can be described by its energy and power boundaries. The energy upper and lower boundaries, which respectively correspond to the fastest and the slowest paths of consuming energy, describe the charging flexibility of this PEV during its parking horizon. Power boundaries are used to limit the instantaneous charging power of this PEV, which has to be strictly no greater than the rated power of the charging port. The aggregations of these two types of boundaries of all PEVs are used to present the collective charging load boundaries of the aggregator.

Before we present the way to compute the energy and power limits, we first compute the maximum possible departure $SOC_{ijk}^{D,a}(t)$ of each PEV as follows.

$$SOC_{ijk}^{D,a}(t) = \min\left(SOC_{ijk}^{D}, SOC_{ijk}(t) + \frac{H_{ijk}(t) \times \rho \times P_{ijk}^{\max} \times \Delta}{B_{ijk}}\right), \forall k \in K_{ij}^{t}, \forall j \in J_i \tag{3.21}$$

where $H_{ijk}(t)$ is the planning horizon of PEV k, $\forall k \in K_{ij}^{t}$, which is normally selected as its remaining parking duration.

The energy upper limit and lower limit of a single PEV from time step t to time step $t + H_{ij}(t) - 1$ are then computed by recursion. Mathematically,

$$\begin{aligned} e_{ijk}^{\max}(t+l) = e_{ijk}^{\min}(t+l) = SOC_{ijk}^{D,a}(t) \times B_{ijk}, \\ l = H_{ijk}(t), \ \ldots, H_{ij}(t) - 1, \forall k \in K_{ij}^{t}, \forall j \in J_i \end{aligned} \tag{3.22}$$

$$\begin{aligned} e_{ijk}^{\min}(t+l) = \max\left(e_{ijk}^{\min}(t+l+1) - \rho \times P_{ijk}^{\max} \times \Delta, SOC_{ijk}(t) \times B_{ijk}\right), \\ l = 0, \ \ldots, H_{ijk}(t) - 1, \forall k \in K_{ij}^{t}, \forall j \in J_i \end{aligned} \tag{3.23}$$

$$e_{ijk}^{\max}(t) = SOC_{ijk}(t) B_{ijk}, \forall k \in K_{ij}^{t}, \forall j \in J_i \tag{3.24}$$

$$e_{ijk}^{\max}(t+l) = \min\left(e_{ijk}^{\max}(t+l-1) + \rho \times P_{ijk}^{\max} \times \Delta, SOC_{ijk}^{D,a}(t) \times B_{ijk}\right),\\ l = 1, \ldots, H_{ijk} - 1, \forall k \in K_{ij}^t, \forall j \in J_i \tag{3.25}$$

where $e_{ijk}^{\min}(t)$ and $e_{ijk}^{\max}(t)$ are respectively the energy lower and upper bounds of PEV k at time t. Equations (3.22) constrain the energy state of a PEV after its departure, which has to be fixed at the requested departure state. Equations (3.23) specify the minimum energy state of PEV at the $(t+l)$-th interval could be at most $\rho \times P_{ijk}^{\max} \times \Delta$ lower than its energy state at time $(t+l+1)$, $(l = 0, \ldots, H_{ijk}(t) - 1)$ but cannot be lower than its initial energy state. Equations (3.24) imply the initial energy state and Eq. (3.25) make sure that the maximum energy state of a PEV at one period later can be as much as $\rho \times P_{ijk}^{\max} \times \Delta$ larger than the energy state of the adjacent previous time period and should also be strictly no larger than $SOC_{ijk}^{D,a}(t) \times B_{ijk}$.

The charging power upper limits of a single PEV from time step t to time step $t + H_{ij}(t) - 1$ are determined by the rated power of the charging port. Specifically, if the PEV is connected to a charging port, its charging power can be no larger than the rated power, otherwise its charging power should be zero. Analytically,

$$p_{ijk}^{\max}(t+l) = P_{ijk}^{\max}, l = 0, \ldots, H_{ijk}(t) - 1, \forall k \in K_{ij}^t, \forall j \in J_i \tag{3.26}$$

$$p_{ijk}^{\max}(t+l) = 0, l = H_{ijk}(t), \ldots, H_{ij}(t) - 1, \forall k \in K_{ij}^t, \forall j \in J_i \tag{3.27}$$

where $p_{ijk}^{\max}(t)$ is the power upper limit of PEV k at time step t.

Based on the energy and power boundaries of each PEV in aggregator j, the collective energy and power boundaries of this aggregator are computed by simple summation. Additionally, the aggregate charging power should also not lead to the overloading of the local distribution transformer as in (3.30). Each aggregator then reports the aggregate charging load boundaries (3.28)–(3.30) to the DSO for further centralized coordination.

$$E_{ij}^{\min}(t+l) = \sum_{k \in K_{ij}^t} e_{ijk}^{\min}(t+l), l = 0, \ldots, H_{ij}(t) - 1, \forall j \in J_i \tag{3.28}$$

$$E_{ij}^{\max}(t+l) = \sum_{k \in K_{ij}^t} e_{ijk}^{\max}(t+l), l = 0, \ldots, H_{ij}(t) - 1, \forall j \in J_i \tag{3.29}$$

$$P_{ij}^{\max}(t+l) = \min\left(\sum_{k \in K_{ij}^t} p_{ijk}^{\max}(t+l), A_{ij} \times \xi_{ij}(t+l)\right), l = 0, \ldots, H_{ij}(t) - 1, \forall j \in J_i \tag{3.30}$$

3.5.4 Centralized Aggregate Charging Load Optimization at the DSO

With the aggregate charging load boundaries, the DSO attempts to coordinate the charging of all aggregators to minimize energy costs and to control peak demand. Preferred charging curves for each aggregator are centrally determined via a linear optimization model, where the aggregate charging requirements of each aggregator and system load profile are explicitly considered [22].

Specifically, we propose the centralized charging coordination model to be solved at t as follows.

$$\begin{aligned}\min_{\theta, p_{ij}^r} \quad W_i(t) = &\sum_{j \in J_i} \sum_{l=0}^{H_{ij}(t)-1} c_{ij}(t+l) \times p_{ij}^r(t+l) \times \Delta \\ &+ \mu \times \sum_{l=0}^{H_i(t)-1} \theta(t+l) - \kappa \times \sum_{l=0}^{H_i(t)-1} (H_i(t) - l) \times p_{ij}^r(t+l)\end{aligned} \tag{3.31}$$

$$\text{s.t. } p_{ij}^r(t+l) \leq P_{ij}^{\max}(t+l), l = 0, \ldots, H_{ij}(t) - 1, \forall j \in J_i \tag{3.32}$$

$$p_{ij}^r(t+l) = 0, l = H_{ij}(t), \ldots, H_i(t) - 1, \forall j \in J_i \tag{3.33}$$

$$\begin{aligned}&E_{ij}^{\min}(t+l) \leq \sum_{\tau=0}^{l-1} \rho \times p_{ij}^r(t+\tau) \times \Delta + E_{ij}^{\max}(t) \leq E_{ij}^{\max}(t+l), \\ &l = 1, \ldots, H_{ij}(t), \forall j \in J_i\end{aligned} \tag{3.34}$$

$$\sum_{j \in J_i} p_{ij}^r(t+l) \leq A_i \times \xi_i(t+l) + \theta(t+l), l = 0, \ldots H_i(t) - 1 \tag{3.35}$$

where $p_{ij}^r(t)$ is the preferred aggregate charging power for aggregator j at time t; μ is a large positive penalty factor for positive slack variables $\theta(t)$, which is introduced as slack variable to ensure the feasibility of the optimization problem in cases of excessive charging demands or limited charging load margins; κ is a small positive factor related to charging earliness considerations and will be explained in detail later. $H_i(t)$ is the planning horizon of the DSO i, which is selected as the maximum value of the planning horizons of all aggregators. A_i is the capacity of the primary distribution transformer and $\xi_i(t)$ is similarly defined as the proportion of available capacity of the primary distribution transformer that can be used for PEV charging at time step t.

The first term of the objective function quantifies the electricity purchase costs of all aggregators over the planning horizon. The second term penalizes positive slack variables $\theta(t)$ so as to keep the planned power at each time interval from violating the distribution transformer capacity limit to the greatest extent but also helps ensure the problem feasibility in case of excess charging demand. The last term implies our preference to early charging. As the weight factor of charging power declines associated with time, the charging system tends to plan PEVs charging power early so as to prepare for unexpected early departure of PEVs or unanticipated mass arrivals for charging.

Following the linear objective function, constraints (3.32) imply power upper limits on the preferred power curves. At any interval, the preferred power for each aggregator should not exceed its maximum charging power boundary. Constraints (3.33) constrain the preferred power at intervals out of the range of the aggregator's planning horizon, which should be strictly fixed at 0. Constraints (3.34) ensure the preferred charging curves within aggregator's energy boundaries. Constraints (3.35) are capacity limits. The introduction of slack variables keeps the model from infeasibility.

The computational burden of the above linear programming model is directly related to the number of aggregators controlled and the length of planning horizon but has little to do with the number of PEVs connected. Therefore, the prevalent algorithms, such as Interior Point Method or Simplex Algorithm can effectively solve this linear programming problem.

Note that with abundant available capacity for PEVs charging during off-peak periods in practices, values of $\theta(t)$ can be well restricted at zero under most conditions. In cases when the value of $\theta(t)$ is positive, we proportionally derate the preferred charging power $p_{ij}^{r}(t+l)$ to ensure system reliable operation. With derated $p_{ij}^{r}(t+l)$, undesirable sacrifice of customer charging requirement might occur. Mathematically, $p_{ij}^{r}(t+l)$ are adjusted by following equations:

$$p_{ij}^{r}(t+l) = p_{ij}^{r}(t+l) \times \frac{A_i \times \xi_i(t+l)}{A_i \times \xi_i(t+l) + \theta(t+l)}, l = 0, \ldots H_i(t) - 1, \forall j \in J_i \tag{3.36}$$

Then the DSO sends the updated preferred charging power $p_{ij}^{r}(t)$ at time t to corresponding aggregator j for further power allocation.

3.5.5 Charging Coordination Within Aggregator

With the preferred power dictated by the DSO, the objective of each aggregator at this step is to allocate the planned power $p_{ij}^{r}(t)$ to its controlled PEVs. Specifically, we design a fast, completeness value based scheduling algorithm [22]. We define the completeness value of a PEV charging task as follows. It is jointly determined by PEV's current SOC, departure SOC requirement and remaining parking duration.

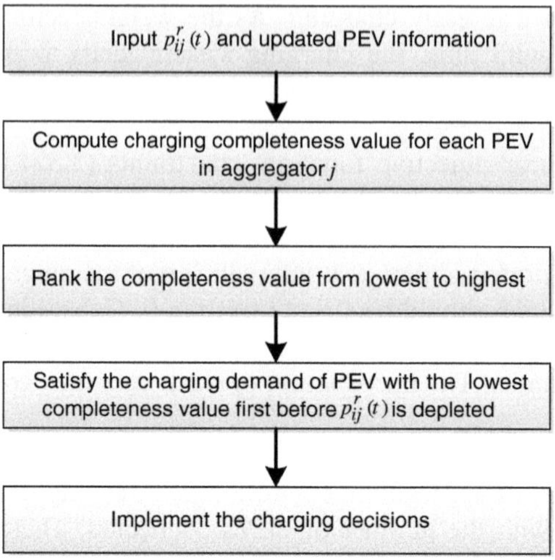

Fig. 3.4 Power allocation algorithm for aggregator i

$$\Psi_{ijk} = \frac{H_{ijk}(t) - C_{ijk}(t)}{\left(SOC_{ijk}^{D,a}(t) - SOC_{ijk}(t)\right) \times B_{n_i}} \tag{3.37}$$

where $C_{ijk}(t) = \left(SOC_{ijk}^{D,a}(t) - SOC_{ijk}(t)\right) \times B_{ijk}/(\rho \times P_{ijk}^{\max} \times \Delta)$, i.e. the minimum remained charging time needed for this PEV to charge to its required SOC level.

In the scheduling algorithm, we design the aggregator always choose to allocate the power to vehicles with the smallest completeness value first. Specifically, the fast, completeness value based scheduling power allocation algorithm can be schematically illustrated in Fig. 3.4.

Each aggregator then implements the charging schedule based on the results of the presented scheduling algorithm at the beginning of time step t. When the time proceeds to the beginning of next time interval $t + 1$, the above three-step procedure is repeated to determine the charging schedule for interval $(t + 1, t + 2]$ based on updated system information.

3.6 Three Level Coordinated Charging for Large Scale of PEVs

In the previous part, we investigated the problem of coordinating PEV charging across multiple aggregators at the distribution system level. When large scale of PEVs integrates into the power networks, their charging flexibilities could be potentially exploited further at the transmission level. Especially for regions with vertically regulated utilities, where the transmission and distribution sectors are operated by a single company, the charging coordination of large scale of PEVs are more appropriate to be implemented in a hierarchical way. In this subsection, a three-level hierarchical framework for coordinated PEV charging is presented [26].

3.6.1 Three Level Hierarchical Coordinated Charging Framework

We demonstrate the proposed three level hierarchical framework for coordinated charging of a large scale of PEVs in Fig. 3.5. In general, this hierarchical control framework includes three levels: transmission level control, distribution level control and charging station level control [26]. We assume the system operation data can be communicated between the transmission level operator and the distribution level operator directly. Stations including parking decks with charging points, charging stations and battery swapping stations are at the lowest level in the coordinated control framework. By arbitraging the TOU prices, the electricity purchase costs are minimized under the constraints of customer charging requirements. Meanwhile, we assume the station level operators in this framework comply with the regulation signal sent by the distribution operator under the pre-specified incentive and penalty terms.

Specifically, at the transmission level, the day-ahead forecast of the aggregated charging demand of PEVs and the base load profiles are carried out. A day-ahead reference aggregate charging load curve for each distribution system operator is then decided with the objectives of minimizing system peak load, load fluctuation and total charging costs while respecting the aggregated PEV charging demand flexibility of each distribution system operator and various generation and transmission constraints.

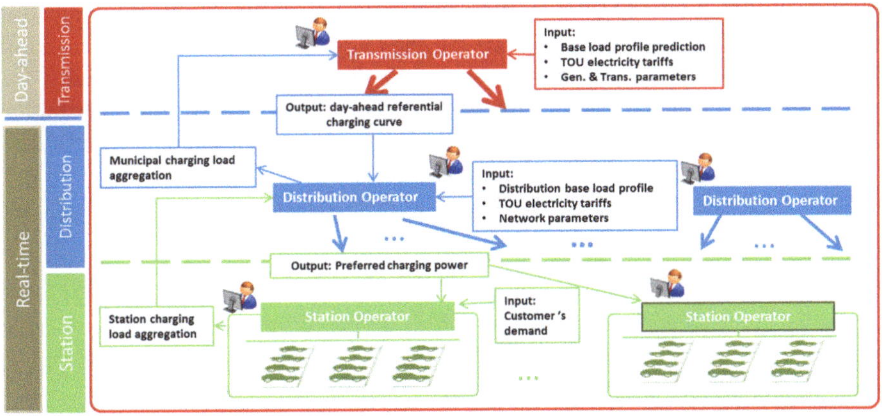

Fig. 3.5 Three-level hierarchical framework for coordinated charging of PEVs

In real time, at the distribution level, based on the day-ahead referential PEV charging curves sent by the transmission-level operator and real time aggregated charging needs of charging stations, the distribution operators dynamically allocate charging power to each station to satisfy their charging requirements and to achieve distribution system load control objectives at the same time. The decision process is similar to Sect. 3.5.

At the station level, the assumptions of arrival and departure processes of PEVs at each station are similar to those as described in Sect. 3.5. For battery swapping stations, we also assume the information of the size and initial SOC of the depleted batteries could also be conveniently obtained, and the expected time for next usage of this battery with $SOC_{ijk}^{D} = 1$ is estimated in advance. With charging requirements, the station level operator dynamically communicates and negotiates with the distribution level operator, follows the preferred charging power specified by the distribution system operator and flexibly determines the charging schedule of PEVs or batteries to satisfy customer charging preferences while minimizing its charging costs under TOU tariffs.

As the coordination process between the distribution operator and the station operator in real time is almost the same as described in Sect. 3.5. In this subsection, we mainly focus on introducing the problem formulation at the transmission level operator and the formulation at the distribution level operator, which also seeks to follow the day-ahead reference charging curve.

3.6.2 Problem Formulation

3.6.2.1 Charging Load Aggregations at the Distribution Level

Similar to the charging aggregation process at the station level as in Sect. 3.5.3, the charging load can be further aggregated at the distribution level [26]. We forecast the charging need of PEVs in each aggregator and describe each charging load through its energy and power bounds. We then aggregate the charging demand of PEVs both at the charging station level and the distribution level. Specifically, the forecasts of the aggregation of charging demands requested by the distribution operator i are calculated based on the following equations in the day-ahead.

$$\hat{E}_{i}^{\max/(\min)}(t) = \sum_{j \in J_i} \sum_{k \in K_{ij}} e_{ijk}^{\max/(\min)}(t), \forall t \in T \tag{3.38}$$

$$\hat{P}_{i}^{\max}(t) = \sum_{j \in J_i} \sum_{k \in K_{ij}} p_{ijk}^{\max}(t), \forall t \in T \tag{3.39}$$

Note that the realized charging requirement of each PEV or battery in real time operation does not necessarily match with the day-ahead forecasts precisely. However, the pooled charging needs at the distribution level can still be reasonably estimated when each distribution operator provides charging services to sufficient number of PEVs. Therefore, we apply the day-ahead forecasted aggregated charging requirements of each distribution operator for day-ahead charging coordination at the transmission level.

3.6.2.2 Day-Ahead Charging Coordination at the Transmission Level

With the aggregated charging demand information of each distribution operator, the transmission operator seeks to find optimal referential charging load profiles for each distribution operator to optimize the transmission level objectives [26]. Day-ahead unit commitment and economic dispatch considering both network and flexible charging load could be implemented at this level. For illustrative purposes, we only formulate a centralized coordination problem without considering the generation and network constraints. Specifically, the optimization problem that is solved day-ahead by the transmission operator is formulated as follows.

$$\min_{\hat{P}_i^r(t), L_p, L_i} L_p + \lambda \times \sum_{i \in I} L_i + \phi \times \sum_{i \in I} \sum_{t \in T} \hat{P}_i^r(t) \times c_i(t) \times \Delta + \alpha \times \sum_{t \in T} \left(\hat{P}^{BL}(t) + \sum_{i \in I} \hat{P}_i^r(t) \right)^2 + \beta \times \sum_{i \in I} \sum_{t \in T} \left(\hat{P}_i^{BL}(t) + \hat{P}_i^r(t) \right)^2 \tag{3.40}$$

$$\text{s.t. } \hat{P}^{BL}(t) + \sum_{i \in I} \hat{P}_i^r(t) \leq L_p, \forall t \in T \tag{3.41}$$

$$\hat{P}_i^{BL}(t) + \hat{P}_i^r(t) \leq L_i, \forall i \in I, \forall t \in T \tag{3.42}$$

$$0 \leq \hat{P}_i^r(t) \leq \hat{P}_i^{\max}(t), \forall i \in I, \forall t \in T \tag{3.43}$$

$$\hat{E}_i^{\min}(t) \leq \sum_{\tau=1}^{t} \rho \times \hat{P}_i^r(\tau) \times \Delta \leq \hat{E}_i^{\max}(t), \forall i \in I, \forall t \in T \tag{3.44}$$

where L_p is the day-ahead peak demand of the system over the planning horizon. L_i is the day-ahead peak demand of the i-th DSO. $\hat{P}_i^r(t)$ stands for the day-ahead referential charging load trajectory for the DSO i at time t. $c_i(t)$ is time-of-use charging cost per kWh at time t, which is uniform for all aggregators under the

DSO i. $\hat{P}^{BL}(t)$ is the forecasted base load of the system at time t and $\hat{P}_i^{BL}(t)$ is the forecasted base load under the DSO i. λ, ϕ, α and β are weighting coefficients.

The first term of the objective function (3.40) quantifies the system peak load at the transmission level over the planning horizon. Meanwhile, the second term quantifies the peak loads of all distribution operators. The third term indicates system preferences for lower charging costs. Finally, to consider system operator's preferences for smoother load profile, we add the last two terms to penalize projected load variations in the whole system and each distribution operator, respectively. Following the quadratic objective function, constraints (3.41) impose the limit on system peak load. Likewise, we introduce constraints for distribution peak demand in (3.42). Equation (3.43) constrain that the referential charging power of each distribution operator should be well kept below their respective power boundaries. Constraints (3.44) make sure the referential charging power of each distribution operator satisfy their forecasted aggregated cumulative charging energy boundaries.

Note that the above model is a quadratic convex optimization problem and its dimension is mainly related to the number of distribution operators. Hence, it can be solved efficiently. Since the aggregated charging demand of each distribution operator can be reasonably estimated in the day ahead when sufficient number of PEVs are integrated, in real time operations, the distribution operator does not communicate with the transmission operator but only seeks to follow the dictated referential charging power trajectory $\hat{P}_i^r$ as closely as possible by coordinating the charging power of charging stations.

3.6.2.3 Real-Time Charging Coordination at the Distribution Level

Similar to the Sect. 3.5 in real time operations, the charging power of each charging station is dictated by its corresponding DSO [26]. At each time interval, the operator of charging station j under the distribution operator i first computes its aggregated charging load based on (3.21)–(3.30) and sends its cumulative charging energy and power boundaries $(E_{ij}^{\max}, E_{ij}^{\min}, P_{ij}^{\max})$ to the distribution operator i for further coordination. The DSO i then solves the following convex optimization model at time t.

$$\begin{aligned}\min_{P_{ij}^r, D_i} U_i(t) = & \sum_{j\in J_i}\sum_{l=0}^{H_{ij}(t)-1} c_i(t+l)\times P_{ij}^r(t+l)\times\Delta + \gamma\times\sum_{\tau=0}^{H_i(t)-1}\left|\hat{P}_i^r(t+l) - \sum_{j\in J_i} P_{ij}^r(t+l)\right| \\ & - \kappa\times\sum_{j\in J_i}\sum_{l=0}^{H_{ij}(t)-1}(H_i(t)-l)\times P_{ij}^r(t+l) + \mu\times L_i + v \\ & \times\sum_{l=0}^{H_i(t)-1}\left(\hat{P}_i^{BL}(t+l) + \sum_{j\in J_i} P_{ij}^r(t+l)\right)^2\end{aligned} \tag{3.45}$$

$$\text{s.t. } 0 \le P_{ij}^r(t+l) \le P_{ij}^{\max}(t+l), \forall l = 0, \ldots, H_{ij}(t)-1, \forall j \in J_i \tag{3.46}$$

$$E_{ij}^{\min}(t+1) \le \sum_{\tau=0}^{l} \rho \times P_{ij}^{r}(t+\tau) \times \Delta \le E_{ij}^{\min}(t+1), \quad \forall j \in j_i, \\ \forall l = 0, \ldots, H_{ij}(t) - 1 \tag{3.47}$$

$$\hat{P}_i^{BL}(t+l) + \sum_{j \in J_i} P_{ij}^{r}(t+l) \le L_i, \forall l = 1, \ldots, H_i(t) - 1 \tag{3.48}$$

where γ, μ and v are weight coefficients.

The control objective of the distribution operator includes five components. The first term quantifies the projected total charging costs under the TOU tariffs along the real time planning horizon. The second term quantifies the total deviations of the total charging power from the day-ahead referential charging profile dictated by the transmission operator. The third term implies the preferences for early charging. As the weight factor of referential charging power declines in time, the charging system tends to plan PEV charging early so as to prepare for unexpected early departure or unanticipated mass arrivals for charging. The last two terms of the objective function respectively quantify the projected distribution peak load and load variations along the planning horizon. Since the optimization of distribution load profile has been included in day-ahead transmission level coordination model, the choices of parameters of μ and v can be small in real time control. Following the objective function, similarly, the resulting aggregate charging profile of each charging station is supposed to satisfy their respective accumulated energy and power boundaries in constraints (3.46) and (3.47). In addition, the value of peak demand over the real time planning horizon is characterized in constraints (3.48). Though other system constraints such as voltage and line thermal limits are not considered explicitly in this formulation as [8, 27], they can be readily incorporated if needed in practice.

The complexity of this problem is primarily related to the number of charging stations, which makes the formulated optimization problem easy to solve. After solving the above convex optimization model, the distribution operator then sends the resulting referential charging power at time step t to each charging station for further intelligent power allocation.

3.6.2.4 Real-Time Charging Coordination at the Station Level

With the referential charging power dictated by the distribution operator, the objective of the station operator is to intelligently allocate the planned power $P_{ij}^{r}(t)$ to its controlled PEVs or batteries. For the detailed allocation algorithm, please refer to Sect. 3.5.5.

3.6.2.5 Hierarchical Coordinated Charging Control Framework

Based on the above description of computation at each operator level, we can summarize the hierarchical decision processes at different levels of operators in Fig. 3.6 [26].

3.6.3 Case Studies

In this subsection, to understand how the control strategies perform under various circumstances, we carry out numerical simulations on three cases. Each case has three distribution system operators and the number of PEVs in each case is different. All simulations are conducted on a PC with Intel, Core i3 (2.93 GHz) CPU and 4 GB RAM. The formulated optimization problems at the transmission and the distribution level charging coordination are solved via CPLEX [23].

3.6.3.1 Case Specifications

In order to account for the differences in charging behaviors of various types of vehicles, we consider four types of PEVs, i.e. buses, taxis, cars owned by government or public institutions (abbreviated as GIOcar) and private cars (abbreviated as PRIcar) [28]. At the initial stage of PEV adoptions, the development of PEVs is still subject to many uncertainties, in terms of vehicle technology maturity, government subsidies, charging infrastructure availability and public awareness, etc. In many places around the globe, such as China, PEVs are first widely introduced in public transportation, but only have limited adoptions in the private sector. With the maturity of PEV technology and the wide availability of charging infrastructure,

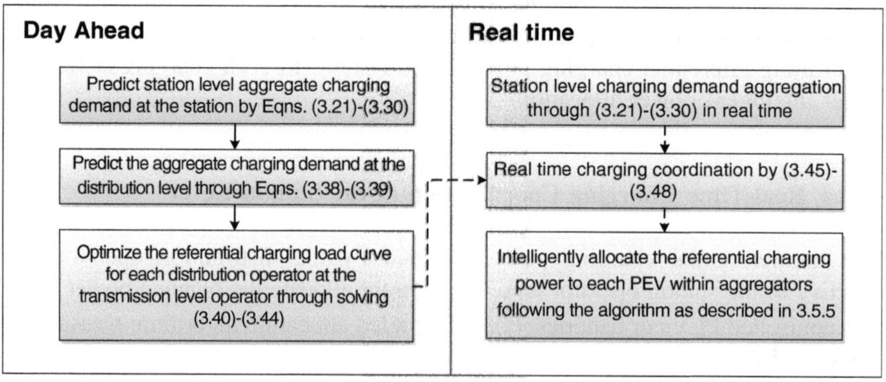

Fig. 3.6 The hierarchical decision processes at different levels of operators

Table 3.1 Scales of PEVs under three different cases (in thousand)

Aggregator	Type	Case A	Case B	Case C
X	Bus	0.1	0.3	0.6
	Taxi	0.4	0.5	1
	GIOcar	0.2	0.5	1
	PRIcar	0.2	10	20
Y	Bus	0.1	0.2	0.4
	Taxi	0.4	0.5	1
	GIOcar	0.2	0.5	1
	PRIcar	0.4	13	26
Z	Bus	0.2	0.6	1.2
	Taxi	0.4	0.5	1
	GIOcar	0.4	1.5	3
	PRIcar	1	45	90

private customers will become more inclined to adopt PEVs. To account for these development trends, in our designed case A, PEVs are widely adopted in public transportation but the adoptions in private sector are still very limited. In our case B and case C, we respectively consider a scenario where PEVs are normally adopted in private sector and a scenario that PEVs are heavily integrated. To be specific, the scales of different types of PEVs in all of three cases are listed in Table 3.1.

Considering that fast charging is more widely applied for electric buses, we assume that 60 % of buses get charged through fast charging and 40 % of buses adopt battery swapping. We further assume other types of PEV get refueled only through charging.

According to our empirical study on PEV charging behavior [29], parameter settings of different categories of PEVs are given in Table 3.2, where $N(a, b^2)$ denotes a normal distribution with mean a and standard deviation b and $U(a, b)$ stands for a uniform distribution with support $[a, b]$.

The electricity rates we adopted in our simulations are Time-of-Use tariffs, which is designed to encourage PEV off-peak charging, being $0.1659/kWh and $0.0411/kWh for 7:00–23:00 and 23:00–7:00 (next morning), respectively. We select three typical base load profiles for each DSO.

We assume that the decision interval $\Delta = 0.25$ hour and charging efficiency $\rho = 0.92$. Other parameter values are tuned as follows: $\lambda = 0.1$, $\phi = 1$, $\alpha = 0.01$, $\beta = 10^{-4}$, $\gamma = 10^{-5}$, $\kappa = 10^{-3}$, $\mu = 5 \times 10^{-6}$ and $v = 5 \times 10^{-9}$. The choices of these weight coefficients can be arbitrarily adjusted to satisfy the control preferences of system operators.

Table 3.2 Parameters for the calculation of charging load of PEVs

Type	Charging times/day	Charging initial time	Constraints on charging time (h)	Initial SOC	Expected SOC upon leaving	Charging (kW)	Battery size (kWh)	Probability of charging
Bus (charging)	4	N(10:00, 2^2)	1	N(0.6, 0.1^2)	0.8	91.7	324	1
		N(14:00, 2^2)	1	N(0.4, 0.1^2)	0.8	91.7		
		N(19:00, 2^2)	1	N(0.4, 0.1^2)	0.8	91.7		
		N(22:00, 2^2)	8	N(0.4, 0.1^2)	1	39.6		
Bus (battery swapping)	4	N(10:00, 2^2)	4	N(0.6, 0.1^2)	1	39.6	324	1
		N(14:00, 2^2)	4					
		N(19:00, 2^2)	10					
		N(22:00, 2^2)	8					
Taxi	2	N(13:00, 2^2)	1.5	N(0.3, 0.1^2)	1	32	60	1
		N(23:00, 2^2)	Till 7:00 am			7		
GIOcar	1	U(18, 24)	Till 8:00 am	N(0.4, 0.1^2)	1	7	60	1
PRIcar	1	N(9:00, 0.5^2)	8	N(0.6, 0.1^2)	1	7	60	0.2
		N(19:00, 1.5^2)	Till 8:00 am	N(0.6, 0.1^2)				0.8

3.6.3.2 Other Charging Strategies

In order to demonstrate the effectiveness of the proposed three-level hierarchical coordinated control framework (abbreviated as ThrC), three other benchmarks are also considered [26].

- The no-control strategy (abbreviated as NoC) or uncoordinated charging strategy is considered. Once a PEV is connected to its charging port, it is charged at its rated charging power until the PEV departs or its battery is full.
- A strategy where charging coordination is only implemented at the station level (abbreviated as OneC) is considered. In other words, each station level operator decides their charging schedule independently with the objective of minimizing total station charging costs and completing charging as early as possible.
- Finally, we consider a strategy where the day-ahead charging coordination at the transmission level is absent (Abbreviated as TwoC). Specifically, each distribution operator independently decides charging reference trajectories for charging stations in real time by solving problem (3.45)–(3.48) without considering the second term in objective function (3.45).

3.6.3.3 Simulation Results

We carry out simulations for the three cases mentioned above, following four different charging strategies. The overall system performances in terms of average charging costs and system peak demand are summarized in Table 3.3. It is shown that cost savings can be effectively achieved by following coordinated charging strategies, i.e. ThrC, TwoC and OneC. Furthermore, even with other load control objectives, the ThrC and the TwoC strategies also realize similar cost savings as the OneC strategy. In case A, when most of the PEVs on roads are used for public transportations, the cost savings are limited. Whereas the average cost savings reach over 40 % both in case B and case C, this is because both in these two cases, the charging schedules of a large number of private PEVs can be flexibly determined and thus the off-peak prices can be better exploited and the reductions in charging costs become significant.

In addition, note that increasing system peak load implies additional generation capacity and network reinforcement, we investigate how different charging strategies impact system peak under all of these three cases. In case A, the impact of PEV charging on system peak is not significant. In case B and C, however, the benefits of coordinated control have been clearly demonstrated. Under the ThrC and the TwoC strategies, the system peak demand is effectively reduced compared to that under the OneC or the NoC strategies. The ThrC outperforms other strategies in terms of reducing the total peak demand whereas the resulting peak demand of DSO turns out to be better under the TwoC strategy, when the coordination at the transmission level is absent in the day-ahead. Moreover, though cost savings can be effectively achieved under the OneC strategy, the absence of coordination across multiple

Table 3.3 System performances following different charging control strategies

Control strategy	Aggregator	Charging costs ($/kWh)			Increase of peak load (MW)		
		A	B	C	A	B	C
ThrC	X	0.11	0.09	0.09	2.3	19.1	49.1
	Y	0.11	0.08	0.08	3.6	6.5	21.9
	Z	0.11	0.08	0.08	7.1	20.2	29.0
	Total	**0.11**	**0.08**	**0.08**	**17.1**	**40.1**	**62.9**
TwoC	X	0.11	0.10	0.10	2.3	10.7	27.0
	Y	0.11	0.09	0.09	3.6	6.5	21.8
	Z	0.11	0.10	0.10	7.1	9.1	10.0
	Total	**0.11**	**0.10**	**0.10**	**17.1**	**43.2**	**63.5**
OneC	X	0.11	0.09	0.09	5.0	20.2	47.2
	Y	0.11	0.08	0.08	4.4	15.0	38.8
	Z	0.11	0.08	0.08	7.4	78.2	14.8
	Total	**0.11**	**0.08**	**0.08**	**17.3**	**125.2**	**236.5**
NoC	X	0.12	0.15	0.15	5.0	20.2	73.8
	Y	0.12	0.15	0.15	4.4	27.5	105.4
	Z	0.13	0.16	0.15	7.4	78.2	204.2
	Total	**0.12**	**0.15**	**0.15**	**17.3**	**125.2**	**405.4**

aggregators leads to the undesirable increase both in the DSO and the transmission system peak demand.

Figures 3.7 and 3.8 show the system load profiles following different charging strategies under case B and case C, respectively. The base load (system load excluding charging load) is referred to as BSL. It is indicated that, the ThrC, the TwoC and the OneC strategies have all successfully delayed the charging demand to off-peak period, which starts from 23:00. Whereas when no charging coordination is implemented under the NoC, the charging process begins instantaneously when massive private PEVs are connected at night. The overlapping of the charging peak and system base load undesirably increases overall system peak demand. We also observe that under the OneC strategy, at the moment of price turning cheaper at 23:00, station operators choose to charge PEVs due to the charging earliness preferences. Though this undesirable rebound effect does not become the major cause of the peak demand increase in both case B and case C, it requires significant amount of system ramping up reserves at 23:00 and may cause system operation stability issues. The system peak demand reduction and valley filling are effectively achieved under the ThrC and the TwoC strategies.

Based on the above observations, we find that through charging coordination, the charging costs can be reduced by exploiting the charging flexibility and shifting the charging loads to off-peak periods. Meanwhile, when the scales of PEVs are not large and most PEVs are public vehicles which only have limited charging flexibilities, the benefits of implementing charging coordination at the distribution and the transmission level turn out to be still insignificant and unnecessary. While in

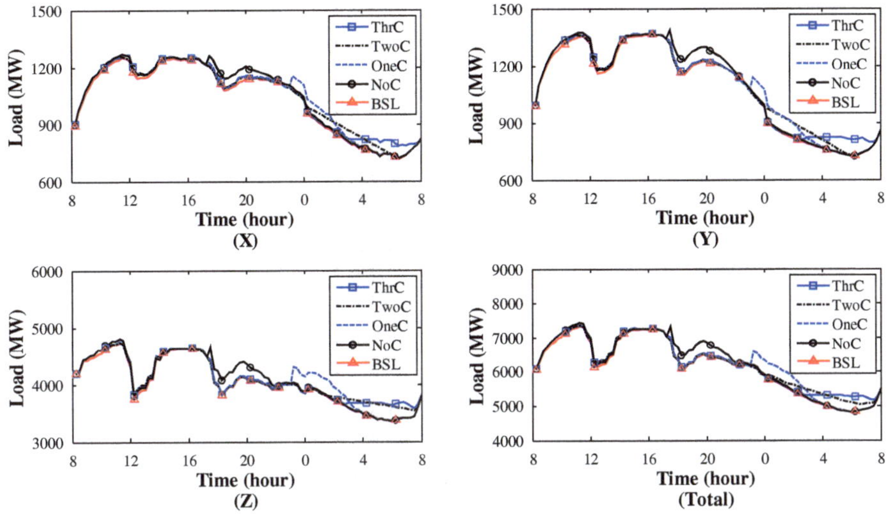

Fig. 3.7 Load profiles following different charging control strategies under case B

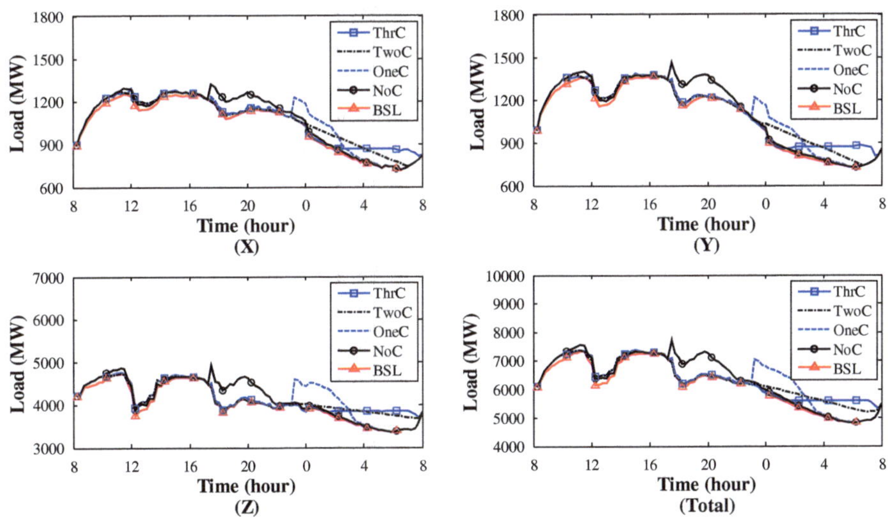

Fig. 3.8 Load profiles following different charging control strategies under case C

case B and case C, charging coordination at the distribution and transmission level seems to become effective in alleviating the rebound effect and thus reducing system requirements on ramping reserves and achieving better system load profiles. System operators can deploy the charging coordination operators step by step depending on the development scale of PEVs and other system control requirements.

3.7 Chapter Summary

In this book chapter, we investigate the hierarchical charging coordination strategies for PEVs. Starting from investigating the problem of charging coordination in the charging station and battery swapping station, we further introduce a two level coordination framework across multiple aggregators. Finally, we present a hierarchical charging coordination framework, which consists of three levels: the transmission level control, the distribution level control and the charging station level control. The respective control objective and constraints of each level operator are specified. Communication/control interfaces between different levels are also clearly presented.

We propose the cumulative energy and power boundaries to present the aggregate flexibility of large scale of PEVs both at the station operator and distribution system operator and effectively reduce the computation burden and communication overhead. The hierarchical control framework with reduced information exchange and computation also makes the whole control system more reliable to system communication and computational failures. Even if the communication is lost between the distribution operator and the station operator, the station level operator can still switch to local coordination strategy and effectively shift the charging load to off-peak periods by dynamically responding to TOU prices. Based on mathematical models of three levels control framework, simulation studies on three case studies are performed. Numerical examples demonstrate the effectiveness of the proposed control framework, and indicate its scalability.

Finally, for simplicity, we neglect the network constraints in our study. While it is worth noting that the hierarchical coordinated charging control framework is flexible to incorporate system constraints, such as power flow constraints, in the transmission and the distribution formulations.

References

1. Lopes JAP, Soares FJ, Almeida PMR (2011) Integration of electric vehicles in the electric power system. Proc IEEE 99(1):168–183
2. Su W, Rahimi-Eichi H, Zeng W, Chow MY (2012) A survey on the electrification of transportation in a smart grid environment. IEEE Trans Ind Inform 8(1):1–10
3. Roe C, Meisel J, Meliopoulos S, Evangelos F, Overbye T (2009) Power system level impacts of PHEVs. In: 42nd Hawaii international conference on system sciences, pp 1−10
4. Fernandez LP, Romn T, Cossent R, Domingo CM, Frłas P (2011) Assessment of the impact of plug-in electric vehicles on distribution networks. IEEE Trans Power Syst 26(1):206–213
5. Clement K, Haesen E, Driesen J (2010) The impact of charging plug-in hybrid electric vehicles on a residential distribution grid. IEEE Trans Power Syst 25(1):371–380
6. Han S, Han S, Sezaki K (2010) Development of an optimal vehicle-to-grid aggregator for frequency regulation. IEEE Trans Smart Grid 1(1):65−72
7. Sundstrom O, Binding C (2012) Flexible charging optimization for electric vehicles considering distribution grid constraints. IEEE Trans Smart Grid 3(1):26–37

8. Richardson P, Flynn D, Keane A (2012) Optimal charging of electric vehicles in low-voltage distribution systems. IEEE Trans Power Syst 27(1):268–279
9. Sortomme E, El-Sharkawi M (2012) Optimal combined bidding of vehicle-to-grid ancillary services. IEEE Trans Smart Grid 3(1):70–79
10. Wu D, Aliprantis DC, Ying L (2012) Load scheduling and dispatch for aggregators of plug-in electric vehicles. IEEE Trans Smart Grid 3(1):368–376
11. Vagropoulos SI, Bakirtzis AG (2013) Optimal bidding strategy for electric vehicle aggregators in electricity markets. IEEE Trans Power Syst 28(4):4031–4041
12. Luo Z, Hu Z, Song Y, Xu Z, Lu H (2013) Optimal coordination of plug-in electric vehicles in power grids with cost-benefit analysis—part I: enabling techniques. IEEE Trans Power Syst 28 (4):3546–3555
13. Yao W, Zhao J, Wen F, Xue Y, Ledwich G (2013) A hierarchical decomposition approach for coordinated dispatch of plug-in electric vehicles. IEEE Trans Power Syst 28(3):2768–2778
14. Qi W, Xu Z, Shen ZJM, Hu Z, Song Y (2014) Hierarchical coordinated control of plug-in electric vehicles charging in multifamily dwellings. IEEE Trans Smart Grid 5(3):1465–1474
15. Wen C, Chen J, Teng J, Ting P (2012) Decentralized plug-in electric vehicle charging selection algorithm in power systems. IEEE Trans Smart Grid 3(4):1779–1789
16. Hamid QR, Barria JA (2013) Distributed recharging rate control for energy demand management of electric vehicles. IEEE Trans Power Syst 28(3):2688−2699
17. Liu H, Hu Z, Song Y, Lin J (2013) Decentralized vehicle-to-grid control for primary frequency regulation considering charging demands. IEEE Trans Power Syst 28(3):3480–3489
18. Ma Z, Callaway DS, Hiskens IA (2013) Decentralized charging control of large populations of plug-in electric vehicles. IEEE Trans Control Syst Technol 21(1):67–78
19. Gan L, Topcu U, Low SH (2013) Optimal decentralized protocol for electric vehicle charging IEEE Trans Power Syst 28(2): 940−951
20. Sheikhi A, Bahrami Sh, Ranjbar AM, Oraee H (2013) Strategic charging method for plugged in hybrid electric vehicles in smart grids: a game theoretic approach. Int J Electr Power Energy Syst 53:499–506
21. Xu S, Chen W (2006) The reform of electricity power sector in the PR of China. Energy Policy 34(16):2455–2465
22. Xu Z, Hu Z, Song Y, Zhao W, Zhang Y (2014) Coordination of PEVs charging across multiple aggregators. Appl Energy 136:582–589
23. IBM ILOG CPLEX Optimization studio 12.5 (2013) IBM ILOG. http://www-947.ibm.com/support/entry/portal/overview/software/websphere/ibmilogcplexoptimizationstudio. Accessed 15 March 2013
24. Bohnsack R, Pinkse J, Kolk A (2014) Business models for sustainable technologies: exploring business model evolution in the case of electric vehicles. Res Policy 43(2):284–300
25. Xu Z, Hu Z, Song Y, Luo Z, Zhan K, Wu J (2010) Coordinated charging strategy for PEVs charging stations. In: Proceedings of power and energy society general meeting, pp 1−8
26. Xu Z, Su W, Hu Z, Song Y, Zhang H (2014) A hierarchical framework for coordinated charging of plug-in electric vehicles in China. Submitted to IEEE Trans Smart Grid (under 3rd round review)
27. Deilami S, Masoum A, Moses P, Masoum M (2011) Real-time coordination of plug-in electric vehicle charging in smart grids to minimize power losses and improve voltage profile. IEEE Trans Smart Grid 2(3):456−467
28. Luo Z, Hu Z, Song Y, Xu Z, Lu H (2013) Optimal coordination of plug-in electric vehicles in power grids with cost-benefit analysis—part II: a cast study in China. IEEE Trans Power Syst 28(4):3556–3565
29. Luo, Z (2012) Formulations, solutions and benefits analyses on coordinated charging/discharging of large scale plug-in electric vehicles. Ph.D. dissertation, Department of Electrical Engineering, Tsinghua University, Beijing

Chapter 4
Impacts of Plug-in Electric Vehicles Integration in Distribution Networks Under Different Charging Strategies

Filipe J. Soares, Pedro N.P. Barbeiro, Clara Gouveia and João A.P. Lopes

Abstract The uncertainties related to when and where Plug-in Electric Vehicles (PEVs) will charge in the future requires the development of stochastic based approaches to identify the corresponding load scenarios. Such tools can be used to enhance existing system operators planning techniques, allowing them to obtain additional knowledge on the impacts of a new type of load, so far unknown or negligible to the power systems, the PEVs battery charging. This chapter presents a tool developed to evaluate the steady state impacts of integrating PEVs in distribution networks. It incorporates several PEV models, allowing estimating their charging impacts in a given network, during a predefined period, when different charging strategies are adopted (non-controlled charging, multiple tariff policies and controlled charging). It uses a stochastic model to simulate PEVs movement in a geographic region and a Monte Carlo method to create different scenarios of PEVs charging. It allows calculating the maximum number of PEVs that can be safely integrated in a given network and the changes provoked by PEVs in the load diagrams, voltage profiles, lines loading and energy losses. Additionally, the tool can also be used to quantify the critical mass (percentage) of PEV owners that need to adhere to controlled charging schemes in order to enable the safe operation of distribution networks.

Keywords Charging strategies · Critical mass · Distribution grid · Markov model · Monte Carlo simulation · Plug-in electric vehicle impacts · Steady-state operation

---

F.J. Soares (✉) · P.N.P. Barbeiro · C. Gouveia · J.A.P. Lopes
Centre for Power and Energy Systems (CPES) of INESC TEC, Porto, Portugal
e-mail: filipe.j.soares@inesctec.pt

P.N.P. Barbeiro
e-mail: pnpb@inesctec.pt

C. Gouveia
e-mail: cstg@inesctec.pt

J.A.P. Lopes
e-mail: jpl@fe.up.pt

S. Rajakaruna et al. (eds.), *Plug In Electric Vehicles in Smart Grids*,
Power Systems, DOI 10.1007/978-981-287-317-0_4

4.1 Introduction

The foreseen rollout of Plug-in Electric Vehicles (PEVs) will considerably affect distribution grids management and operation. The extra amount of power they will demand from the grid will oblige system operators to understand the impacts resulting from PEVs connection into distribution networks.

Several approaches to this problem have been pursued. In [1], for instance, authors follow a deterministic strategy to locate PEVs along the network buses and, consequently, determine PEVs load during an entire day. Conversely, in [2], authors introduced a probabilistic method for determining PEVs load. In [3], Heydt analyzed the changes in the load diagram of a community of about 150–300 thousand people, in the USA, for increasing penetration levels of PEVs in the vehicle fleet. The author concluded that a salient factor to be considered in PEVs deployment is their charging during peak hours and referred that a possible method to alleviate peak loading and temperature rise in distribution transformers is through the use of load management techniques. Lopes et al. [1, 4], studied the impacts of PEVs in distribution grids. These authors evaluated the PEVs charging impact on the grid technical constraints and concluded that PEVs can lead to the violation of statutory voltage and ratings limits, as well as to a significant increase in the energy losses. The authors stressed the need to develop and implement efficient management procedures for coordinating PEVs charging, in order to minimize the need to reinforce the grid infrastructures. Papadopoulos et al. [5], also addressed the technical challenges related with the PEVs integration. Steady state voltage profiles of a typical Low Voltage (LV) network from the UK, under different PEVs penetration scenarios, were investigated and the results obtained showed that the grid voltage profiles are highly dependent on the number of PEVs integrated in the grid. Clement et al. [2, 6], analyzed the PEVs impacts in distribution grids power losses and voltage deviations. The authors concluded that PEVs uncoordinated charging is very likely to lead to voltage problems, even for low PEVs integration levels. Other works, such as [7–10], presented similar studies with analogous conclusions.

Several approaches to this problem have been pursued. In [1], for instance, authors follow a deterministic strategy to locate PEVs along the network buses and, consequently, determine PEVs load during an entire day. Conversely, in [2], authors introduced a probabilistic method for determining PEVs load. In [3], Heydt analyzed the changes in the load diagram of a community of about 150–300 thousand people, in the USA, for increasing penetration levels of PEVs in the vehicle fleet. The author concluded that a salient factor to be considered in PEVs deployment is their charging during peak hours and referred that a possible method to alleviate peak loading and temperature rise in distribution transformers is through the use of load management techniques. Lopes et al. [1, 4], studied the impacts of PEVs in distribution grids. These authors evaluated the PEVs charging impact on the grid technical constraints and concluded that PEVs can lead to the violation of statutory voltage and ratings limits, as well as to a significant increase in the energy losses. The authors stressed the need to develop and implement efficient management procedures for coordinating PEVs charging, in order to minimize the need to reinforce the grid infrastructures. Papadopoulos et al. [5], also addressed the technical challenges related with the PEVs integration. Steady state voltage profiles of a typical Low Voltage (LV) network from the UK, under different PEVs penetration scenarios, were investigated and the results obtained showed that the grid voltage profiles are highly dependent on the number of PEVs integrated in the grid. Clement et al. [2, 6], analyzed the PEVs impacts in distribution grids power losses and voltage deviations. The authors concluded that PEVs uncoordinated charging is very likely to lead to voltage problems, even for low PEVs integration levels. Other works, such as [7–10], presented similar studies with analogous conclusions.

One common point from the studies presented in [1–10] was that the technical problems identified could be easily avoided if adequate PEVs load management techniques were implemented. This proved to be true, as described by several authors in [11–15]. These works were focused on the determination of optimal (or near optimal) PEVs charging schedules. In [11], for instance, Lopes et al. suggested a smart charging scheme based on a hierarchical structure that monitors the grid operating conditions and manages PEVs charging to avoid violations of the grid technical restrictions. In [14], Geth et al. developed an algorithm to determine the optimal charging profiles for fleets of PEVs in Belgium. Sortomme et al. [15], suggested three distinct smart charging schemes that exploited the relationship between feeder losses, load factor and load variance.

Interesting approaches were proposed in these works, though they were only able to reveal the effects of a possible scenario for a given period.

Therefore, it is important to develop tools that allow exploring different scenarios in a coordinated way, which may result in both average scenarios and extreme case scenarios to be used for network steady state evaluation. Such tools can be used to help system operators in planning their operation for the next hours or to enhance existing system operators planning techniques, allowing them to obtain additional knowledge on the impacts of PEVs battery charging. Given the fact that PEVs are mobile loads that may appear in almost any bus of a given electricity network, voltage profiles, lines loading, peak power and energy losses variations need to be properly evaluated for the simulation of the operating conditions or for the planning exercise.

To achieve these objectives, a simulation tool to accurately estimate the PEVs impacts along one typical week (with 336 time intervals of 30 min) in Low and Medium Voltage (LV/MV) networks was developed, considering different PEV charging strategies. It includes a stochastic model to simulate PEVs movement in a geographic region and a Monte Carlo method to create different scenarios of operation. This tool, which uses PSS/E [16] and Python programming language [17] to conduct power flow studies in 30 min time steps, will be described in detail in this chapter, as well as the results of its application to several case studies.

The tool can also be used to quantify the critical mass (percentage) of PEV owners that need to adhere to controlled charging schemes in order to enable the safe operation of the distribution networks.

The charging approaches modelled in the simulation tool will be presented in Sect. 4.2. Section 4.3 presents details regarding the modelling approach, while the case studies used to evaluate the tool performance and the results obtained are presented in 4.4. The main conclusions are drawn in Sect. 4.5.

4.2 PEVs Charging Approaches

Taking into account the expected business models in the PEV field, [18], three different charging approaches were assumed to be the most promising in the near future and modelled in the simulation tool: Non-controlled Charging (commonly referred in the literature as "dumb charging"); Multiple Tariff; and Controlled Charging (commonly referred in the literature as "smart charging"). These approaches are described in the following sections. Although not modelled in the simulation tool, a brief description of Vehicle-to-Grid (V2G) will also be provided.

4.2.1 Dumb Charging

This is a no control mode where PEVs can be freely operated having no restrictions or incentives to modulate their charging. Therefore, PEVs are regarded as normal loads, like any other appliance. In this mode, it is then assumed that PEV owners are completely free to connect and charge their vehicles whenever they want. The charging starts automatically when PEVs plug-in and lasts until its battery is fully charged or charge is interrupted by the PEV owner. In addition, electricity price for these users is assumed to be constant along the day, what means that no economic incentives are provided in order to encourage them to charge their vehicles during the valley hours, when the grid operating conditions are more favorable to an increment in the energy consumption.

For scenarios of large PEV deployment, this approach will provoke technical problems in the generation system and on the grid (potential large voltage drops and lines overloading). The only way to tackle the foreseen problems provoked by PEV is then to reinforce the existing generation system and grid infrastructures and plan new networks in such way that they can fully handle PEV grid integration. Yet this is a somewhat expensive solution that will require high investments in network infrastructures that utilities would like to avoid.

4.2.2 Multiple Tariff Policy

As in the previous approach, the multiple tariff policy assumes that PEV owners are completely free to charge their vehicles whenever they want. However, electricity price is assumed to vary along the day, existing some periods where its cost is lower (during valley hours). However, as this is not an active management strategy, its success depends on the PEV owner willingness to take advantage of this policy, and thus only part of the PEVs load will eventually shift to valley hours.

It should be taken into account that the economic signals provided to PEV owners with the multiple tariff policy might have a perverse effect in scenarios characterized by a high integration level of PEVs. It might happen that a large number of PEVs connects simultaneously in the beginning of the cheaper electricity periods, making the grid reach its technical limits.

4.2.3 Smart Charging

The uncontrolled PEVs charging strategies referred above are more prone to provoke negative impacts for the networks operation. In addition, the non-controllability of the PEVs charging will also impact negatively the profit that the electricity retailers (commonly referred in the literature as PEV Supplier/Aggregators—PEVS/A) might achieve from the markets negotiations. They will not have flexibility to shift the PEVs load towards the lower demand periods, being thus incapable of profiting from lower energy prices. Given the high number of technical restrictions violations that are expected to occur, the Distribution System Operator (DSO) will only have one possible solution to maintain the quality of service levels: make large investments in network reinforcements to solve the problems as they arise.

In this sense, an alternative path that might be followed is to foment the adherence of the PEV owners to controlled slow charging schemes, like the smart charging. The smart charging strategy envisions an active management system, as described in [11], where there are two hierarchical control structures, one headed by an PEVS/A and other by the DSO. The possibility of controlling the PEVs charging will be of great benefit for both PEVS/A and DSO. The PEVS/A will have the possibility of exploiting the PEVs flexibility for charging, namely the PEVs that are parked during large periods of time overnight, thus profiting from lower energy prices. Under these circumstances, the PEVs charging management performed by the PEVS/A will naturally shift a significant amount of the PEVs load from the peak hours towards lower demand periods, contributing to improve the network operating conditions, to reduce the energy losses and to reduce the DSO need to invest in network reinforcements.

As explained in [11], when operating the grid in normal conditions, PEVs will be managed and controlled exclusively by the PEVS/A, whose main functionality will be grouping PEVs, according to their owners' willingness, to exploit business opportunities in the electricity markets. The PEVS/A will monitor all the PEVs connected to the grid and its state, providing power or requesting from them the services that it needs to cope with what was previously defined in the market negotiations. This is accomplished by sending set points to vehicle controllers related with rates of charge or requests for provision of ancillary services. To accomplish successfully such a complex task, it is required that every fixed period (likely to be defined around 30 min), the State of Charge (SoC) of each PEV battery is communicated to the PEVS/A, to assure that, at the end of the charging period, batteries will be charged according to PEV owners requests.

In presence of emergency operating conditions, i.e. when the grid is being operated above its technical limits, the DSO should have the possibility of acting over PEVs charging. In these situations, PEVs might receive simultaneously two different set points, one from the PEVS/A and other from the DSO. To avoid violation of grid operational restrictions, the DSO signals should override the PEVS/A ones. This type of PEVs charging management provides the most efficient usage of the resources available at each moment, enabling congestion prevention and voltage control [19].

Since the smart charging not only contemplates PEVs management performed by the PEVS/A, but also the DSO control over the PEVs charging when required, it offers the possibility to manage the load of the smart charging adherents in the way that best fits the PEVS/A purposes as well as the network needs.

However, it should be stressed that the DSO and PEVS/A flexibility to manage the load of the smart charging adherents is always constrained by the PEV owners' requests, which should be fulfilled at all times. For the purpose of this work, the smart charging performed by the PEVS/A is assumed to be the management of the PEVs load in such a way that flattens the load diagram as much as possible.

4.2.4 Vehicle-to-Grid

This approach is an extension of the previous one where, besides the charging, the PEVS/A controls also the power that PEVs might inject into the grid. In the V2G mode of operation, both PEVs load controllability and storage capabilities are exploited. From the grid perspective, this is the most interesting way of using PEVs capabilities given that besides helping managing lines overloading and voltage related problems in some problematic spots of the grid, PEVs have also the capability of providing regulation services, such as frequency control. Nevertheless, there are also some drawbacks related with the batteries degradation. Batteries have a finite number of charge/discharge cycles and its usage in a V2G mode might represent an aggressive operation regime due to frequent shifts from injecting to absorbing modes. Thus the economic incentive to be provided to PEV owners must be even higher than in the smart charging approach, so that they cover the battery damages owed to its extensive use.

Being the most aggressive mode for charging PEVs, due to possible implications with PEV batteries lifecycle, this option is not likely to be a reality neither in the short run nor in the medium term. Only when battery technology has reached a high maturation stage, this strategy may be adopted. For this reason, it has not been considered in the current implementation of this software tool.

4.3 Simulation Tool

The simulation tool was developed to perform three different types of studies:

1. **Evaluate the impacts of a given number of PEVs in a distribution network.** For this study, a Monte Carlo simulation method was implemented to make the tool capable of simulating different scenarios (for the same PEV integration percentage), providing a reliable characterization of the grid operating conditions regarding voltage profiles, branches loading, grid peak power, energy losses and the networks components that are more likely to be operated near, or even above, their technical limits. An overview of the Monte Carlo simulation method is provided in the light grey area of Fig. 4.1.
2. **Compute the maximum number of PEVs that can be integrated in a given network.** This is achieved by using iteratively the procedure described in **1**, increasing in a stepwise manner the integration of PEVs (in steps of 1 %). The algorithm, whose flowchart is presented in Fig. 4.1, is stopped when there is a violation of the specified voltage limits or a line overloading.
3. **Quantify the critical mass (percentage) of PEV owners that need to adhere to the controlled charging schemes to enable the safe operation of the networks.** For this study, the first step of the algorithm consists in considering a fixed PEV integration percentage, of which one half of the PEVs is assumed to be "non-controlled charging" adherents and the other "time of day tariffs" adherents. Then, if problems are not detected, PEV integration is increased by 10 %. This procedure is repeated until a problem in the network is detected (either a voltage lower limit violation or a line overloading). After, the second step of the algorithm consists in iteratively increasing the percentage of "controlled charging" adherents, in steps of 5 %, while the "non-controlled charging" and the "time of day tariffs" adherents are decreased accordingly. The second step is repeated until the technical problems identified are solved. In the end of the procedure, the percentage of controlled charging adherents that allowed solving the problems is stored.

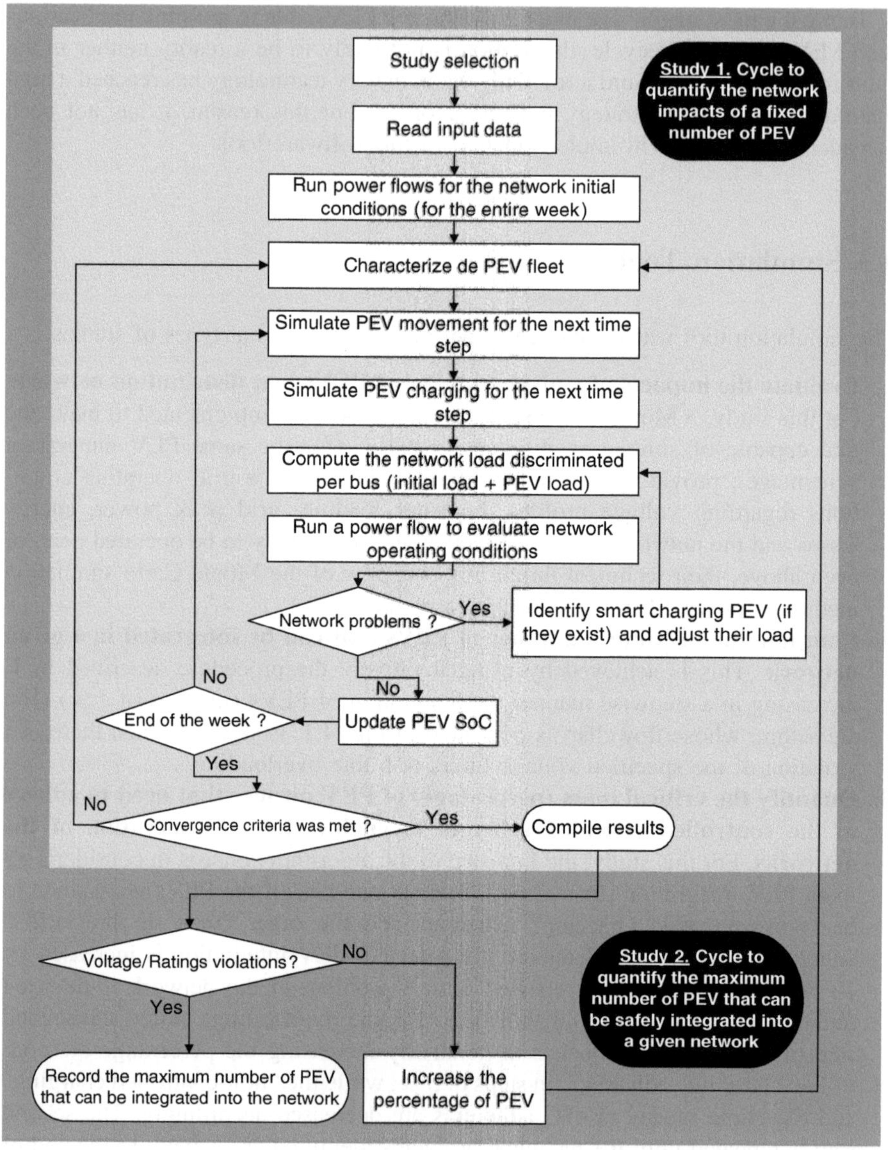

Fig. 4.1 Flowchart of the simulation tool

4.3.1 Impacts of a Given Number of PEVs in a Distribution Network

Concerning the study described in **1**, the cycle to quantify the network impacts of a given number of PEVs runs for each Monte Carlo simulation (light grey area in Fig. 4.1) the steps described next.

Step 1—Read input data

The input data is related with variables defined for each case study (e.g. number of conventional vehicles in the geographical area under study) and technical details of the network components (i.e. data required to run power flows).

Step 2—Run power flows for the initial scenario, without PEVs, and store relevant results

A power flow for each time step (30 min) is run considering only the network base load (no PEVs considered). The network indexes (voltages, lines ratings, losses and peak load) are stored to compare them later with the different PEV integration scenarios evaluated.

Step 3—Characterize the PEV fleet

During this procedure, all the PEV characteristics relevant for the simulation are generated. The PEVs battery capacity, charging power, energy consumption and initial battery SoC (battery SoC in the beginning of the simulation) are defined according to truncated Gaussian probability density functions, whose average, standard deviation, maximum and minimum values allowed are presented in Table 4.1.

While the parameters of the Gaussian density function used for the initial battery SoC were assumed for the purpose of this work, the parameters for battery capacity, slow charging rated power and energy consumption were obtained from the information made available by the manufacturers of 42 different PEVs. In [20–23] are presented some of the Internet sites from where PEV characteristics were obtained for this study. The maximum and minimum values allowed, presented in Table 4.1, were used to confine the values drawn for each PEV within realistic boundaries. A driver behavior was also assigned initially to each PEV. The different behaviors considered in this study were defined according with the findings of a survey made within the framework of the MERGE project [24]. The results

Table 4.1 Gaussian distributions for PEVs characterization

	Average	Standard deviation	Maximum value allowed	Minimum value allowed
Battery capacity (kWh)	29.0	14.5	72.0	10.0
Slow charging rated power (kW)	3.0	1.5	9.0	2.0
Energy consumption (kWh/km)	0.16	0.08	0.25	0.10
Initial battery SoC (%)	90.0	25.0	100.0	50.0

Table 4.2 Drivers' behaviors considered

	Percentage of the responses (%)
PEV charge at the end of the day	33
PEV charge only when it needs	23
PEV charge whenever possible	20
PEV charge whenever is convenient and the driver has time	24

revealed that there are four major types of behaviors regarding PEVs charging, as presented in Table 4.2.

For the purpose of this work, regarding the behaviors modeling and simulation, there was no relevant differences between the drivers that "charge at the end of the day" and those who "charge whenever is convenient and they have time". Therefore, the PEVs to which one of these drivers' behaviors was assigned were assumed to behave equally along the simulations.

For the drivers who charge their PEV only when it needs, it was assumed that the minimum battery SoC that triggers the need for charging was 30 %.

Step 4—Simulate PEVs movement and charging

The PEVs movement was simulated using a discrete-state, discrete-time Markov chain to define the states of all the PEVs at each time step of 30 min. A detailed description of the Markov chain can be found in [25]. It was assumed that, at every unit of time, PEVs can be in one of the following states: "in movement", "parked in a residential area", "parked in a commercial area" and "parked in an industrial area".

If a PEV is in the state "in movement", there is no need to define its location. However, if it is in a "parked" state and connected to the grid for charging purposes, it is crucial to know the PEV location to allocate its load to a specific network bus. Thus, for each time instant, a bus location was attributed to parked PEVs.

In the beginning of the simulation, a draw was made using Eqs. 4.1 and 4.3, to define the network nodes where each PEV stays parked when they are in the states "parked in a residential area" and "parked in an industrial area". These nodes represent the location of the household and of the workplace of each PEV and they were kept fixed during the simulations to emulate daily home-workplace commuting. Thus, every time a PEV was in "parked in a residential area" and "parked in an industrial area" states, it was automatically assumed to be parked in the nodes initially defined.

The procedure followed for "parked in a commercial area" was different. Every time a PEV was in this state, a new draw is made, using Eq. 4.2, to define the network bus where the PEV is parked. This means that PEVs can be in different places of the network when they are in the "parked in a commercial area" state.

The draw of the PEVs location was made taking into consideration the real nature of the loads connected to each network bus, as it can be observed in Eqs. 4.1–4.3. Thus, for the network under study, all the existing loads were classified as industrial, commercial or residential loads.

$$P^R(Bus\,b) = \frac{Load^R_{Bus\,b}}{\sum Load^R} \tag{4.1}$$

$$P^C(Bus\,b) = \frac{Load^C_{Bus\,b}}{\sum Load^C} \tag{4.2}$$

$$P^I(Bus\,b) = \frac{Load^I_{Bus\,b}}{\sum Load^I} \tag{4.3}$$

where $P^{R/C/I}(Bus\,b)$ is the probability of a PEV be located in bus b, if "parked in a residential/commercial/industrial area", $Load^{R/C/I}_{Bus\,b}$ is the residential/commercial/industrial load installed in bus b and $\sum Load^{R/C/I}$ is the network total residential/commercial/industrial load.

For the PEVs in movement, a procedure was developed to account for their energy consumption and the respective reduction in the battery SoC. First a Gaussian probability density function was used to draw the travelled distances for all the PEVs in movement. Therefore, if a PEV was in movement in time instant t and its battery SoC went below a predefined threshold (assumed to be 15 %) in time instant $t+1$, it was considered that the PEV would make a short detour to a fast charging station for recharging purposes. The travelled distance during the detour was obtained using also a Gaussian probability density function, whose parameters are presented in Table 4.3.

The fast charging was assumed to be made during 15 min with a power of 40 kW [26].

The average of the Gaussian distribution used to characterize the travelled distance in common journeys was obtained by dividing the average daily mileage in Europe by the average number of journeys per day [27]. The standard deviation was assumed to be 50 % of the average.

The values of the Gaussian function for the travelled distance to the fast charging station, were obtained by assuming that they were 25 % of those used in the travelled distance in common journeys distribution.

For the parked PEVs, an optimization procedure is used by the PEVS/A to define which smart charging adherents should charge at each time step to minimize the deviations between the energy bought in the market by the PEVS/A and the energy consumed by PEVs. It should be stressed that it was assumed that the power

Table 4.3 Gaussian distributions for travelled distances

	Average	Standard deviation	Maximum value allowed	Minimum value allowed
Travelled distance in common journeys (km)	40	20	200	10
Travelled distance to fast charging station (km)	10	5	50	2.5

charging rate for smart charging adherents could be controlled between 0 and the slow charging rated power presented in Table 4.1. To achieve the intended objective, it is required to find a set of n load values, being n the number of smart charging adherents, which can be defined as optimal in the sense that they allow minimizing the deviations referred above. This problem can be formulated as an optimization problem, as shown next.

$$\min \left| EBA_t - TIPEVL_t - \sum_{i=1}^{n} FPEVL_t^i \right| \tag{4.4}$$

subject to

$$0 \le SOCR_{td}^i - SOC_t^i \le \frac{\left(FPEVL_t^i + (td - (t+1)) \times MCR^i\right) \times 1/2 \times PEV_{ce}}{PEV_i^{bc}} \times 100 \tag{4.5}$$

$$0 \le FPEVL_t^i \le MCR^i \tag{4.6}$$

$$0 \le SOCR_{td}^i \le 100 \tag{4.7}$$

$$0 \le SOC_t^i \le 100 \tag{4.8}$$

$$t + 1 \le td \tag{4.9}$$

where:

- i—represents the "flexible PEV" index; "Flexible PEV" are the PEV whose owners adhered to the smart charging;
- t—represents the time index;
- n—is the nr. of "flexible PEV" under the PEVS/A control;
- EBA_t—represents the average power during ½ h, in kW, related with the energy bought in the day-ahead market by the PEVS/A for time period between t and $t+1$; $EBA_t(\text{kW}) = energy\,bought_{t \to t+1}(\text{kWh})/1/2\,\text{h}$;
- $TIPEVL_t$—is the "inflexible PEV" load, in kW, in time step t; "Inflexible PEV" are the PEV whose owners adhered to the dumb charging or multiple tariff schemes;
- $FPEVL_t^i$—is the power absorbed by "flexible PEV" i, in kW, in time step t; the $nFPEVL_t^i$ are the decision variables of the optimization problem; they can assume continuous values in the interval $[0, PEV\,maximum\,charging\,rate]$;
- td—represents the time step at which a given "flexible PEV" disconnects from the grid;
- SOC_t^i—are the PEV i battery SoC, in percentage, in time step t;
- $SOCR_{td}^i$—represents the battery SoC required by the owner of PEV i, in percentage, in time step td;

- PEV_i^{bc}—represents the battery capacity, in kWh, of PEV i;
- PEV_{ce}—represents the efficiency of the PEV charging process.

Equation 4.5 is used to assure that the PEV battery SoC, required by the PEV owners at the moment of disconnection, is possible to attain when considering the PEV maximum charging rate.

Equation 4.6 assures that only charging rates between $[0, PEV\,maximum\,charging\,rate]$ kW will be attributed to "flexible PEV".

Equations 4.7 and 4.8 are used to guarantee that the required battery SoC and battery SoC in the time step t are always within the interval $[0, 100]\%$.

Equation 4.9 assures that the time of disconnection always takes place after time step $t + 1$.

The objective of this optimization problem is then to minimize the sum of the absolute value of the deviations. It is a linear optimization problem, which is suitable for quasi-real-time applications since it is very fast to solve and does not require any type of forecasted data. It is only needed to know, for the current time step (t), the energy bought by the PEVS/A, the power consumed by the "inflexible PEV", the moment of disconnection of the "flexible PEV" that are plugged-in and the energy required by their owners during the connection period.

At each time instant, the PEVs battery SoC is updated according to the energy spent travelling, using Eq. 4.10 or according to the energy absorbed in slow charging mode or in fast charging stations, using Eq. 4.11.

It was assumed that PEVs "parked in a residential area" and "parked in an industrial area" charge at their nominal charging rate while "parked in a commercial area" PEVs have the capability of charging at the double of their nominal charging rate.

$$PEV_{SoC}^{t+1}(\%) = PEV_{SoC}^{t} - \frac{PEV_{\text{kWh/km}\,consumption} \times PEV_{Travelled\,distance}}{PEV_{Battery\,capacity}} \times 100 \quad (4.10)$$

$$PEV_{SoC}^{t+1}(\%) = PEV_{SoC}^{t} + \frac{PEV_{Charging\,efficiency} \times PEV_{Charging\,power} \times {}^{1}\!/_{2}}{PEV_{Battery\,capacity}} \times 100 \quad (4.11)$$

where:

- PEV_{SoC}^{t+1}—represents the battery SoC in time step $t + 1$;
- PEV_{SoC}^{t}—represents the battery SoC in time step t;
- $PEV_{\text{kWh/km}\,consumption}$—is the PEV energy consumption in kWh/km;
- $PEV_{Travelled\,distance}$—is the distance that the PEV travels in time step t;
- $PEV_{Battery\,capacity}$—is the capacity of the PEV battery in kWh;
- $PEV_{Charging\,efficiency}$—is the efficiency of the PEV charging process; $PEV_{\text{Charging efficiency}} \in [0, 1]$;
- $PEV_{Charging\,power}$—is the PEV charging power in kW.

Step 5—Compute the network load discriminated per bus (base load + PEVs load)

The total load in each bus is obtained by summing the network initial load to the PEVs load (in each bus).

Step 6—Sample analysis

The evaluation of the samples is made by running a power flow for each time instant to gather information regarding voltage profiles, power flows in branches and energy losses.

During the simulations, the day where the highest peak load occurs is recorded, in order to provide an idea of the worst situation that might occur when a percentage of conventional vehicles are replaced by PEVs.

In order to keep track of the most problematic buses and branches within the grid, the number of out of limit voltages and lines overloading occurrences are recorded along the simulations. According with [28], voltages must be kept with the interval 0.90–1.10 p.u. during 95 % of the time, in a weekly basis. It was considered in this work that a voltage violation occurs when the values are outside the referred interval. Then, the probability of having voltages below the imposed limit and branches overloading is computed using Eqs. 4.12 and 4.13.

$$P^{Bus\,b}_{Voltage\ violation} = \frac{Voltage\ violation^{Bus\,b}}{Nr.\,iterations \times 336} \times 100 \tag{4.12}$$

$$P^{Line\,l}_{Overloading} = \frac{Overloading^{Line\,l}}{Nr.\,iterations \times 336} \times 100 \tag{4.13}$$

where:

- $P^{Bus\,b}_{Voltage\ violation}$—represents the probability of a voltage violation occur in bus b;
- $P^{Line\,l}_{Overloading}$—represents the probability of a line overloading occur in line l;
- $Voltage\ violation^{Bus\,b}$—represents a voltage violation occurred in bus b;
- $Overloading^{Line\,l}$—represents a line overloading in line l;
- $Nr.\,iterations$—is the number of iterations run until reaching the Monte Carlo convergence;
- 336—represents the number of 30 min time intervals within a week.

Step 7—In case of network problems, adjust the load of the PEVs that adhered to the smart charging and that are contributing for the network problem identified

The indexes recorded in previous step are evaluated. If a network problem occurs and if there are smart charging adherents in the scenario being analyzed, a load reduction signal is sent to those that are contributing for the problem. A 10 % load reduction is requested.

Step 8—Compute new network total load

The total load in each bus is updated, taking into account the load reduction of the smart charging adherents.

Step 9—Sample reevaluation

The procedure described in Step 6 is used to update network indexes.

Step 10—In case of network problems, readjust the load of the PEVs that adhered to the smart charging

The network indexes stored in the previous step are evaluated. If the network problem persists, a new load reduction of 10 % is requested to the smart charging PEVs, following the same procedure as in Step 7. Steps 8–10 are repeated until the problem is solved or until the smart charging adherents load is reduced to zero.

Step 11—Check convergence criteria

To terminate the Monte Carlo process, two criteria are used: number of iterations and the variances of the aggregated network load of each one of the 336 time instants. The latter means that one variance value is computed for the total network load per time instant. The process is set to perform 500 iterations (500 weeks) and check, in the end, if the variation of all the 336 variances in the last 5 iterations is lower than 1e^{-3}. If at least one of the 336 variances did not meet this convergence criterion, the process is kept running more iterations until all the variances variations are lower than the predefined value. The variances variation is calculated using Eq. 4.14.

$$\Delta Variance = \left| Variance_h^t - Variance_{h-5}^t \right| < 1e^{-3} \tag{4.14}$$

where $Variance_h^t$ is the variance of the network load at time instant t, $t \in [1, 336]$, in the hth iteration.

4.3.2 Maximum Number of PEVs That Can Be Integrated in a Given Network

Regarding the study to compute the maximum number of PEVs that can be integrated in a given network, the algorithm stops if any technical problems are detected and the maximum number of PEVs that can be integrated in the network is computed. Otherwise, the PEV integration percentage is increased and the steps from 3 onwards are repeated.

4.3.3 Critical Mass of PEVs

For the critical mass study, the first step of the procedure followed consists in the consideration of a fixed PEV integration percentage, of which one half of the PEVs are assumed to be dumb charging adherents and the other half multiple tariff adherents.

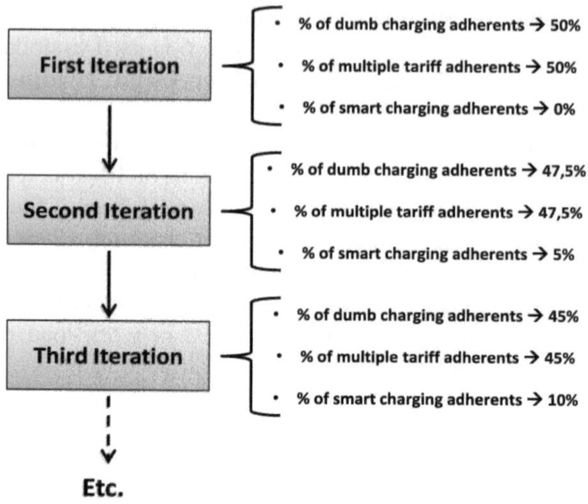

Fig. 4.2 Flowchart of the steps followed for the critical mass estimation

Then, if problems are not detected, the PEV integration percentage is increased by 10 %, assuming the same proportion of dumb charging and multiple tariff adherents (50 % of each).

This procedure is repeated until a problem in the network is detected (either a voltage lower limit violation or a branch overloading).

After detecting a technical problem, the second step of the procedure consists in iteratively increasing the percentage of smart charging adherents, in steps of 5 %, while the dumb charging and multiple tariff adherents percentage is decreased accordingly, as explained in Fig. 4.2.

The second step of the procedure is repeated until the technical problems previously identified are solved. In the end of the procedure, the percentage of smart charging adherents that allowed solving the problems detected (the critical mass of smart charging adherents) is recorded.

4.4 Simulations and Impacts Evaluation

In order to test the tool developed, several test cases were considered: a LV network and five MV networks. For each of the networks, three PEV charging scenarios were considered:

- All PEVs in dumb charging mode;
- All PEVs in multiple tariff mode;
- All PEVs in smart charging mode.

The maximum allowable PEV integration was computed by increasing in a stepwise manner the integration of PEVs in the network, until a violation of the voltage limits or a branch overloading occur. By considering these three extreme scenarios, it is possible to evaluate separately the effectiveness of the implemented algorithm for every charging strategy. Combinations of charging strategies would also be feasible, but in a first instance, for the purpose of validation, would not be as meaningful.

In a second stage, two distinct approaches were followed for the LV and the MV networks.

In the LV network case, four more simulations were performed in order to evaluate the effectiveness of the smart charging strategy when compared with the dumb charging and with two distinct multiple tariff policies. These simulations allow evaluating the network operating conditions when the number of PEVs that can be integrated with the smart charging behave as:

- Dumb charging adherents;
- Multiple tariff (22–8 h) adherents;
- Multiple tariff (1–7 h) adherents;
- Smart charging adherents.

For the MV networks, only three more simulations were performed considering the maximum allowable PEV integration with the dumb charging, multiple tariff (1–7 h) and smart charging.

4.4.1 Networks Used as Case Studies

The networks used as case studies were carefully chosen in order to evaluate systems with different characteristics, like their topology (rural or urban) and their type of consumers (industrial, commercial or residential).

4.4.1.1 Low Voltage Grid

Figure 4.3 shows the single line diagram of the LV network from an urban area (400 V) used as test case. It is composed essentially by residential loads, having only a small share of commercial clients.

The power factor assumed for the conventional load is 0.96, whereas the specified voltage in the feeding point is 1.00 p.u. There are a total of 125 conventional vehicles enclosed in the geographical area covered by this network. As this is a LV network, no fast charging station was assumed to be available within the grid boundaries, as this type of infrastructures will be most likely connected at the MV level.

The network's typical weekly load diagram used in the simulations is presented in Fig. 4.4. It was obtained by aggregating the load diagrams of the residential and commercial consumers within the network. The residential and commercial consumers' diagrams were combined taking into account the proportion of installed

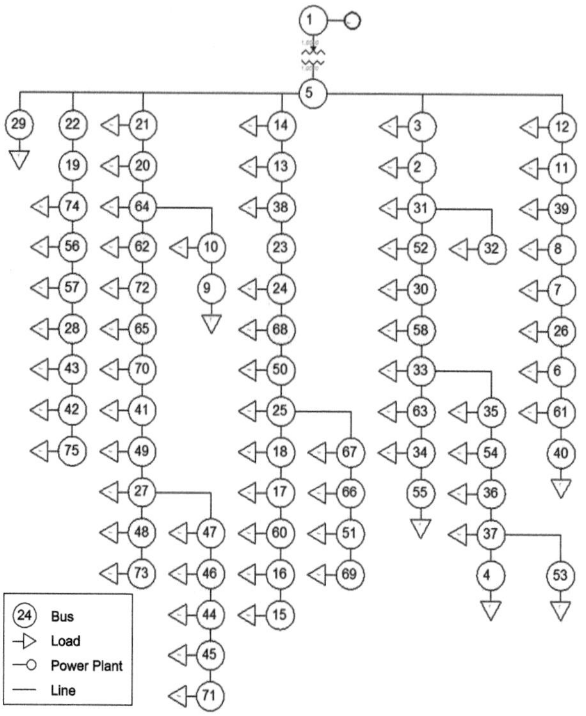

Fig. 4.3 Single line diagram of the network

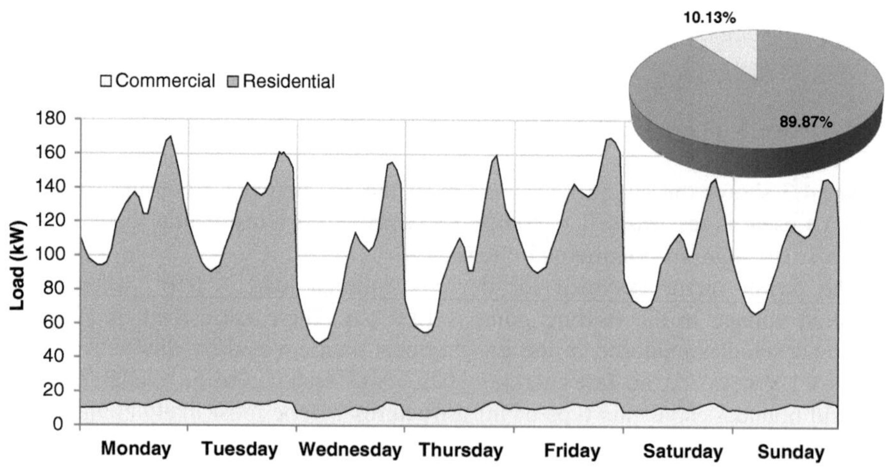

Fig. 4.4 Load diagram of a typical week (the pie chart shows the energy consumption per sector)

power related with each type of these consumers. As shown in the pie chart of Fig. 4.4, the final diagram has a contribution of ca. 90 % of the residential sector and ca. 10 % of the commercial sector.

The peak of conventional load is 170 kW, distributed over 69 of the 74 network buses, and the energy consumption during a typical week assumes the value of 18.7 MWh.

As the clients of this type of network usually leave their homes in the morning to go to work and only return at night, it is assumed that PEVs only plug-into one of the network nodes for charging when they are "parked in a residential area". When PEVs are "parked in a commercial area" or "parked in an industrial area", it is assumed that they are outside of the network boundaries and thus their load is not assigned to any network node.

4.4.1.2 Medium Voltage Grids

The set of five real MV networks analyzed cover a wide spectrum of systems, where rural and urban environments were enclosed, as well as different types of consumers: industrial, commercial and residential. A detailed description of these networks can be found in [29]. The networks are the following:

- MV Network 1—urban network from a historic city zone, mainly with commercial consumers;
- MV Network 2—rural network;
- MV Network 3—urban network from a recently built residential area in a city center;
- MV Network 4—network from a residential area in a city surroundings;
- MV Network 5—network from a touristic area.

The load profiles of each network, during a typical week, are presented in Fig. 4.5. As it can be seen, there is a significant variety of load profiles, which may be explained by the different climate, social-cultural and economic conditions of each area. Despite the differences, well defined daily patterns are easily identified for all the networks except the rural, where the load consumption along the week is more irregular.

The most relevant characteristics of the tested networks are presented in Table 4.4.

4.4.2 Maximum Allowable PEV Integration

4.4.2.1 Low Voltage Grid

The maximum allowable PEV integration percentages in the LV network are depicted in Fig. 4.6. The percentages are relative to the total number of

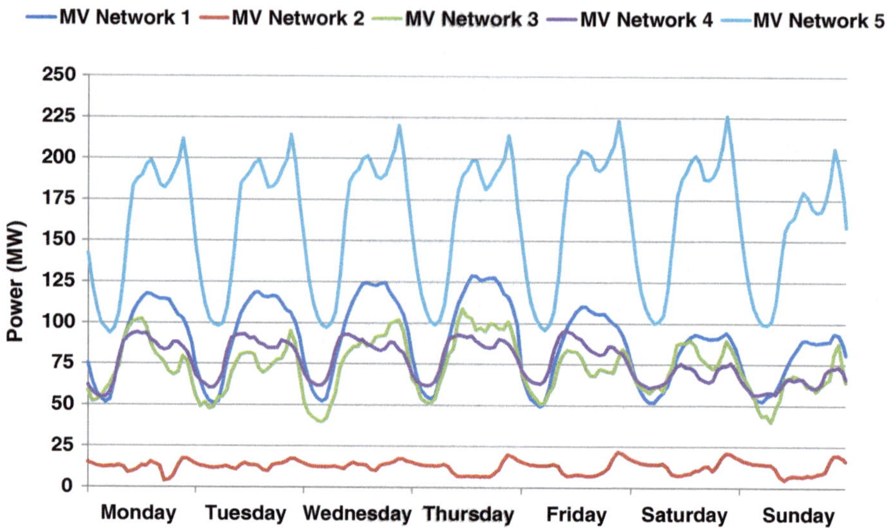

Fig. 4.5 MV networks load profiles for a typical week

Table 4.4 Networks characteristics

	MV Network 1	MV Network 2	MV Network 3	MV Network 4	MV Network 5
Type of network	Urban network from a historic city zone, mainly with commercial consumers	Rural network	Urban network from a recently built residential area in the city center	Network from a residential area in the city surroundings	Network from a touristic area
Nr. of HV/MV substations	9	5	18	8	8
Voltage level (kV)	15/20	13/13.2/20	13.2/20	20	20
Nr. of buses	930	5,292	2,323	4,598	15,077
Nr. of branches	1,150	5,355	2,462	4,680	15,353
Peak power (MW)	128.5	21.5	108.3	94.9	226.1
Weekly energy consumption (GWh)	14.9	2.1	12.2	12.8	27.0
Power factor of the conventional load	0.93	0.93	0.93	0.93	0.93
Nr. of conventional vehicles	21,135	5,203	109,641	21,749	34,155
Bus where the fast charging station is installed	25	4,614	1,419	31	12,524
Voltage lower limit (p.u.)	0.93	0.93	0.93	0.93	0.93
Period of lower energy price (h)	1–7	1–7	1–7	1–7	1–7

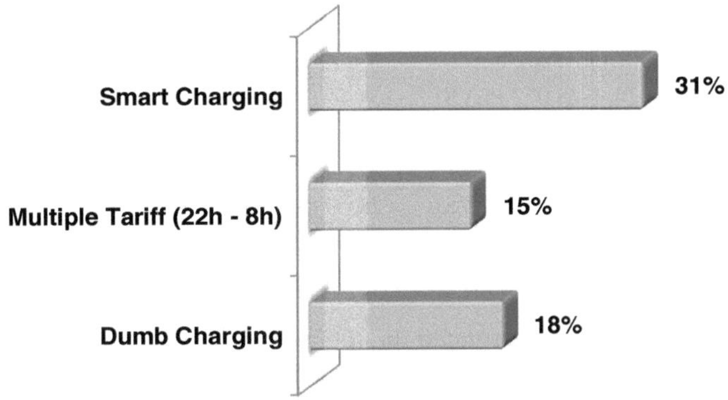

Fig. 4.6 Maximum allowable PEV integration in the LV network

conventional vehicles enclosed in the geographical area covered by this network, which is, as referred previously, ca. 125 vehicles. For the dumb charging, multiple tariff and smart charging, the number of PEVs that can be safely integrated in this network is therefore 23, 19 and 39, respectively. For this network, it was assumed that the period of lower energy prices is between 22 and 8 h, every day of the week, and that PEV owners that adhere to this charging approach will always put their vehicles charging during this period.

As it will be further discussed ahead, both under voltage problems and branches overloading were the factors that limited the PEV integration in all the charging modes studied.

In this network, the multiple tariff (22–8 h) is the charging strategy that leads to the lowest PEV integration percentage. Despite the maximum allowable PEV integration being the lowest with the multiple tariff, this is the charging strategy that accounts for the highest peak load, as shown in Fig. 4.10. The instantaneous increase of the PEVs load verified around 22 h, due to a large number of multiple tariff adherents starting their charging almost simultaneously, is the reason why the maximum allowable PEV integration with the multiple tariff is lower than with the dumb charging. This load increase occurred in specific locations of the grid, where some grid components were already operating very near their limits, provoking the occurrence of voltage limits violations and lines overloading.

4.4.2.2 Medium Voltage Grids

An overview of the maximum PEV integration percentage and the correspondent absolute value of PEVs allowed in each of the MV networks studied are presented in Table 4.5.

From the results obtained, it can be observed that the analyzed systems can handle, up to a certain level, the penetration of PEVs without concerns to the network infrastructures. However, it was verified that the maximum number of

Table 4.5 Maximum allowable PEV integration in the MV networks

	Dumb charging	Multiple tariff	Smart charging
MV Network 1	24 % (5,072 PEVS)	34 % (7,186 PEVS)	56 % (11,836 PEVS)
MV Network 2	40 % (2,081 PEVS)	57 % (2,965 PEVS)	74 % (3,850 PEVS)
MV Network 3	2 % (2,193 PEVS)	4 % (4,386 PEVS)	8 % (8,771 PEVS)
MV Network 4	28 % (6,090 PEVS)	24 % (5,220 PEVS)	42 % (9,135 PEVS)
MV Network 5	10 % (3,416 PEVS)	5 % (1,708 PEVS)	24 % (8,197 PEVS)

PEVs that can be safely integrated in the networks depends on the charging schemes adopted by the PEV owners. From the three strategies analyzed, smart charging yielded better results in all the case studies addressed, as with it was possible to reach higher PEV integration levels without violating the network technical restrictions, meaning that higher investments deferral can be obtained. The multiple tariff can be classified as the second best strategy, as in three of the five networks it attained better results than the dumb charging. It should be noted that for the MV networks, it was assumed that the period of lower energy prices is between 1 and 7 h, every day of the week, and that PEV owners that adhere to this charging approach will always put their vehicles charging in this period.

The fact of the dumb charging yielding better results than the multiple tariff in some of the networks can be explained by the instantaneous increase of the PEVs load verified around 1 h. This occurs due to a large number of multiple tariff adherents' start charging almost simultaneously. This load increase might occur in specific locations of the grid, where some grid components are already operating very near their limits, provoking the occurrence of technical violations.

4.4.3 Power Demand

4.4.3.1 Low Voltage Grid

Although an entire week has been simulated, only the load diagrams changes during Wednesday are shown in Fig. 4.7, with the dumb charging (upper right), multiple tariff (22–8 h) (lower left), multiple tariff (1–7 h) (lower right) and with the three referred scenarios plus the aggregators' smart charging (upper left). The results presented were obtained considering a PEV integration percentage of 31 % in all the charging strategies, which is the maximum PEV integration possible with the smart charging strategy, without any network reinforcements.

With the dumb charging strategy, the PEVs tend to charge mostly at the end of the day, which is the time period when people arrive home from work. The amount of power requested by the PEVs provokes a very large increase in the peak load, leading to the violation of the technical limits of several network components. In order to avoid these violations, the DSO would have to curtail 62.3 kWh of the

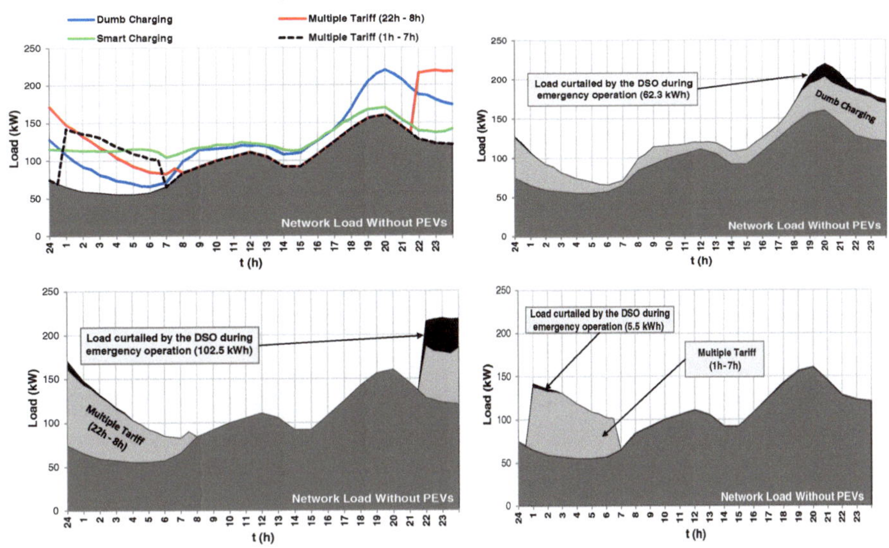

Fig. 4.7 Changes in the load diagram with the dumb charging (*upper right*), multiple tariff (22–8 h) (*lower left*), multiple tariff (1–7 h) (*lower right*) and with the three referred scenarios plus the EVS/A smart charging (*upper left*)

energy demanded by PEVs during the day analyzed (black areas in the upper right chart).

In the multiple tariff strategy, it was assumed that PEVs will only charge between 22 and 8 h, since this is the period when the energy prices are assumed to be lower. For these reasons, there are a high number of PEVs connecting to the grid for charging at 22 h and the amount of power requested provokes the violation of the technical limits of several network components. In order to avoid these violations, the DSO would have to curtail 102.5 kWh of the energy demanded by PEVs (black areas in the lower left chart).

As it can be observed, the multiple tariff policy adopted is not the most adequate for the PEVs charging, as it does not allow shifting a substantial part of the PEVs load to the valley periods. For this reason, a different multiple tariff policy was tested, where the energy prices were assumed to be lower between 1 and 7 h. The results obtained are presented in the lower right chart of Fig. 4.7. This multiple tariff strategy allows shifting a very significant part of the PEVs load to the valley hours. As a result, the occurrence of technical limits violations is greatly reduced. The DSO would only have to curtail 5.5 kWh of the energy demanded by PEVs, against the 102.5 kWh with the previously discussed multiple tariff strategy (22–8 h). It should be noted, however, that the amount of power consumed by PEVs in the multiple tariff (1–7 h) scenario is lower than in the multiple tariff (22–8 h). This happens because in the former PEVs only have a 7 h period to charge their batteries, while in the latter they have a 10 h period. The consequence will probably be

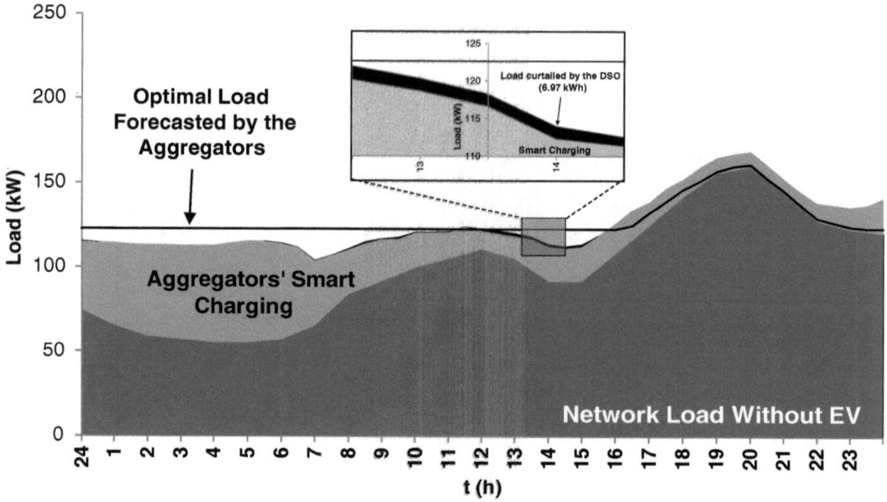

Fig. 4.8 Load diagram in the smart charging scenario (wednesday zoomed-in)

having more PEVs using to the fast charging station during the remaining hours of the day in the multiple tariff (1–7 h) scenario.

The optimal curve for the PEVS/A smart charging is represented by the black line in Fig. 4.8 and is basically the result of an optimization problem run to flatten as much as possible the load diagram. Thus, the PEVS/A will try to manage the smart charging adherents to minimize the deviations between their optimal load curve and the energy effectively consumed by PEVs (light grey area in Fig. 4.8, referred to as smart charging). The deviations between the energy bought by the PEVS/A and the energy effectively consumed by PEVs would probably be greatly reduced if more smart charging adherents were available in the network.

The load curtailed by the DSO when the system enters in the emergency operating state is presented in the zoomed area of Fig. 4.8 (black areas). As it can be observed, in the period between 13:30 and 14:30 h, a load curtailment of *ca.* 7 kWh (black area) was performed by the DSO. This load reduction allowed the DSO to solve some low voltage problems and lines overloading that were detected in some areas of the network.

To provide a clear picture of the load diagram changes provoked by the PEV integration in this LV network, the load diagrams, discriminated per bus, obtained in the scenario without PEVs (upper left), dumb charging (upper right), multiple tariff (22–8 h) (lower left) and PEVS/A smart charging (lower right) are presented in Fig. 4.9.

Differently from the load diagrams presented in Figs. 4.7 and 4.8, which represent the average load values obtained along the 500 iterations of the Monte Carlo run for each scenario, the results presented in Fig. 4.9 are referred only to the load values obtained in the last iteration.

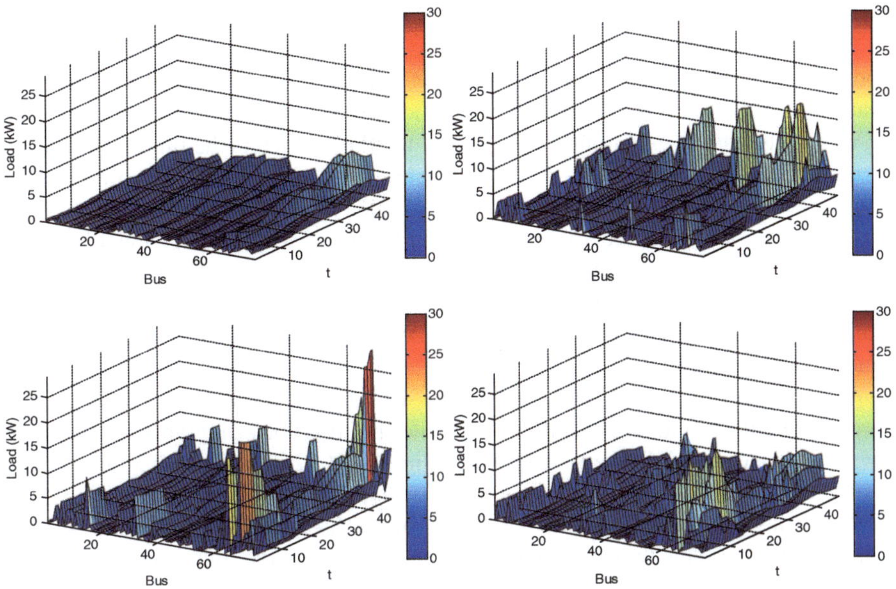

Fig. 4.9 Load diagrams, discriminated per bus, obtained in the last iteration of the simulations performed for the scenario without PEVs (*upper left*), dumb charging (*upper right*), multiple tariff (22–8 h) (*lower left*) and PEVS/A smart charging (*lower right*)

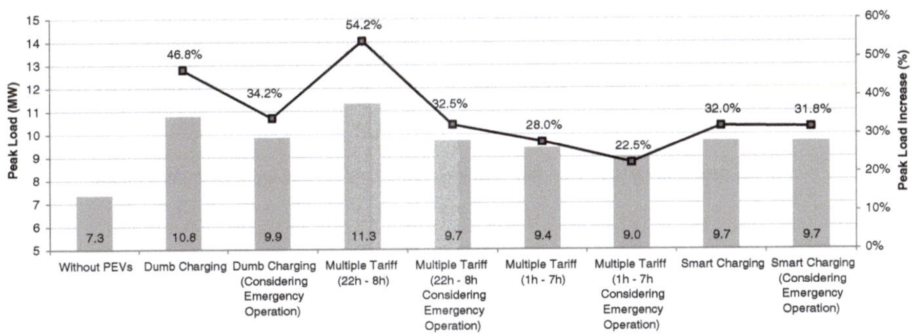

Fig. 4.10 Highest peak load recorded, in all the scenarios studied, and the respective peak load increase when compared with the scenario without PEVs

In the scenario without PEVs, this network has a peak load of 170.0 kW, which is incremented to 239.6 kW with the dumb charging, to 260.2 kW with the multiple tariff (22–8 h), to 188.5 kW with the multiple tariff (1–7 h) and to 180.4 kW with the PEVS/A smart charging.

Figure 4.10 shows the highest peak load recorded in all the scenarios studied, as well as the respective increase when compared with the scenario without PEVs.

4.4.3.2 Medium Voltage Grids

The PEVs power demand profile in the MV network 1, for the three charging strategies, is shown in Fig. 4.11. When considering the dumb charging strategy, the PEVs tend to charge essentially at the end of the day, which is the time period when people arrive home from work. In the multiple tariff scenario, the PEV owners tend to charge their vehicles between 1 and 7 h, which is the period of time when the energy prices are assumed to be lower. With the smart charging, the PEVs are charged mostly during the night, as this is the period when the PEVs availability is higher and the demand is lower. These two facts combined, make it possible to integrate a large number of PEVs in this grid without causing any technical constraints violations.

Adding the PEVs load depicted in Fig. 4.11 to the conventional load of this network clients, makes it possible to compute the total load diagrams for the three charging strategies addressed, as presented in Fig. 4.12. The load diagram for the scenario without PEVs reveals a relatively constant pattern during the week and the weekend days. A significantly large valley period is notorious during the nights, while during the days two small peaks are easily identifiable, one occurring during lunch time and the other during the evening.

In the scenario without PEVs, this network has a peak load of 128.5 MW, which is incremented to 135.6 MW using the dumb charging, to 133.9 MW using the multiple tariff and to 132.1 MW using the smart charging. The latter can be considered an outstanding achievement, since the peak load only increased 3.6 MW with a PEV integration of 57 %, representing *ca.* 12047 PEVs.

It is interesting to notice that the PEVs charging, for the dumb charging and the multiple tariff provokes changes in the hour at which the network peak load occurs. In the particular case of this network, the peak load occurrence changes from 14 to

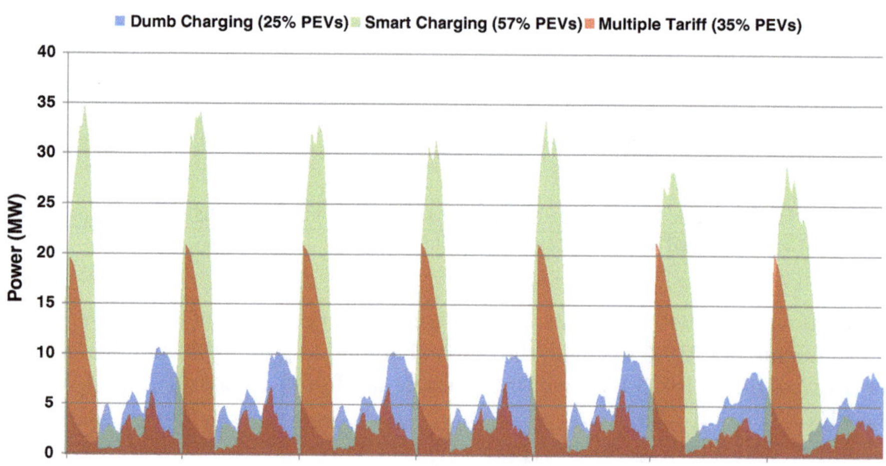

Fig. 4.11 PEV load demand profiles in the MV Network 1

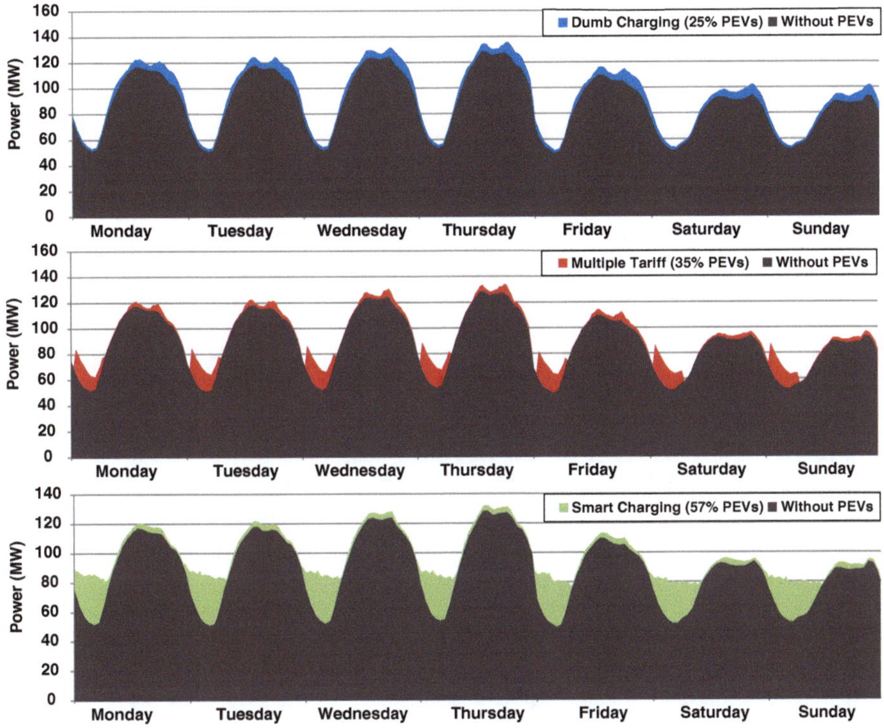

Fig. 4.12 Load profiles: differences between charging strategies (MV Network 1)

19 h of Thursday. For the smart charging, the hour at which the peak load occurs remains unchanged.

In MV Network 2, as shown in Fig. 4.13, the week considered is a typical summer week, where several farmland irrigation pumps are in operation during the night. This fact has a curious impact on the load diagram: it provokes a significant load increase during the night, making the valley periods occur during the afternoons.

When the smart charging is adopted, as PEVs will tend to be charged during the valley hours, when the grid is far from reaching its operational limits, a significant number of PEVs will be charged during the afternoons. However, as during the day the PEVs availability is lower, it is impossible to put all PEVs charging during the afternoons. The PEVs that are unavailable to charge during the day are usually available during the night. This fact explains the reason why there is a significant amount of PEVs that charge during the night and why it is impossible to obtain a "smoother" load diagram.

In the scenario without PEVs, this network has a peak load of 21.5 MW, which is incremented to 25.2 MW using the dumb charging, to 24.5 MW using the multiple tariff and to 23.2 MW using the smart charging.

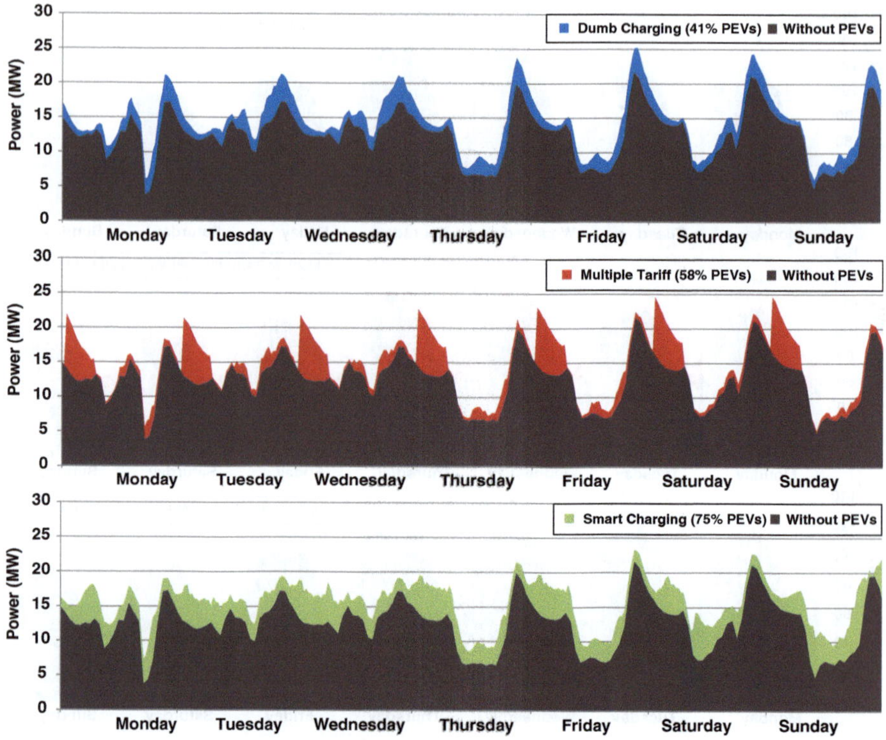

Fig. 4.13 Load profiles: differences between charging strategies (MV Network 2)

It is interesting to notice that the PEVs charging, for the multiple tariff, provokes a shift in the hour at which the daily peak load occurs, from around 22 to 1 h. For both the smart and the dumb charging, the hour at which the daily peak load occurs remains unchanged.

The load diagram of MV Network 3 for the scenario without PEVs is depicted in the dark grey area of Fig. 4.14. It reveals that the hour at which the daily peak load occurs changes from day to day. Conversely, the valley periods that occur during the night are considerably less volatile, what allows charging a high number of PEVs in these periods, as it is observed with the multiple tariff and the smart charging.

In the scenario without PEVs, this network has a peak load of 108.3 MW, which is incremented to 110.1 MW using the dumb charging, to 108.6 MW using the multiple tariff and to 112.5 MW using the smart charging. The latter can be considered an outstanding achievement, since the peak load only increased 4.2 MW with a PEV integration of 9 %, representing *ca.* 9868 PEVs.

As the PEVs consumption in this grid is not very relevant when compared to the total consumption of the conventional loads, the daily peak hours are kept almost unchanged.

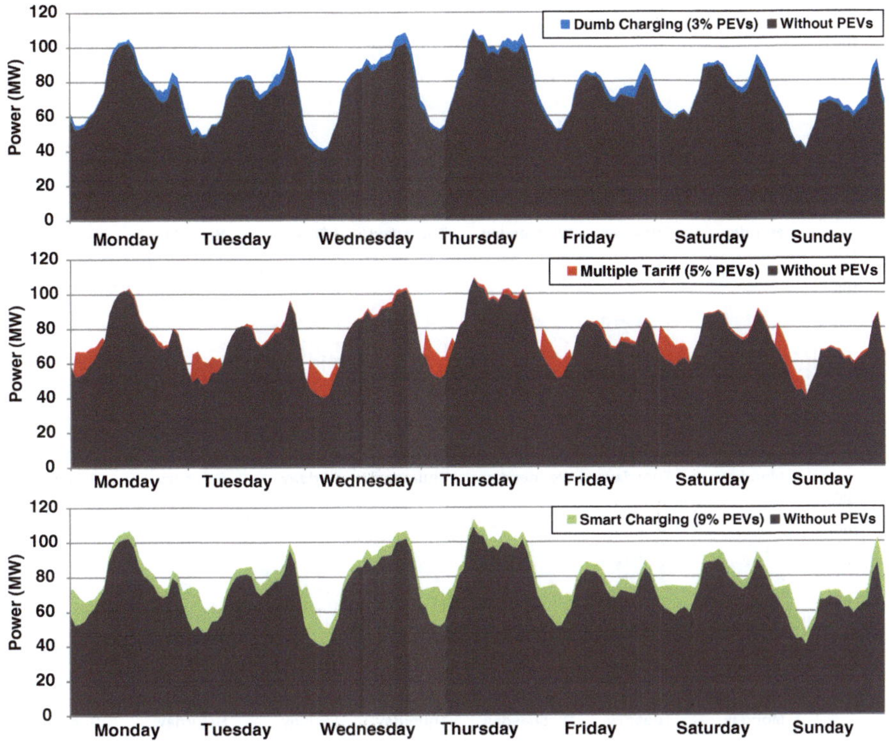

Fig. 4.14 Load profiles: differences between charging strategies (MV Network 3)

Given the particular characteristics of this network, it happens that the peak power in some days of the week increases when the smart charging is implemented. This fact is explained, on the one hand, by the higher number of PEVs that are being considered in the smart charging scenario (9,868 against 5,482 with the multiple tariff and 3,289 with the dumb charging). On the other hand, the smart charging allows increasing the network load in the non-problematic buses, while it postpones PEVs charging in the problematic ones (always taking into account the PEVs availability). This provokes a non-uniform increase of the load in the network that might result in having higher peak loads with smart charging. Nevertheless, it should be stressed that "higher peak load" does not necessarily mean "worst network operating conditions".

The load diagrams of the MV Network 4 are presented in Fig. 4.15. The shape of load diagrams is similar in all the week days, as well as during the weekend days.

The multiple tariff is the charging strategy that accounts for the highest peaks of PEV consumption in all the days of the week, despite the number of PEVs in the grid be the lowest. The instantaneous increase of the PEVs load verified around 1 h, is probably the reason why the maximum allowable PEV integration with the multiple tariff is lower than with the dumb charging. This load increase might occur

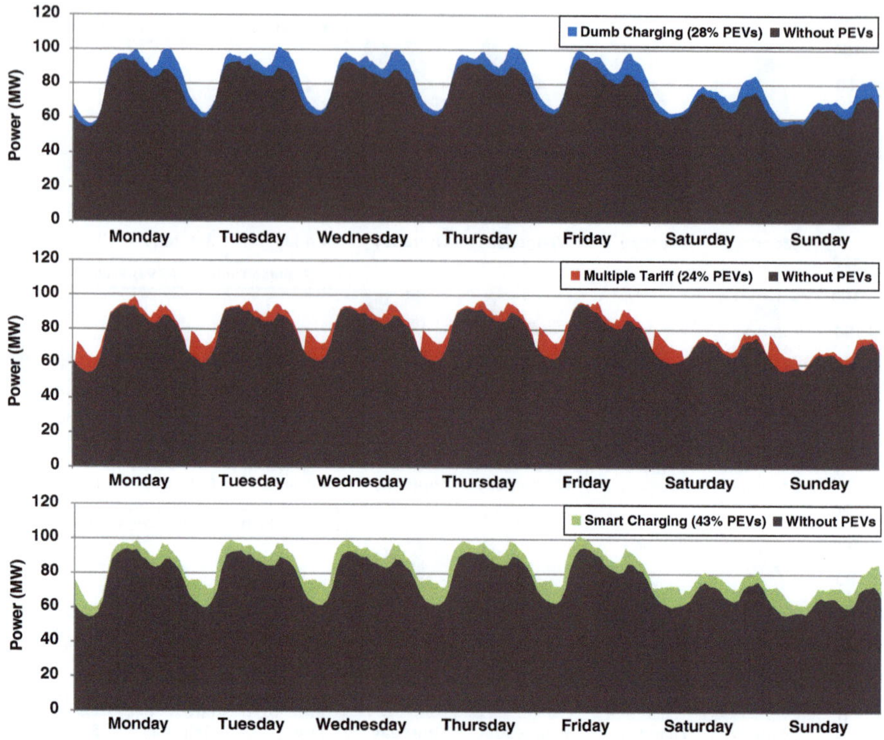

Fig. 4.15 Load profiles: differences between charging strategies (MV Network 4)

in specific locations of the grid, where some grid components are already operating very near their limits, provoking the occurrence of technical violations.

Given the particular characteristics of this network, it happens that the peak power in some days of the week increases when the smart charging is adopted, similarly to what happened in the previous case study.

In the scenario without PEVs, this network has a peak load of 94.9 MW, which is incremented to 101.5 MW using the dumb charging, to 98.4 MW using the multiple tariff and to 102.2 MW using the smart charging.

In MV Network 5, the changes in the load diagram owed to the PEVs charging is almost imperceptible in all the charging approaches, since the PEVs load is almost insignificant when compared with the total load of the conventional clients. For this reason, as shown in Fig. 4.16, the hour at which the peak load occurs did not change in all the scenarios simulated. The load diagram for the scenario without PEVs reveals a relatively constant pattern during the week and the weekend days. A significantly large valley period is notorious during the nights, while during the days two small peaks are easily identifiable, one occurring during lunch time and the other during the evening. The multiple tariff is the charging strategy that accounts for the highest peaks of PEV consumption in all the days of the week, despite the number of PEVs in the grid be the lowest.

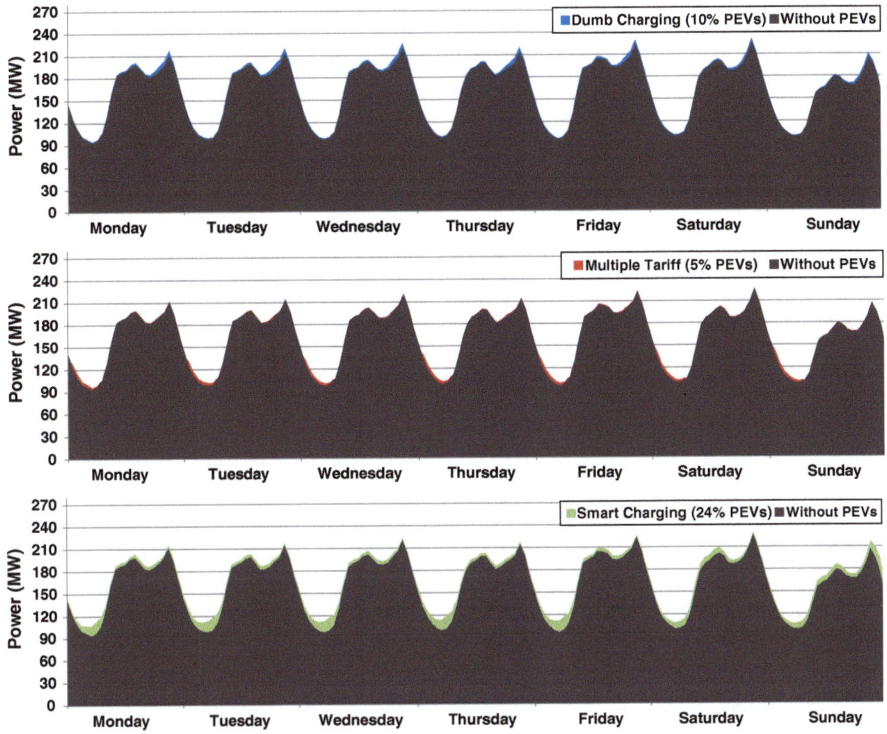

Fig. 4.16 Load profiles: differences between charging strategies (MV Network 5)

In the scenario without PEVs, this network has a peak load of 226.1 MW, which is incremented to 231.1 MW using the dumb charging, to 226.7 MW using the multiple tariff and to 229.3 MW using the smart charging

As for the MV Network 4, the instantaneous increase of the PEVs load verified around 1 h, due to a large number of multiple tariff adherents starting their charging almost simultaneously, might be the reason why the maximum allowable PEV integration with the multiple tariff is lower than with the dumb charging. This load increase might occur in specific locations of the grid, where the components are already operating near their limits, provoking the occurrence of technical violations.

4.4.4 Voltage Profiles

4.4.4.1 Low Voltage Grid

In order to evaluate the impacts that a PEV integration level of 31 % might provoke in what regards voltages in the LV network analyzed, the highest peak load registered along the 500 iterations performed for each scenario was analyzed, and the

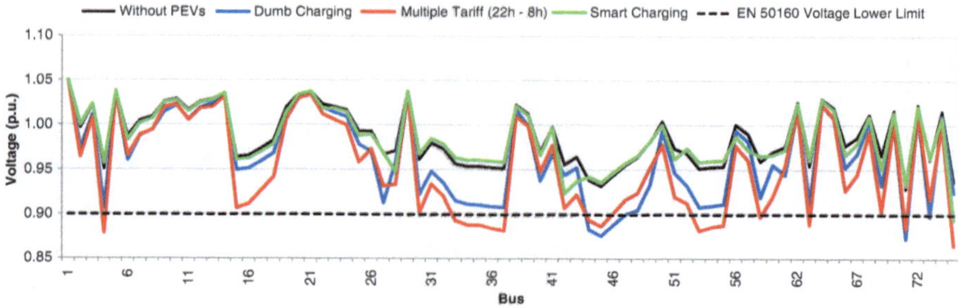

Fig. 4.17 Voltage profiles during the peak hour in the dumb charging, multiple tariff (22–8 h) and smart charging scenarios

corresponding voltage values were plotted in Fig. 4.17. A significant voltage drop occurs along the grid, namely during the periods when the demand is higher, which as Fig. 4.17 shows, causes some violations of the limit of 0.90 p.u. in the dumb charging and multiple tariff (22–8 h) scenarios. These are some of the violations that trigger the emergency operating state and that obliges the DSO to curtail some of the PEVs load. In the PEVS/A smart charging some minor voltage violations were also registered. The network voltage profile for the peak load in the scenario without PEVs is also presented in Fig. 4.17, for comparison purposes.

Figure 4.18 shows the impact of the PEVs charging in the voltage profile of one feeder, during Wednesday. As these charts show, the multiple tariff (22–8 h) is the

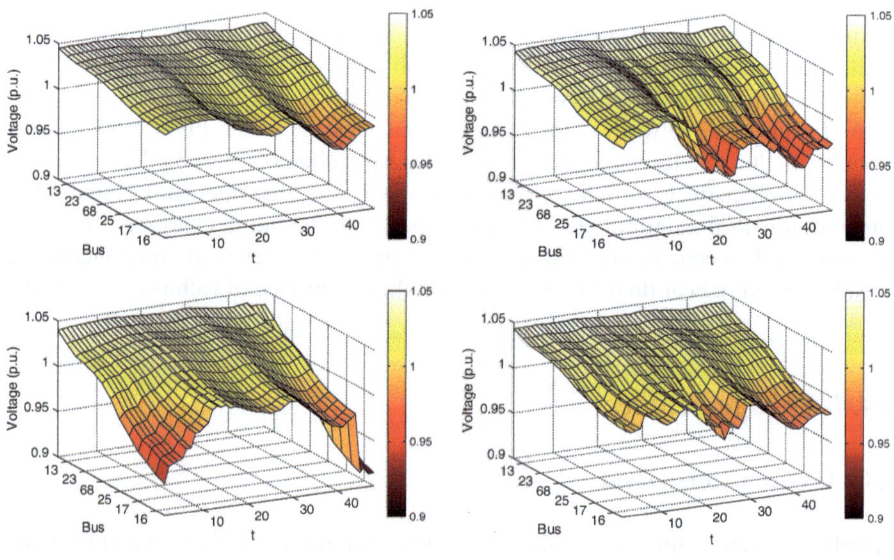

Fig. 4.18 Voltages downstream bus 14 without PEVs (*upper left*), dumb charging (*upper right*), multiple tariff (22–8 h) (*lower left*) and PEVS/A smart charging (*lower right*)

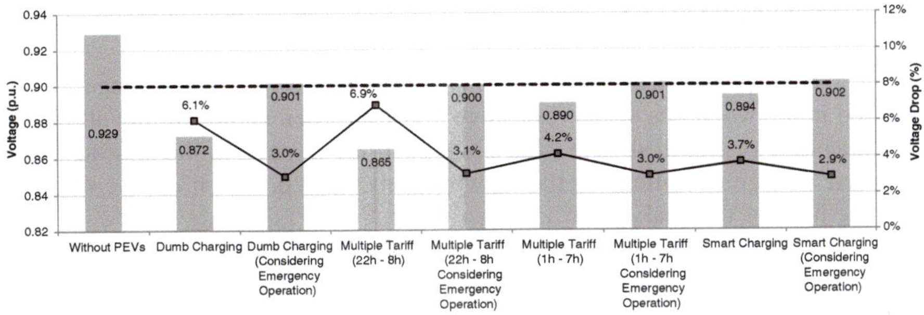

Fig. 4.19 Worst voltage recorded, in all the scenarios studied, and the respective voltage drop when compared with the scenario without PEVs

worst scenario in what regards voltages, as the extra power demanded by PEVs provokes a pronounced voltage drop along this feeder, namely at the beginning and at the end of the day. Nevertheless, no violations of the voltage lower limit were detected for the buses represented. As shown in the lower right chart, the voltage drop is greatly reduced in the PEVS/A smart charging scenario. It should be mentioned that the results presented are referred to the average voltage values obtained from the 500 iterations performed for each scenario.

The worst voltages recorded in all the scenarios studied, as well as the respective voltage drop when compared with the scenario without PEVs, are shown in Fig. 4.19.

All the voltage limits violations were recorded along the simulations, in order to keep track of the most problematic areas of the network. With the results obtained and using Eq. 4.12, the probabilities of having voltages below the imposed limit were calculated for each bus and plotted in Fig. 4.20.

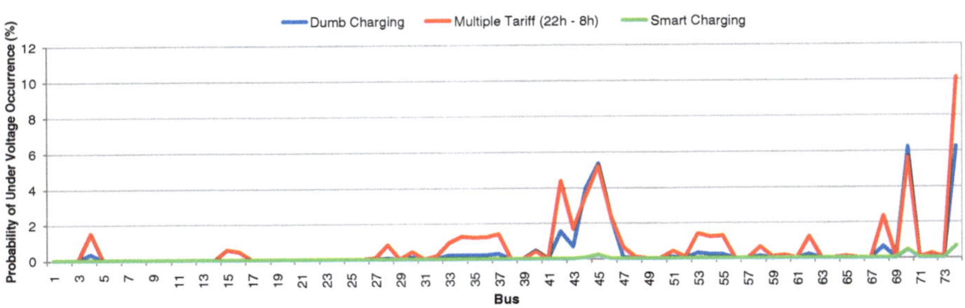

Fig. 4.20 Probability of under voltage occurrence in the dumb charging, multiple tariff (22–8 h) and smart charging

4.4.4.2 Medium Voltage Grids

Figure 4.21, depicts for each of the MV networks analyzed, the voltage values obtained in the worst bus of the respective network, when the maximum allowable PEV integration is reached. The values presented are referred to the hour at which the worst voltage conditions in the networks are verified, which can be different from the hour of the peak load.

As it can be observed, with the exception of the MV Network 2, the PEVs extra demand provokes almost insignificant decreases in the voltage values with relation to the initial value (with no PEVs present in the grids). It is important to recall that in MV networks the R/X ratio is low, contrarily to LV networks, what makes the impacts of the active power consumed by PEVs less relevant regarding voltage drops. In addition, as the majority of the MV networks studied are from urban areas, they are more prone to overloading problems than under voltage issues.

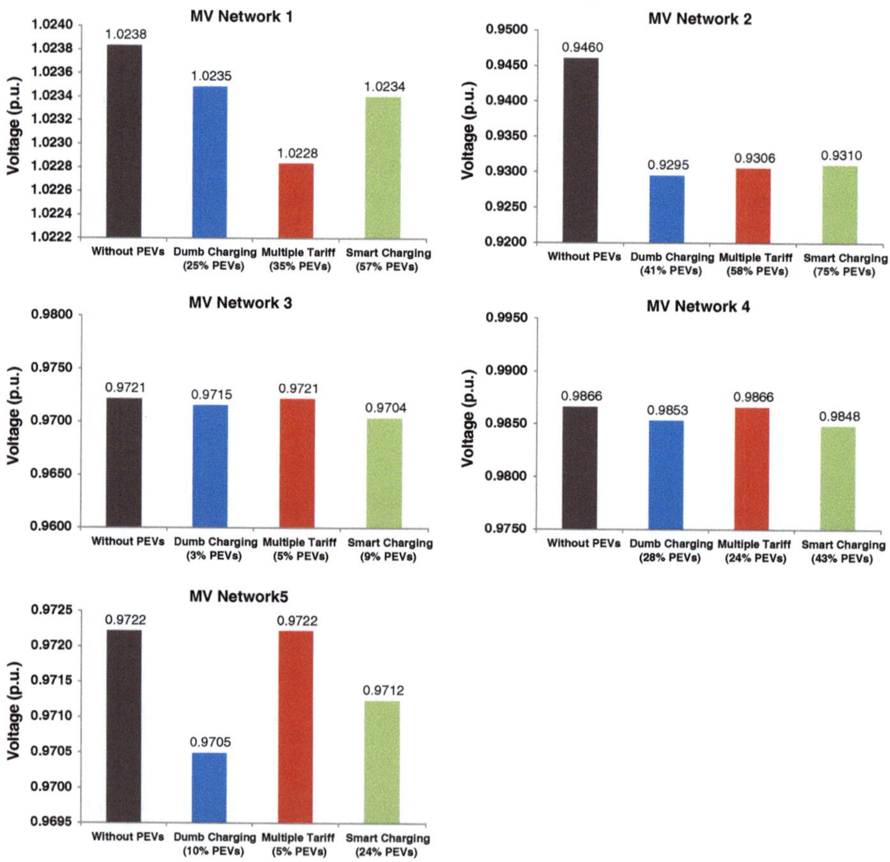

Fig. 4.21 Voltage in the worst bus

Thus, as expected, the voltage values attained for the MV Network 1, MV Network 3, MV Network 4 and MV Network 5 are within acceptable values, while for the MV Network 2 they reach values near or even below the minimum limit allowed (defined for these networks as 0.93 p.u.).

From these results, it is possible to conclude that the voltage lower limit is probably the technical constraint that impedes a higher PEV integration level in the MV Network 2.

Although the voltage values regarding the use of different charging strategies are presented in the same figure, for each network, it should be stressed that they are referred to different scenarios of PEV integration. Thus, the only possible fact that can be concluded from the figures presented is that the smart charging provides better results, as it is the charging strategy that allows the safe integration of a larger number of PEVs in all the case studies evaluated.

4.4.5 Lines Overloading

4.4.5.1 Low Voltage Grid

As urban networks are usually composed by short lines and are subjected to high power demand levels, they are prone to face lines overloading problems. As shown in Fig. 4.23, this is exactly what happens in the case of the LV grid under analysis, where ratings above 100 % in several lines of the network were detected in all the scenarios analyzed, including the PEVS/A smart charging scenario. Figure 4.22 provides an overview of the impact provoked by PEVs in the network lines loading, for the peak load demand in the scenario without PEVs (upper left), dumb charging (upper right), multiple tariff (22–8 h) (lower left) and smart charging (lower right). The color grading between blue and red stands for increasing line loading values, ranging from 0 to 100 %. In Fig. 4.23 the worst ratings recorded in all the scenarios studied are presented, as well as the respective increase when compared with the scenario without PEVs.

A considerable number of violations were registered for some lines, as shown in Fig. 4.24. The results obtained for the worst line reveal that there is a probability of overloading occurrence of *ca.* 12 % in both the dumb charging and multiple tariff (22–8 h) scenarios. The probabilities presented in Fig. 4.24 were obtained using Eq. 4.13.

4.4.5.2 Medium Voltage Grids

In the MV, differently to what was verified for the voltage profiles, lines overloading was the most critical aspect in the generality of the studied grids, with especially emphasis in the networks with urban characteristics.

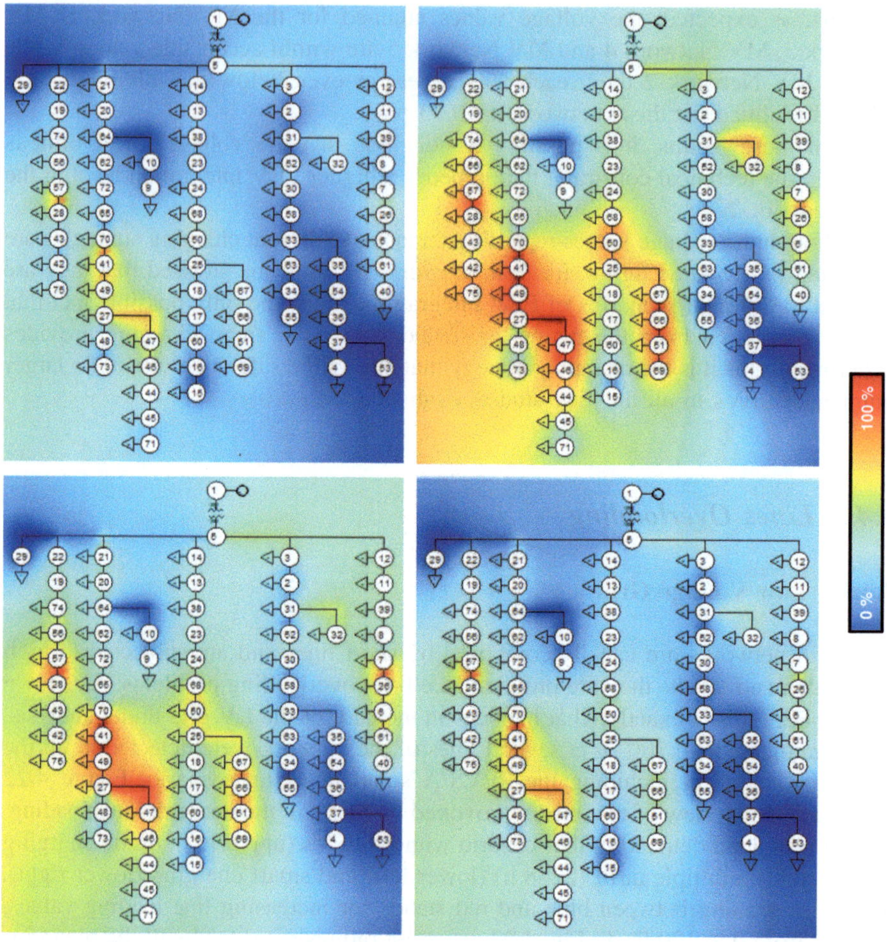

Fig. 4.22 Lines loading during the peak hour in the scenario without PEVs (*upper left*), dumb charging (*upper right*), multiple tariff (22–8 h) (*lower left*) and smart charging (*lower right*)

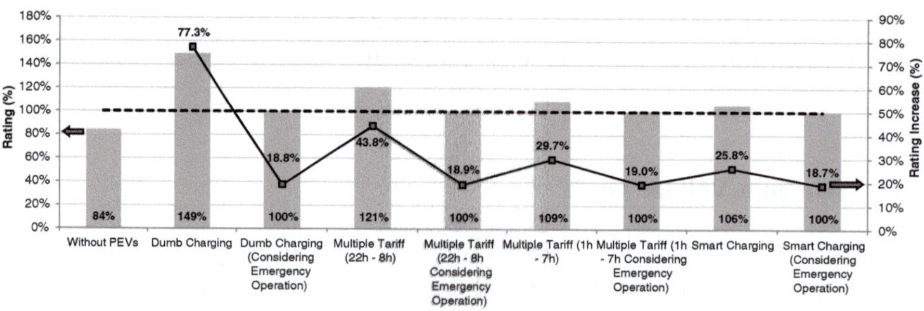

Fig. 4.23 Worst rating recorded and the respective increase when compared with the scenario without PEVs

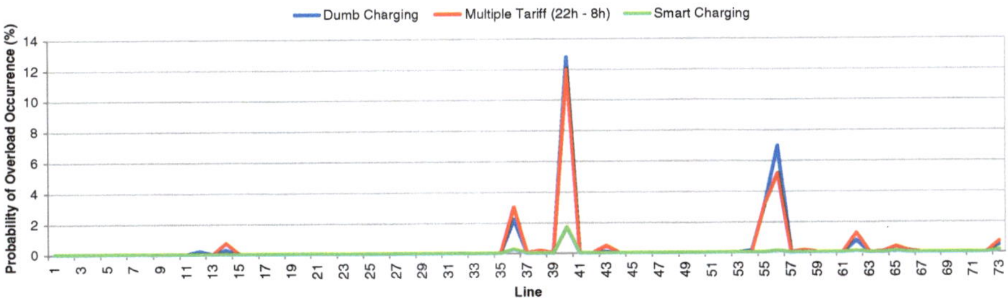

Fig. 4.24 Probability of overload occurrence in the dumb charging, multiple tariff (22–8 h) and smart charging scenarios

Looking at Fig. 4.25, where the rating percentage of the most loaded branch during the hour at which the worst overloading occurs is depicted, it is possible to observe the effects of the PEVs charging when the three different charging methods are applied. The maximum rating allowed was assumed to be 100 %.

The results obtained show, in all the networks, that the branches loading levels considerably worsen with the growth of the number of PEVs present in the grids. In fact, lines loading are the factor limiting a further PEV integration in the MV Network 1, MV Network 3, MV Network 4 and MV Network 5. The MV Network 2, besides having low voltage problems, has also overloading issues.

Likewise to the stated in the analyses performed for voltage profiles, the rating values presented in Fig. 4.25 for the different networks are referred to different scenarios of PEV integration. Thus, as for the voltages results, the only possible fact that can be concluded is that the smart charging provides better results, as it is the charging strategy that allows safely integrating a larger number of PEVs in all the case studies evaluated. If it was considered a fixed number of PEVs in the grids, the worst rating percentage obtained with the smart charging would be significantly lower than the value obtained with the dumb charging and the multiple tariff.

The dumb charging strategy is the charging scheme that accounts for the worst results in the MV Network 1, MV Network 2 and MV Network 3, while multiple tariff strategy accounts for the worst results in the MV Network 4 and MV Network 5.

As referred previously, the worst results of the multiple tariff obtained in the MV Network 4 and MV Network 5, in comparison with the dumb charging approach, might be explained by the specific characteristics of these grids or by the location of the fast charging station, which might be connected in a more fragile place of the network.

The location of the fast charging station is, in fact, a very important variable in what regards branches overloading, as the large amount of power absorbed by these facilities might overload the branches upstream. For this reason, it is advisable that the installation of a fast charging station should always be preceded by a detailed impact study.

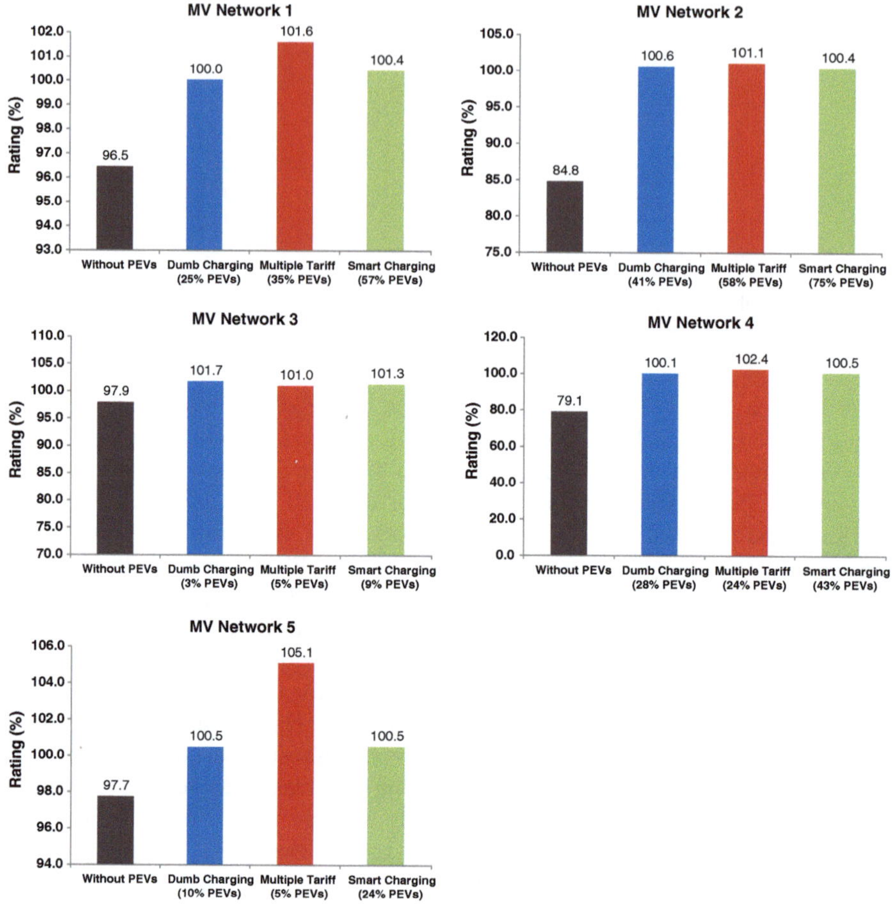

Fig. 4.25 Rating in the worst branch

4.4.6 Energy Losses

4.4.6.1 Low Voltage Grid

In Fig. 4.26, the average value of the weekly energy losses is depicted, obtained along the 500 iterations performed for each scenario, as well as the percentage of the consumption they represent.

The weekly energy losses grow 72 % from the scenario without PEVs to the dumb charging, 65 % to the multiple tariff (22–8 h), 55 % to the multiple tariff (1–7 h) and 54 % to the PEVS/A smart charging scenario.

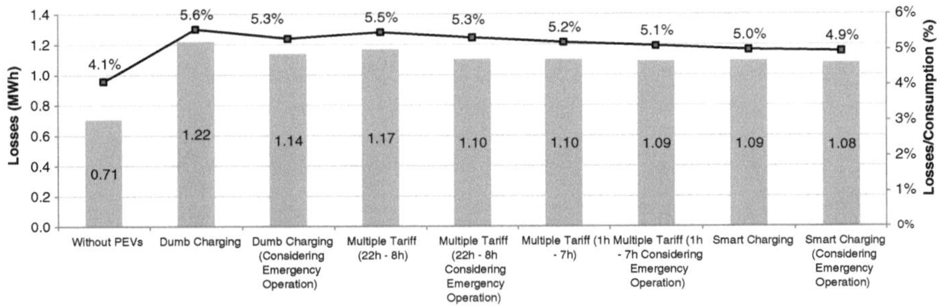

Fig. 4.26 Energy losses in all the scenarios studied (during the entire week)

4.4.6.2 Medium Voltage Grids

Looking at Fig. 4.27, it is possible to evaluate the effects of the PEVs charging in the weekly energy losses of the networks analyzed.

Each chart presents the absolute value of the losses (bars), referred to the left vertical axis, and their value relative to the overall energy consumption (circles), referred to the right vertical axis.

In all the networks a significantly increase in the absolute value of the weekly losses can be observed when comparing the scenarios with and without PEVs.

As the energy losses are directly proportional to the square of the current, when the demand increases, due to the PEVs charging, the current flowing along the grid branches rises as well, provoking an increase in the losses.

Although the absolute value of energy losses increases with the smart charging, when comparing with all the other scenarios, its relative value reveals that this charging strategy yields some benefits in the majority of the cases studied, namely in the MV Network 1, MV Network 3 and MV Network 4.

The adoption of the multiple tariff strategy could also lead to some positive results. As it can be observed, when comparing this strategy with the dumb charging, it is possible to decrease losses relative value in four of the analyzed networks (MV Network 1, MV Network 3, MV Network 4 and MV Network 5), mainly due to the load valleys in the load diagrams that occur between 1 and 7 h. The exception is the MV Network 2 because valley hours, occurring in the late afternoon, do not coincide with the period when the majority of the multiple tariff adherents charge their PEV: between 1 and 7 h.

Generally, the charging method that yields worst results is the dumb charging, since it leads to the occurrence of the highest peak loads, which, expectably, lead to the higher increases in the energy losses.

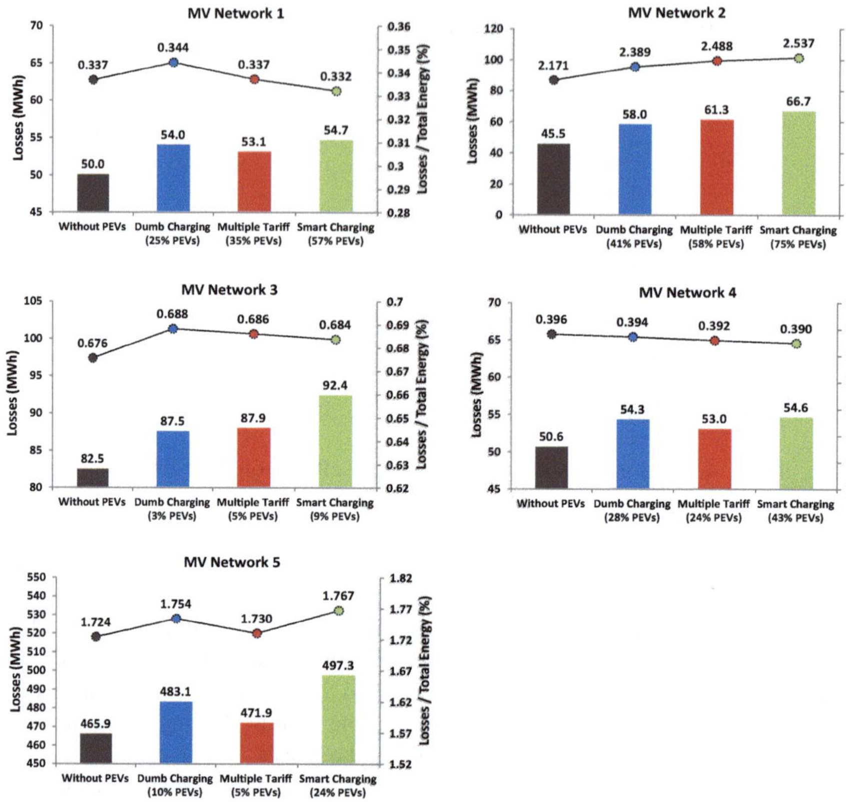

Fig. 4.27 Weekly energy losses

4.4.7 Critical Mass Analysis

The main goal of this section is to identify the percentage of PEV owners that need to adhere to the smart charging in order to safely integrate a given number of PEVs.

The first step of the methodology implemented consisted in the consideration of an initial PEV integration percentage, of which one half of the PEVs were assumed to be dumb charging adherents and the other half multiple tariff adherents.

Then, if problems were not detected in the network, the PEV integration percentage was increased by 10 until a problem in the network was detected. For the MV Network 1, used as test case, the initial PEV integration percentage assumed was of 10 % and the first technical violation was detected with a 30 % PEV integration.

The second step of the methodology consisted in iteratively increasing the percentage of smart charging adherents, in steps of 5 %, while decreasing the dumb charging and multiple tariff adherents accordingly. This procedure was repeated until the technical problems identified were solved.

The problems identified were assumed to be solved when the voltage values in all the network buses and the line ratings in all the branches are within the pre-defined limits (voltage lower limit—0.93 p.u.; voltage upper limit—1.07 p.u.; maximum rating allowed—100 %).

For the case study under analysis, the percentage of smart charging adherents that allowed solving the problems detected—the critical mass—was of 45 %.

A rather obvious assumption about the critical mass is that its value is expected to increase as the number of PEVs connected to the grid increases. In order to demonstrate it, a second scenario with a higher PEV integration (40 %) was analyzed.

4.4.7.1 Case I—MV Network 1 with 30 % of PEVs

As it can be seen in Fig. 4.31, when considering 30 % of PEV integration, with 50 % dumb charging and 50 % multiple tariff, there are already some lines overloading. The worst branch is 1.6 % above its maximum rated capacity.

By incrementing the share of smart charging adherents to 45 % (critical mass value), while decreasing both dumb and multiple tariff adherents to 27.5%, the worst branch rating decreases to 98.4 %, value within the allowed limits. The differences between both scenarios referred have a direct influence on the PEVs load profiles, as presented in Fig. 4.28. In the first scenario (in blue), the PEVs power consumption has two daily peaks: one in the late afternoon (due to dumb charging adherents) and other during the first hours of the night (due to multiple

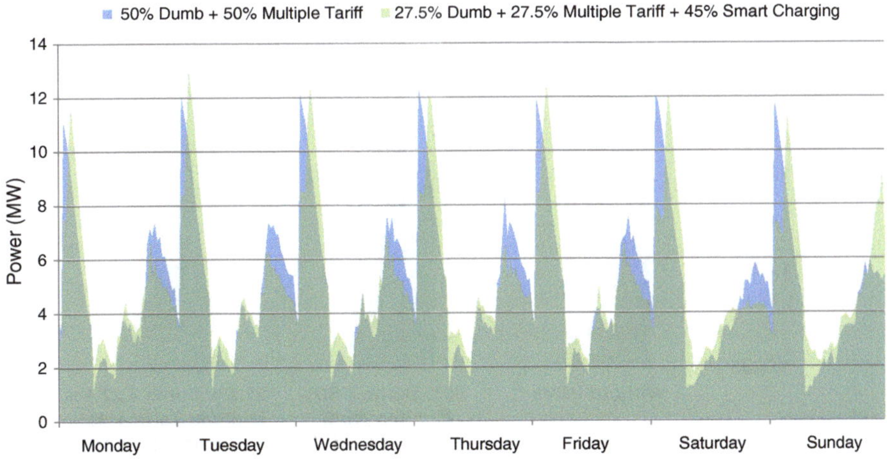

Fig. 4.28 PEV load demand profiles in the MV Network 1 (30 % PEVs)

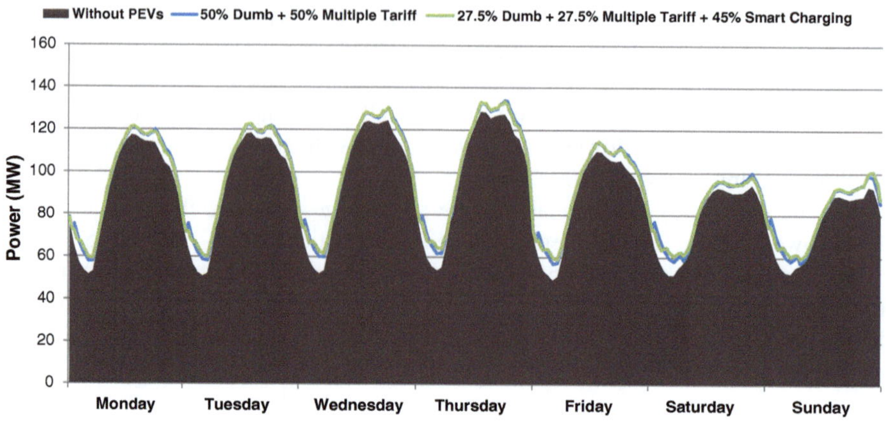

Fig. 4.29 Load profiles without and with PEVs (MV Network 1, 30 % PEVs)

tariff adherents). When the value of the smart charging adherents is incremented to 45 %, a decrease in PEVs power during the late afternoon peak can be noticed.

In Fig. 4.29 the load diagrams for both cases studied are presented. The peak load in the scenario with 45 % of smart charging adherents slightly decreases, in comparison with the scenario with 50 % dumb charging and 50 % multiple tariff.

Figures 4.30, 4.31 and 4.32 present, respectively, the voltages in the worst bus, the rating in the worst branch and the weekly energy losses for both scenarios simulated. As it can be noticed, the increase in the number of smart charging adherents yields benefits in all the indexes analyzed.

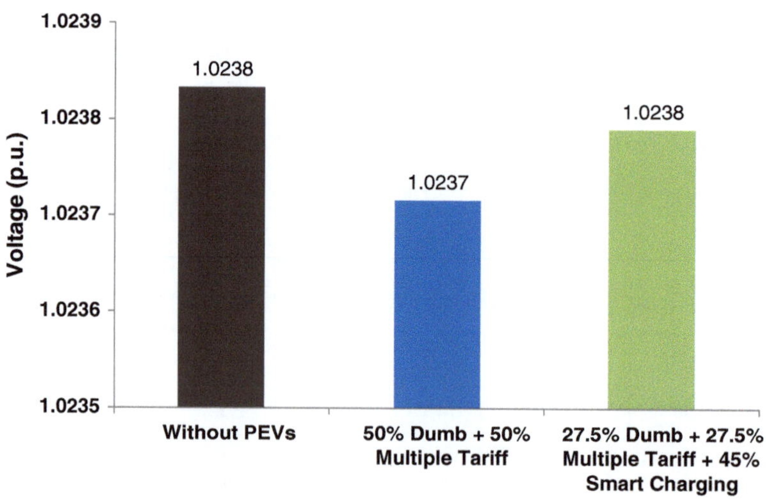

Fig. 4.30 Voltage in the worst bus (30 % PEVs)

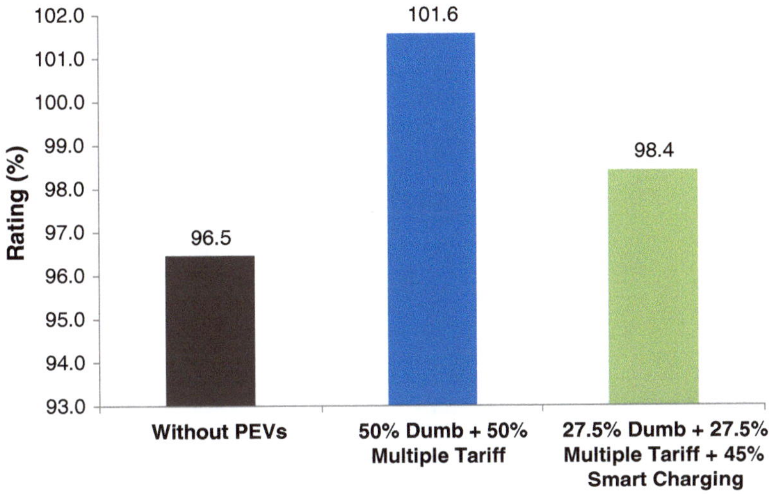

Fig. 4.31 Rating in the worst branch (30 % PEVs)

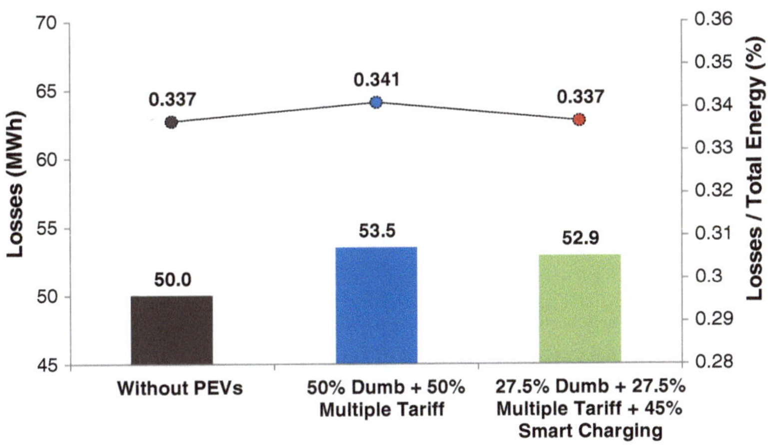

Fig. 4.32 Weekly losses (30 % PEVs)

4.4.7.2 Case II—MV Network 1 with 40 % of PEVs

As referred previously, in Case II it was considered that 40 % of total conventional vehicles were electric. As it can be seen in Fig. 4.36, when considering 40 % of PEV integration, the branches overloading are considerably aggravated, when comparing with Case I. The worst branch is *ca.* 10 % above its maximum rated capacity, against the 1.6 % verified in the previous case. Under these conditions, the worst branch rating can only be decreased to acceptable values if the smart charging

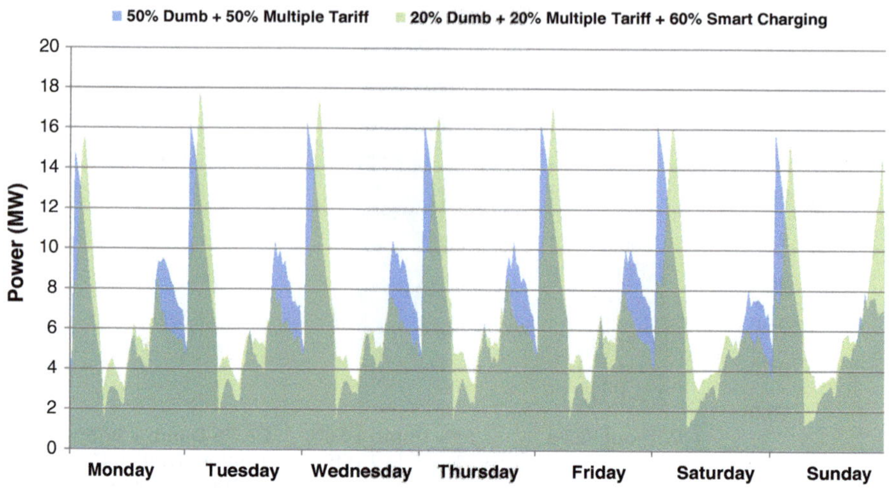

Fig. 4.33 PEV load demand profiles in the MV Network 1 (40 % PEVs)

adherents' percentage reaches 60 % (critical mass value). The differences between both scenarios referred have a direct influence on the PEV load profiles, as presented in Fig. 4.33. When the share of the smart charging adherents is incremented to 60 %, a decrease in PEVs power during the late afternoon peak can be noticed. This reduction is even more evident than the one verified in Case I, mostly due to the higher share of controllable PEVs present in the network.

In Fig. 4.34 the load diagrams for both cases studied are depicted. As in Case I, the peak load in the scenario with 60 % of smart charging adherents decreases

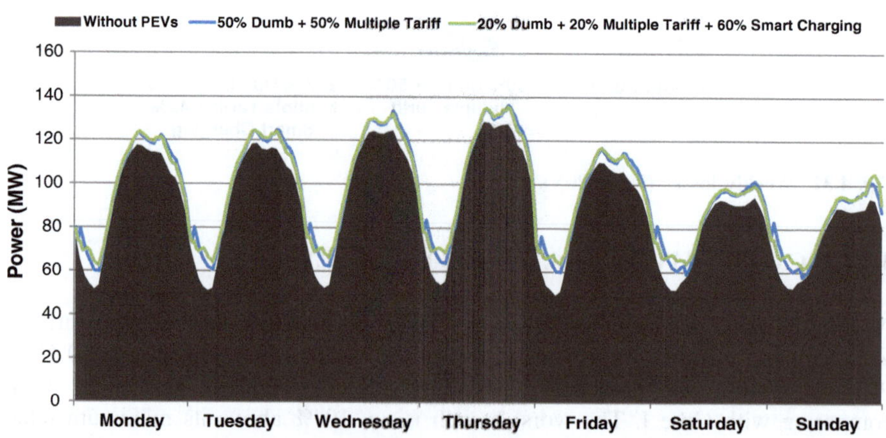

Fig. 4.34 Load profiles without and with PEVs (MV Network 1, 40 % PEVs)

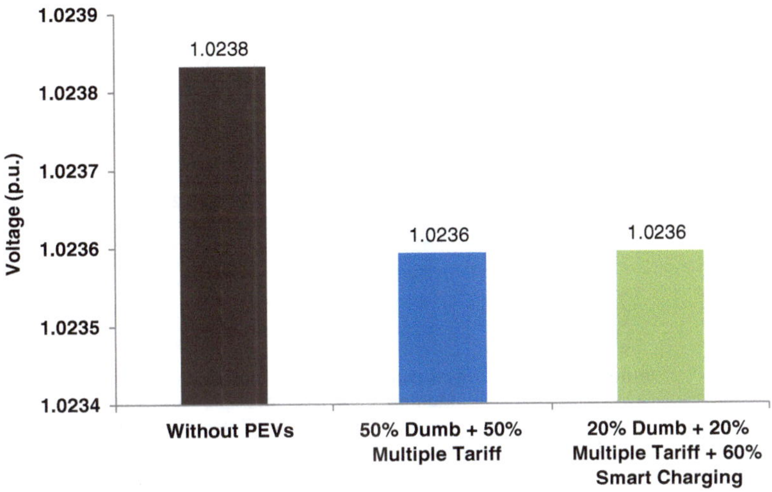

Fig. 4.35 Voltage in the worst bus (40 % PEVs)

slightly, in comparison with the scenario with 50 % dumb charging and 50 % multiple tariff.

Figures 4.35, 4.36 and 4.37 present, respectively, the voltages in the worst bus, the rating in the worst branch and the weekly energy losses for both scenarios simulated. As it can be noticed, the increase in the number of smart charging adherents yields benefits in all the indexes analyzed.

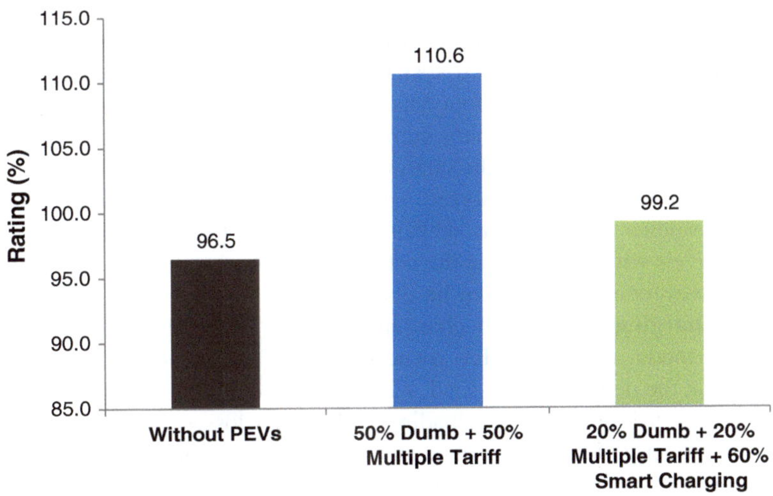

Fig. 4.36 Rating in the worst branch (40 % PEVs)

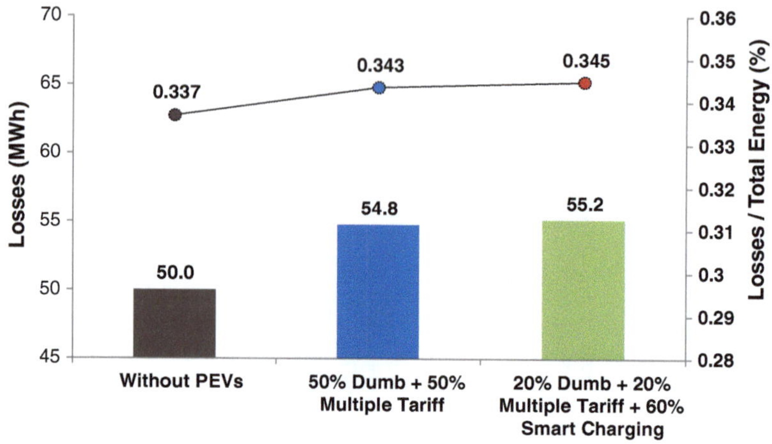

Fig. 4.37 Weekly losses (40 % PEVs)

4.5 Conclusions

By analyzing the results presented along this chapter, it is easy to understand that the magnitude of the PEV impacts in distribution networks are influenced by several factors, like the PEV integration level, the PEV owners' behavior, mobility patterns, the networks load profiles and technical characteristics, the number and location of fast charging stations in the grid, the PEV charging modes, among others. These factors have been carefully analyzed during the simulations performed, as reported along the chapter, being possible to reach the following conclusions:

- **Maximum allowable PEV integration levels (without considering network reinforcements)**: The analyzed systems can handle, up to a certain level, the penetration of PEVs without concerns to the networks infrastructures. However, it was verified that the maximum number of PEVs that can be safely integrated in the networks depends on the charging schemes adopted by the PEV owners. From the three strategies analyzed, smart charging yielded better results in all the case studies addressed, as with it was possible to reach higher PEV integration levels without violating the networks technical restrictions, meaning that higher investments deferral can be obtained.
- **Dumb charging**: The dumb charging revealed to be the most problematic charging mode, as letting PEV owners charge freely leads to a considerable increase in the networks peak load, with negative consequences in what regards voltage profiles, branches overloading and energy losses. This happens due to the simultaneity that exists between the PEVs home arrival and plug-in for charging and the periods when the households consumption is higher.

- **Multiple tariff**: As referred, the multiple tariff strategy can be an effective charging strategy for some networks, provided that pronounced valley periods exist in the daily load diagrams and that they occur more or less during the same daily periods. Nevertheless, it should be stressed that the instantaneous increase of the PEVs load verified in the beginning of the lower energy price period, due to a large number of multiple tariff adherents starting their charging almost simultaneously, might cause several technical problems in some networks, namely in those operating in more strained conditions.
- **Smart charging**: The smart charging is the charging mode that allows obtaining better results, as the envisaged mechanisms to manage the PEVs charging enable a better exploitation of the resources available at each moment, preventing the occurrence of voltage problems and branches overloading. The smart charging also proved to be very effective in reducing the energy losses, and consequently the CO_2 emissions inherent to the electricity generation sector, as it prevents the occurrence of high peak loads. Nevertheless, it should be stressed that even with smart charging, there is a moment when it is impossible to integrate further PEVs in the networks and investments in reinforcements are then unavoidable.
- **Urban networks**: As these networks are usually composed by short lines and are subjected to high power demand levels, they are very likely to face branch/transformer overloading problems faster than voltage drop issues. The results presented in this report prove this fact, as congestion problems were identified in all networks studied with urban topology: LV network, MV Network 1, MV Network 3, MV Network 4 and MV Network 5.
- **Rural networks**: Differently from urban networks, rural networks have usually long radial lines, which provoke considerable voltage drops. Thus, low voltage problems are expected in this type of networks, namely in the buses farthest from the feeding points. The results presented in this report prove this fact, as low voltage problems were only registered in the rural network analyzed (MV Network 2).
- **Changes in the load diagrams**: The extra power demanded by PEVs provokes several changes in the networks load diagrams, which are more pronounced as the PEV integration level rises. Nevertheless, the analysis performed allows concluding that it is impossible to generalize results in a rigorous manner, as the changes induced in the load diagrams depend of several factors that are different from network to network, like the PEV integration level, the PEV owners' behavior, mobility patterns, the networks load profiles and technical characteristics, the number and location of fast charging stations, the PEV charging modes, etc. Even so, some simplistic statements can be made, which should interpreted by the reader as case dependent information: (1) the increase in the power demand, with the dumb charging, is highly related with the journeys distribution along the day; (2) the increase in the power demand, with the multiple tariff, is highly related with the period during which the energy prices are lower; (3) the smart charging allows, to a certain extent, filling the valleys in

the load diagrams, resulting in relatively more uniform power demand profiles along the day.

- **Fast charging stations**: The location of the fast charging stations should be carefully analyzed, as they might provoke severe voltage violations or branches overloading, due to the large amount of power that they consume (*ca.* 40 kW × nr. PEVs charging).
- **Critical mass (i.e. percentage of PEV owners that need to adhere to the smart charging, for a given PEV integration level, in order to enable the safe operation of the networks)**: The methodology proposed for this purpose proved to be an effective procedure to calculate the critical mass, as with it was possible to identify the percentage of smart charging adherents that are required to solve the network problems identified for a given PEV integration level. Nevertheless, as it happened with the load diagrams, the simulations performed allow concluding that it is impossible to generalize results in a rigorous manner. From the analysis of the results obtained, it is only possible to conclude that the critical mass, besides being dependent of the network considered, increases with the PEV integration level.

From what was stated in the previous points, it is clear that the path to safely integrate large quantities of PEVs in distribution networks, without making large investments in components reinforcements, is to implemented mechanisms that allow managing the PEVs charging not only taking into account their owners' requests, but also the networks technical conditions. Nevertheless, it should be remarked that the adherence to these controlled charging schemes will ultimately be always a decision of the PEV owners. Thus, it is of utmost importance to timely define and implement adequate incentives policies, attractive enough to make PEV owners willing to participate in such controlled charging schemes.

References

1. Peças Lopes JA, Soares FJ, Almeida PMR (2009) Identifying management procedures to deal with connection of electric vehicles in the grid. In: 2009 IEEE Bucharest PowerTech, pp 1–8, June 2009
2. Clement-Nyns K, Haesen E, Driesen J (2010) The impact of charging plug-in hybrid electric vehicles on a residential distribution grid. IEEE Trans Power Syst 25(1):371–380
3. Heydt GT (1983) The impact of electric vehicle deployment on load management strategies. IEEE Trans Power Appar Syst 102(5):1253–1259
4. Peças Lopes JA, Soares FJ, Almeida PMR (2009) Smart charging strategies for electric vehicles: enhancing grid performance and maximizing the use of variable renewable energy resources. In: EVS24: electric vehicle symposium, May 2009
5. Papadopoulos P, Cipcigan LM, Jenkins N, Grau I (2009) Distribution networks with electric vehicles. In: Proceedings of the 44th international universities power engineering conference (UPEC), pp 1–5, Sept 2009
6. Clement K, Haesen E, Driesen J (2009) Coordinated charging of multiple plug-in hybrid electric vehicles in residential distribution grids. In: Power systems conference and exposition, PSCE'09, pp 1–7

7. Raghavan SS, Khaligh A (2012) Impact of plug-in hybrid electric vehicle charging on a distribution network in a smart grid environment. In: Innovative smart grid technologies (ISGT), 2012 IEEE PES, pp 1–7
8. Helmschrott T, Godde M, Szczechowicz E, Matrose C, Schnettler A (2012) Methodical approach for analyzing the impact of a mass introduction of electric vehicles on the electricity networks in Europe. In: Power and energy conference at Illinois (PECI), 2012 IEEE, pp 1–6
9. Richardson P, Flynn D, Keane A (2010) Impact assessment of varying penetrations of electric vehicles on low voltage distribution systems. In: IEEE power and energy society general meeting, pp 1–6
10. Pieltain Fernández L, Román TGS, Cossent R, Domingo CM, Frias P (2011) Assessment of the impact of plug-in electric vehicles on distribution networks. IEEE Trans Power Syst 26 (1):206–213
11. Lopes JA, Soares FJ, Almeida PMR (2011) Integration of electric vehicles in the electric power system. Proc IEEE 99(1):168–183
12. Galus MD, Andersson G (2008) Demand management of grid connected plug-in hybrid electric vehicles (PHEV). In: IEEE energy 2030 conference, pp 1–8
13. Galus MD, Andersson G (2009) Integration of plug-in hybrid electric vehicles into energy networks. In: IEEE Bucharest PowerTech 2009, pp 1–8
14. Geth F, Willekens K, Clement K, Driesen J, De Breucker S (2010) Impact-analysis of the charging of plug-in hybrid vehicles on the production park in Belgium. In: MELECON 2010—2010 15th IEEE Mediterranean electrotechnical conference, pp 26–28, Apr 2010
15. Sortomme E, Hindi MM, MacPherson SDJ, Venkata SS (2011) Coordinated charging of plug-in hybrid electric vehicles to minimize distribution system losses. IEEE Trans Smart Grid 2 (1):198–205
16. Siemens PTI. http://w3.siemens.com/smartgrid/global/en/products-systems-solutions/software-solutions/planning-data-management-software/planning-simulation/pages/pss-e.aspx. Accessed 20 Feb 2007
17. Python Software Foundation. Python language reference. http://www.python.org
18. Moreira CL, Rua D, Bessa RJ et al (2010) Extend concepts of MG by identifying several EV smart control approaches to be embedded in the smart grid concept to manage EV individually or in clusters. Research report, deliverable D1.2, EU MERGE project
19. Soares FJ, Almeida PMR, Lopes JAP (2014) Quasi-real-time management of electric vehicles charging. Electr Power Syst Res 108:293–303
20. VENTURI FORMULA-E TEAM website. http://www.venturi.fr. Accessed 12 June 2014
21. MICRO-VETT website. http://www.micro-vett.it/. Accessed 12 June 2014
22. Mitsubishi Motors website. http://www.mitsubishi-cars.co.uk/imiev/. Accessed 12 June 2014
23. Peugeot website. http://www.peugeot.co.uk/showroom/ion/5-door/. Accessed 12 June 2014
24. Downing N, Ferdowsi M (2010) Identification of traffic patterns and human behaviours. Research report, deliverable D1.1, EU MERGE project
25. Bessa RJ, Soares FJ, Pecas Lopes JA, Matos MA (2011) Models for the EV aggregation agent business. In: 2011 IEEE Trondheim PowerTech, pp 1–8
26. Decreto-Lei n.º 39/2010, Diário da República, 1.ª série — N.º 80 — 26 Apr 2010 (Decree-Law in Portuguese)
27. Moreira C, Ribeiro P, Almeida PMR et al (2011) Functional specification for tools to assess steady state and dynamic behavior impacts, impact on electricity markets and impact of high penetration of EV on the reserve levels. Research report, deliverable D2.2, EU MERGE project
28. EN 50160:2007. Voltage characteristics of electricity supplied by public distribution systems. European committee for electrotechnical standardization—CENELEC
29. Sánchez C, Gonzalez A, Ferreira R et al (2010) Scenarios for the evolution of generation system and transmission, distribution grid evolution requirements for different scenarios of EV penetration in different countries. Research report, deliverable D3.1, EU MERGE project

Chapter 5
Smart Management of PEV Charging Enhanced by PEV Load Forecasting

E. Xydas, C. Marmaras, L.M. Cipcigan and N. Jenkins

Abstract According to the U.K. Department for Transport, the 97 % of transport energy consumption comes from the usage of oil. Therefore, a fuel diversification is needed to improve the energy security, and plug-in electric vehicles (PEVs) seem promising in giving an alternative solution. However, PEV owners need electric power from the grid in order to recharge the batteries of their vehicles. PEV charging load is a new type of demand, influenced by additional factors such as travel and driving patterns. Average travel distance within a day, the connection and disconnection time and the PEV's power consumption will directly affect the daily load curve. This chapter proposes a decentralized control algorithm to manage the PEV charging requests. The aim of the control algorithm is to achieve a valley-filling effect on the demand curve, avoiding a potential increase in the peak demand. The proposed model includes an algorithm for PEV short term load forecasting. This forecast contributes to the effectiveness of the control model. Through different case studies, the performance of the proposed model is evaluated and the value of the PEV load forecasting as part of the PEV load management process is illustrated.

Keywords Plug-in electric vehicle (PEV) · Data mining · PEV load forecast · Energy management

5.1 Introduction

Due to environmental concerns and energy security issues, the proportion of the PEVs in the car sales is anticipated to increase in the following years. A large deployment of PEVs will lead to lower greenhouse gas emissions, fuel efficiency, oil independency and increased penetration of renewable energy. Road transport today is dominated by oil-delivered fuels and internal combustion engines and such

E. Xydas (✉) · C. Marmaras · L.M. Cipcigan · N. Jenkins
Department of Electrical Engineering, Cardiff University, Queen's Building, The Parade, Cardiff CF24 3AA, Wales, UK
e-mail: XydasE@cardiff.ac.uk

S. Rajakaruna et al. (eds.), *Plug In Electric Vehicles in Smart Grids*,
Power Systems, DOI 10.1007/978-981-287-317-0_5

a high level of dependence on one single source of primary energy carries strategic, climatic and economic risks [1, 2]. Electric mobility offers an opportunity for diversification of the primary energy sources used in transport, but also brings new risks, technological challenges and commercial imperatives. Depending on the location and the times the PEVs are plugged in, they could cause local constraints on the grid [3]. According to [4], for the extreme scenario of the PEV penetration in Great Britain in 2030, it is estimated that the electricity demand will increase by 59.6 %. In order to maintain the normal operation of the power grid, the generation capacity must be increased to meet this new additional demand of PEV charging. Equipment, especially in the existing distribution and transmission networks, will be overloaded and this may affect the stability and reliability of the power system. It is anticipated that the system may face voltage-drops, power losses increase and overloading of distribution transformers [4]. The impact of PEVs is significant for the Distribution Network Operators (DNO) as there is a need to manage the line congestion and voltage drops [3]. The future electricity networks will also have to integrate distributed generators, as well as energy storage and adapt to new types of demand in addition to the need to power electric vehicles [5]. Network reinforcement is one solution to cope with the large deployment of PEVs, however this solution is expensive. An alternative way is to integrate smart grid control techniques which avoid large investments on the electricity grid [5].

As the penetration of electric vehicles grows, the number of recharging stations where the PEVs can replenish their energy needs is increasing. These charging stations are divided in three main categories according to their location and technical specifications: private residential, private non-residential and public charging stations [6]. The private residential charging stations are installed mainly at home, and have a slow charging rate. Private non-residential chargers are usually installed in the parking lot of a company, accommodating the PEVs energy needs of its employees. Local authorities install publicly available recharging infrastructure on the streets, mainly located at the city centers. The majority of the charging stations have data collection capabilities, keeping records of the PEV charging events.

With the number of PEVs and charging stations gradually rising, charging events are going to occur in various locations and times. This creates a large volume of data, recorded and stored by the individual charging stations or back up offices [7]. According to [8], collecting and managing the dispersed data in a central point is impractical. Therefore, distributed data collection centers are proposed to manage the data from a group of charging stations. The main role of these centers is to aggregate the data from many charging stations offering data reduction services.

According to the recorded charging events, the databases contain information related to (i) the time and place of each charging event, (ii) the amount of requested energy, and in some cases (iii) the ID of the PEV and/or the charging station. This information is used for understanding the charging and travel patterns of the PEV owners, as well as the activity of each charging station. The value of the collected data is useful in many different fields. Various actors are using this information according to their targets. For example a Distribution System Operator (DSO) uses the temporal and spatial information of the charging events to plan future investments in network upgrades. In addition, these data can also be used to determine the PEV charging load profiles [9]. The flexibility of PEV charging load will allow new market entities like PEV aggregators to develop the suitable business models for demand side management in order to provide ancillary services to grid operators [10–14]. PEVs are entities that live in both electricity and transport networks. Charging a PEV in public or street locations requires at least a parking space per charging point. Due to the finite number of parking spaces in a city, especially in the city center, the number of PEVs that are charging at the same time is limited. This will affect the road transport networks particularly the daily travel patterns and the congestion parameters [15]. Authorities should take into consideration this effect and utilize proper mechanisms and parking schemes for the PEV deployment [16]. Different data collected from a typical charging event are used for different applications. More specifically, the analysis of real charging data assists in creating typical PEV charging profiles, information which is important to the planning of the future PEV charging infrastructure. The appropriate number and the charging rate of the public charging stations of an area are defined by understanding the trend of the charging data. Moreover, the extracted charging load patterns are used to explore opportunities for possible ancillary services to the grid operators (load management, frequency response). Using these real data is also important to develop appropriate business models to promote the mass deployment of PEVs. Finally, PEV charging is not only affecting the electric power systems but also the transport networks.

For this reason, authorities are considering possible impacts of PEV charging on the traffic condition or parking spaces of a geographic area.

Due to the variety of charging data, a generic data analysis methodology is needed for extracting the relevant information for each application. The complexity of this process and the large amount of data, make data mining techniques a promising solution in extracting information from charging events records [17, 18].

The chapter is structured as follows: Sect. 5.2 describes the stages of the data mining process and the development of a generic framework for the PEV load forecast methodology based on the data mining principles. Section 5.3 presents the importance of PEV load forecast to the management of the PEV charging. Also, the integrated proposed model is illustrated in detail as well as the operation of the main entities of the model. Section 5.4 presents the simulation results and the effectiveness of the integrated model is evaluated. Finally, conclusions are drawn in Sect. 5.5.

5.2 PEV Load Forecasting Using Data Mining Methods

5.2.1 Data Mining

Data mining is an interdisciplinary process combining different techniques like machine learning, pattern recognition and statistics in order to extract information from large datasets [19]. It is the process of discovering hidden patterns, associations, anomalies and significant structures in large amounts of data. Data mining is a step in the procedure known as Knowledge-Discovery-from-Databases (KDD) [20]. Data pre-processing, data formatting, and data mining actions constitute the KDD process, as presented in Fig. 5.1.

The Data Pre-processing stage includes data selection and data clearance actions. Once the data are collected from a database, a preliminary analysis takes place in order to understand and select the useful data. This selection is critical for the extraction of information as the unnecessary data create noise, and lead to incorrect conclusions. Furthermore, the data selection is reducing the size of the dataset, resulting in lower storage and computational requirements, as well as in reduced processing time. The selected data are then forwarded into a sequence of clearing actions, where missing values are either removed or forecasted whenever it is possible. In addition, outliers like unrealistic charging durations are detected and eliminated so that the extracted conclusions are not distorted.

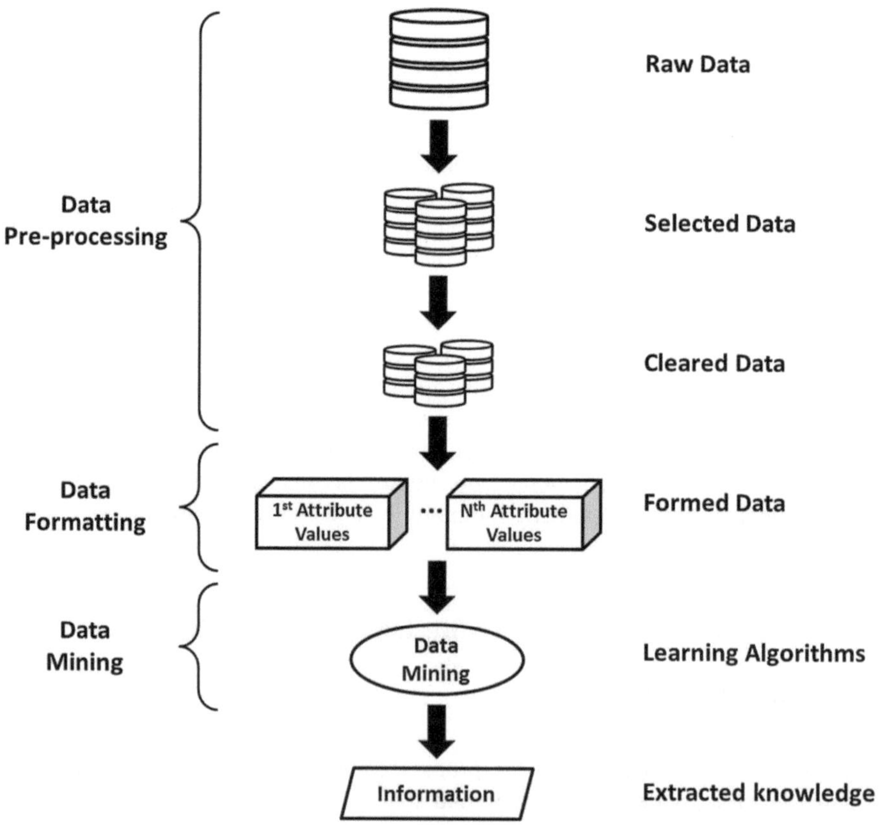

Fig. 5.1 The flowchart of the KDD process

The Data Formatting stage is a stage between the Data Pre-processing and Data Mining processes. In this stage the data are transformed and formatted according to the Data Mining technique of the next stage. According to the available data, attributes are defined to express the different features in the dataset. The data are then organized in attribute groups that express the same type of information. This arrangement is essential for the KDD procedure, and a potential error in the Data Formatting stage will influence the outcome.

The Data Mining stage is the final and the most important stage of the KDD procedure. This stage includes data processing with one or more algorithms, defined accordingly to the goal of the analysis. Two main types of algorithms exist based on the applied learning procedure: unsupervised and supervised learning algorithms. Unsupervised learning algorithms include clustering procedures, often useful for an initial understanding of a dataset, as well as (depending on the application) data partitioning and pattern recognition. In supervised learning algorithms each data string is a pair of an attribute vector and a target (desired) value. Because of this formulation the model is forced to learn the correlation between the attributes and the target values, and they are widely used for classification and forecasting tasks. In order to evaluate the learning capability of a data mining method, the initial dataset is divided in the training and the testing dataset. The training dataset is provided to the KDD procedure to learn the correlations among the attributes and create a trained model. This process is called "training process". The testing dataset is then forwarded to the trained model to evaluate its performance ("testing process"). In case the trained model fails to provide the desired output (within a confidence interval), a reconfiguration of the data mining method is applied and the training-testing sequence is repeated. This iterative process is terminated once the desired output is reached.

5.2.2 PEV Load Forecasting Methodology

In order to develop a PEV load forecasting model, all the stages of the KDD process are considered. Recorded datasets of PEV charging events contain both useful and irrelevant information to the purpose of the particular analysis. For example, in case the purpose is a behavioral analysis of the PEV owners, information regarding the time, the location and the User ID of each charging event are most relevant in contrast to data regarding the charging station manufacturer. On the other hand, in case the purpose is to calculate the utility of a particular company's charging stations, information about the charging station manufacturer becomes more important than data regarding the User ID. Therefore, according to the target of the particular analysis and the availability of the data, an appropriate data selection process is important to be applied. Data regarding a charging event is recorded from the charging station and then forwarded to one or more data collection centers. This process involves a number of components and communication links increasing the risk of a potential failure in this chain. Corrupted or missing data are not a rare phenomenon in such complex communication networks. Therefore, data clearing processes are important to remove the noisy data and the outliers. For example, charging events with zero or negative energy are removed from the dataset. However, a careful analysis at this stage is beneficial from another point of view. By keeping track of the location or the station's ID from where the corrupted data come, an indication of the normal/abnormal operation is obtained.

Due to the amount of charging events and variety of additional data, the Data Pre-processing stage of the KDD process is time consuming without an automated way of processing this volume of information. Furthermore, it is highly possible that additional information about charging events (Stations info or User Info) may be stored in different files. Thus, in order to effectively cope with the data, a script for integrating all sources of information in one dataset is developed. Then, another script is executed in order to select the appropriate data for the ongoing analysis as well as to check the data for mistaken values or outliers within the dataset. The steps of the Data Pre-processing stage are illustrated in Fig. 5.2.

In the Data Formatting stage, a transformation script is applied to the Cleared Data in order to change their structure. For forecasting applications, the structure of the data follows the template presented in Table 5.1. This structure is important for training the model to decode the correlations among the attributes. The time horizon of the forecast defines the time difference between the target and the attribute

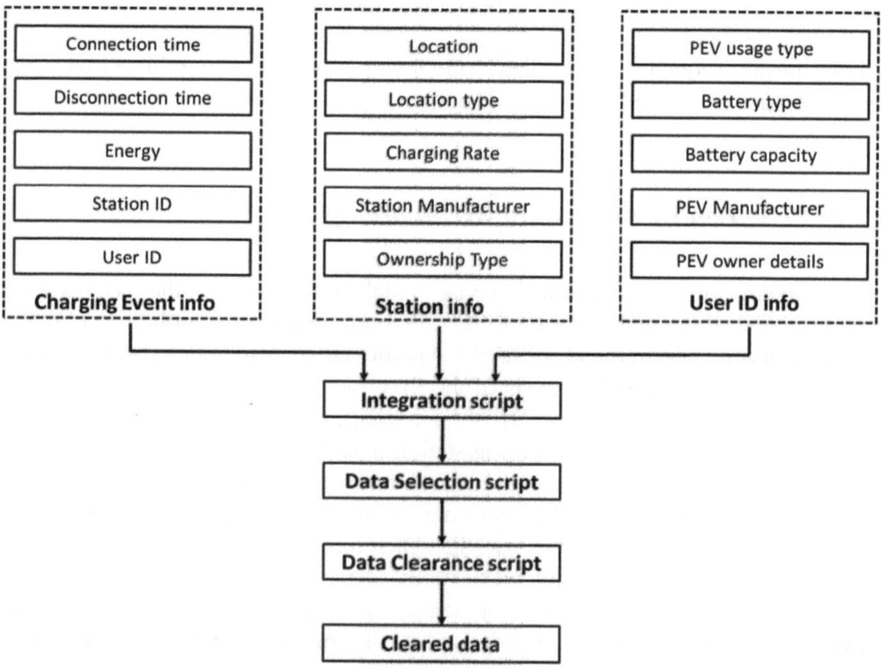

Fig. 5.2 The data pre-processing stage

Table 5.1 Structure of the formed data

Target Title	Attribute_1 Title	…	Attribute_M Title
Target Value-1	Attribute_1 Value-1	…	Attribute_M Value-1
Target Value-2	Attribute_1 Value-2	…	Attribute_M Value-2
…	…	…	…
Target Value-N	Attribute_1 Value-N	…	Attribute_M Value-N

values. For a day ahead PEV charging demand forecast for example, the target values are related to the charging demand of N day while the attribute values refer to N − 1 day. Moreover, the resolution of the forecast defines the time difference between consecutive rows. In the day ahead PEV charging demand forecasting case for instance, assuming a half hour resolution, each row is related to a specific half hour of a day. The new data structure is presented in Table 5.2.

Once the Data Formatting stage is complete, the formed data are used to train the forecasting model. The training process of the forecasting model is shown in Fig. 5.3. An appropriate data mining technique is selected for the forecasting model depending on the characteristics of the PEV charging events. Factors like randomness can make one data mining technique more suitable than another. For example, advanced data mining techniques are needed for an accurate forecast, if high fluctuating data are coming from a public charging station. On the other hand, charging events from a residential charging point are more periodic and easier to predict. A simple method like linear regression can be used for less complicated forecasting models while powerful methods like Support Vector Machines (SVM), Artificial Neural Networks and Trees [21, 22] are used by advanced forecasting models. Regardless the fluctuation of data, a proper configuration of the selected data mining method is also important for the accuracy of the forecasting model (parameters tuning process).

SVM was selected as the appropriate data mining method for the PEV load forecast model due to its high performance. SVM is a machine learning method associated with classification, regression and other learning tasks and was developed by Vapnic, Guyon and Boser [23–27]. SVM tries to find linear separations

Table 5.2 Data structure for a day-ahead PEV charging demand forecast with half hourly resolution

PEV charging demand	Attribute_1 Title	…	Attribute_M Title
PEV charging demand for 1st half hour of N day	Attribute_1 Value for 1st half hour of N − 1 day	…	Attribute_M Value for 1st half hour of N − 1 day
PEV charging demand for 2nd half hour of N day	Attribute_1 Value for 2nd half hour of N − 1 day	…	Attribute_M Value for 2nd half hour of N − 1 day
…	…	…	…
PEV charging demand for 48th half hour of N day	Attribute_1 Value for 48th half hour of N − 1 day	…	Attribute_M Value for 48th half hour of N − 1 day

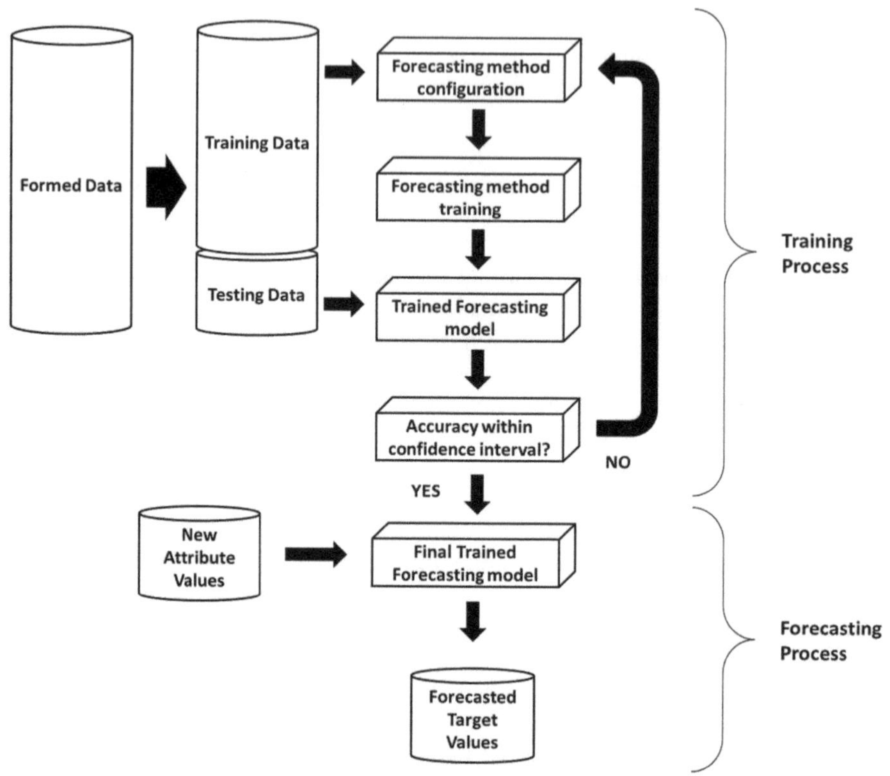

Fig. 5.3 The training and forecasting process

between the data ("decision boundaries" for separating one class from another). Assuming data with two attributes, SVM depicts them into a two dimensional space and search for possible separating lines. If the data are depended on three attributes, they are projected on a three dimensional space, and SVM searches for the possible separating planes. Generalizing for n-attributes, the depiction is on an n-dimension space and SVM search for separating hyperplanes. SVM will find many different lines or hyperplanes which divide the data. The optimal line/hyperplane is selected based on the maximization of the separating distance. When SVM cannot find linear separations in the initial data, they transform these data into new spaces using the kernel functions. For each kernel type, there are different variables that need to be tuned in order to perform effectively [28–33]. The Gaussian Radial Basis Function (RBF) described in Eq. 5.1 is found to outperform in many cases of learning tasks, and thus this kernel type was used in the PEV load forecast model [34]. Thus, for the PEV load forecast model the parameters γ, C and ε are considered in the tuning process. Parameter γ expresses the width in the kernel function [30], parameter C represents the trade-off between the training error and the maximum number of data points that can be separated in all possible ways [35], while parameter ε influences the number of support vectors and consequently the generalization capability of the model [36].

$$K(x,y) = \exp\left(-\gamma(\|x-y\|)^2\right), \quad \gamma > 0 \tag{5.1}$$

where x and y express samples of different attribute vectors.

The formed data are separated in two parts, the Training and Testing dataset. Since the appropriate data mining method and its parameters are selected, the model is trained based on the training dataset. Once the model is trained, the testing dataset is used to evaluate the performance of the forecasting model. In the evaluation process, only the attribute values of the testing dataset are supplied to the trained model in order to forecast the corresponding target values of the testing dataset. By comparing the forecasted values with the actual target values of the testing dataset, the performance of the model is evaluated. If the accuracy of the model is not sufficient, a reconfiguration of the parameters of the data mining method is required and then the model is retrained. Subsequently, the performance of the new trained model is evaluated and this iterative process terminates when the accuracy level is reached.

Once this procedure is completed, the forecasting model can be used on unknown data. The new dataset includes values in the attribute columns, while the target values are missing (unknown). The forecasting model based on the correlations learned from the training process and the supplied attribute values will predict the unknown target values. Note that the time difference between the attribute and target values of the new dataset will match the one of the training dataset. If the model was trained for a day-ahead forecast for example, this will be the time horizon of the forecast, and the target values of the next day constitute the output.

5.2.3 Case Study

The proposed methodology is applied on a dataset coming from real charging events recorded from public charging stations. The data are from a pilot project in Paris involving 71 PEVs. The PEVs' charging activities were recorded for 1 year. The charging events were classified according to the ID of each PEV, and examined individually. For each PEV, charging patterns like the connection/disconnection time and the energy demand per charging event were analyzed in order to produce weekly distributions. An example of the charging demand distribution of four random PEVs for one week is shown in Fig. 5.4.

These distributions are important for analyzing the charging demand profile of each PEV owner. Useful information is also extracted analyzing the distributions for the times the PEV owners connect and disconnect their vehicles to a public charging station. Due to the small size of this sample, a generalization was

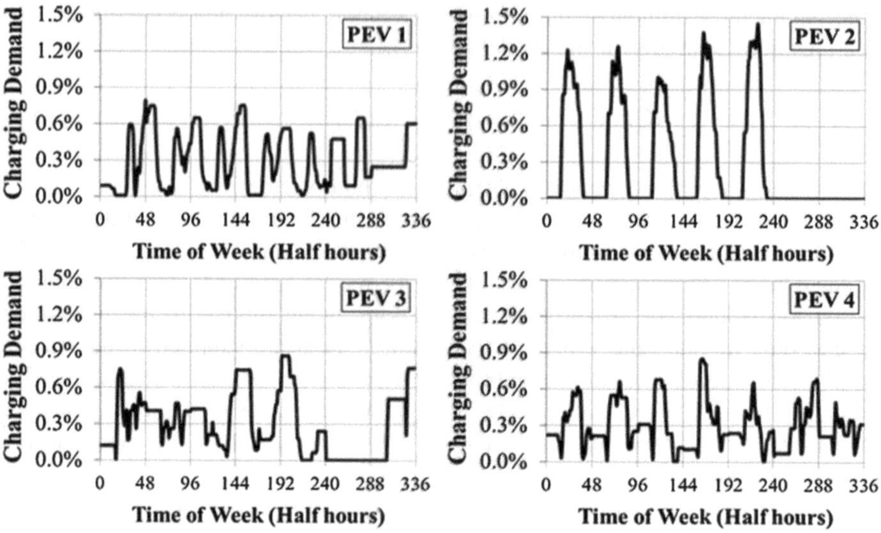

Fig. 5.4 Charging demand distribution for 1 week for four random PEVs

necessary in order to build a larger PEV fleet. The distributions of the analyzed features were used to create PEVs with similar charging demand profiles. In this scenario 2,130 PEVs were created, and the total charging demand of this fleet was calculated for 1 year.

Based on the available information in the initial dataset, the attributes used for the training and testing procedures are:

1. *Previous Day Load*: The charging demand of the previous day for each half hour.
2. *Previous Week Load*: The charging demand of the same day of previous week for each half hour.
3. *Week*: Number of the week (1–53).
4. *Day*: Number of the day (1–7) starting with Monday.
5. *Half Hour*: 1–48 half hour parts of each day.
6. *Number of New Connections*: The new PEV plug-in connections for every half hour.

Once the dataset is properly formed, the last day is considered "unknown" and constitutes the target of the forecast. The rest of the data are split in training and testing datasets, and the procedure presented in Fig. 5.3 is followed. SVM was applied as the data mining method. There are various performance indexes which are used to assess the effectiveness of a model.

Mean Absolute Percentage Error (MAPE) is an accepted industry standard for measuring the forecasting accuracy of model while Root Mean Square Error (RMSE) penalizes large absolute differences between actual and forecasted values.

The r-correlations show the general performance of a model. These performance indexes are calculated with the following formulas:

$$MAPE = \frac{\sum_{i=1}^{N}\left(\frac{|X_i - Y_i|}{X_i}\right)}{N} \times 100\,\% \tag{5.2}$$

$$RMSE = \sqrt{\frac{\sum_{i=1}^{N}(X_i - Y_i)^2}{N}} \tag{5.3}$$

$$r = \frac{\sum_{i=1}^{N}\left(X_i - \overline{X}\right)\left(Y_i - \overline{Y}\right)}{\sqrt{\sum_{i=1}^{N}\left(X_i - \overline{X}\right)^2}\sqrt{\sum_{i=1}^{N}\left(Y_i - \overline{Y}\right)^2}} \tag{5.4}$$

where:

N is the number of the forecasted values
X is the actual values
Y is the forecasted values
$\bar{X}$ is the mean of the actual values
$\bar{Y}$ is the mean of the forecasted values

In this case study the termination criterion of training the model was used to reach a MAPE with less than 5 %. After training the model, the "unknown" last week is supplied for forecast. The results are shown in Fig. 5.5. The performance was evaluated using Eqs. 5.2–5.4, and the results are summarized in Table 5.3.

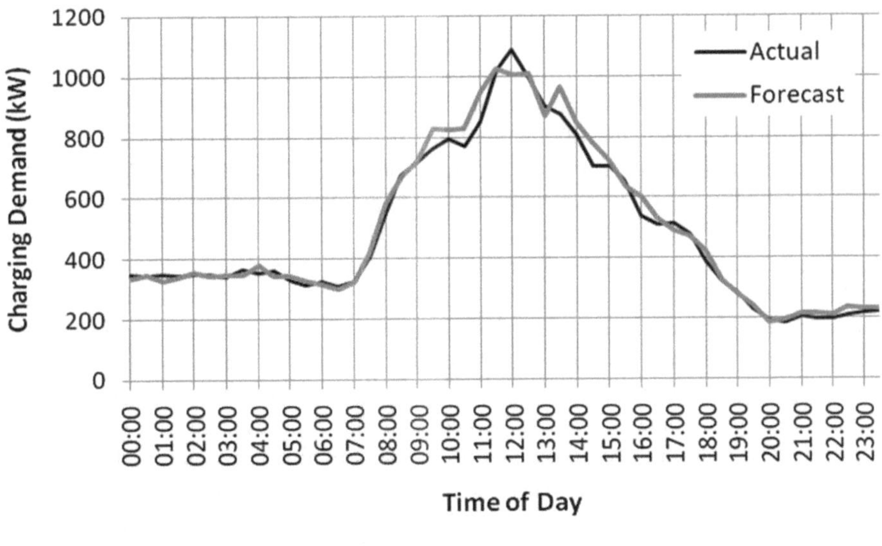

Fig. 5.5 Day-ahead forecasted charging demand

Table 5.3 Performance indices of the forecasting model

Performance index	Value
MAPE (%)	4.79
RMSE	33.87
r-Correlation (%)	99.26

Considering that the charging events were recorded from public stations and present high fluctuation, the performance of SVM was considered accurate enough. Additional information and attributes may increase the accuracy of the forecast. Weather data when available can be used to reduce the errors of a forecast. However, adding more attributes could increase the risk of finding irrelevant connections between the data and reduce the learning capability of the model. Therefore, several trials are necessary to achieve the best results involving different datasets.

5.3 Incorporating PEV Load Forecast in the PEV Load Management

5.3.1 The Importance of PEV Load Forecast in the Charging Control Model

The coordination of PEV charging belongs to demand side management (DSM) or demand response (DR) applications. In those applications, the philosophy of adapting power demand to power generation is applied to maintain the normal operation of the electricity grid. Coordinating PEV charging is an effective and low cost solution to reduce the impacts of this additional electricity demand on the electric power systems. The majority of PEV owners are expected to plug in their vehicles in the evening hours when they return home after work. They would like to have their vehicles fully charged the next day morning when they go to work. Considering the fact that no less than 90 % of the cars are parked during a day, there is opportunity to shift the electricity consumption from PEV charging to times with lower demand [4]. Smart charging control algorithms make use of this flexibility in order to reduce peak loads or charge PEVs preferential from renewable energy sources. These algorithms decide the charging schedules of PEVs according to their objective (e.g. valley-fill, peak shaving, and frequency regulation). Several approaches for the optimal coordination of PEV charging have been proposed in literature. Two main categories are predominantly in the literature: centralized control [37–39] and decentralized control [40–47]. Centralized control approaches are found to perform well for a limited number of PEVs. While the number of PEVs is increasing, the interactions between PEVs and the central aggregator become more complicated. This bi-directional communication requires a large amount of data to be acquired and processed from a central unit, increasing the minimum requirements of the computational resources [46]. In contrast, the effectiveness of the distributed control techniques is independent to the number of PEVs, as each PEV solves the scheduling problem individually.

PEV charging load is a specific type of demand associated with the travel patterns of the PEV owners [21, 22]. Their daily trips can determine their energy requirements for recharging the PEV batteries as well as the times they connect and disconnect their vehicles to the charging stations. The information of where and when the PEV owners will recharge their vehicles will lead to a more effective algorithm for coordinating the PEV charging schedules. In the future smart grids there will be a bi-directional flow of information allowing the network operators to collect data of the charging events within a geographical area. The PEV Aggregator, an entity that will be responsible to coordinate the PEV charging schedules providing ancillary services to network operators will make use of the knowledge recorded in the historical charging data. Data mining techniques are applied to extract useful information that will enhance the effective coordination of the PEV charging schedules [48]. In [43] forecasting actions were found to play an important role in scheduling of the PEV charging. These forecasts are related to probability distributions of arrival/departure times and the initial/desired battery state of charge (SOC) of the upcoming PEV. The model estimates the aggregated charging load and plans the scheduling process of PEV charging accordingly.

To our knowledge, the majority of charging control models assumes that all PEVs are participating in the control scheme. However, this is not a realistic scenario for the future composition of the PEV fleet. In a realistic case, the PEV fleet is separated in responsive and non-responsive PEV to control signals coming from the aggregator. Responsive PEVs are the ones that participate in coordination process responding to control signals from the PEV aggregators or other central management entities. On the contrary, non-responsive PEVs are not willing to participate in the control scheme. This willingness to participate in the control scheme is defined by the PEV owners and their routine. For example, if the daily routine of a PEV owner is affected due to an event, this may also influence the flexibility in charging the vehicle. Note also that some PEVs can be responsive to control signals in most cases. However, this does not mean that abnormal charging events are not happening occasionally from the same PEV. Forecasting the demand from unresponsive PEVs is critical for the effectiveness of the control scheme. Historical charging events are used to extract information about the abnormal charging demand from unresponsive PEVs. The value of PEV load forecast to the control of PEV charging is illustrated through an example.

In this example a mixture of responsive and unresponsive PEVs is assumed. The arrival and departures times of both types of PEVs are shown in Fig. 5.6a. An abnormal event occurs at 10:00, when a number of unresponsive PEVs are connected to the charging stations requiring charging for a short period of time. Despite other PEVs (the responsive ones) having a level of flexibility for the connection time, the inflexible demand from the unresponsive PEVs is critical for the effectiveness of the control algorithm. In this example, a control algorithm was applied to coordinate all PEVs without having future knowledge of the demand from the unresponsive PEVs. The objective of this control algorithm is to have a valley-filling effect on the demand curve of the assumed network. Figure 5.6b shows the final demand from the responsive PEVs. Based on the control model, a number of

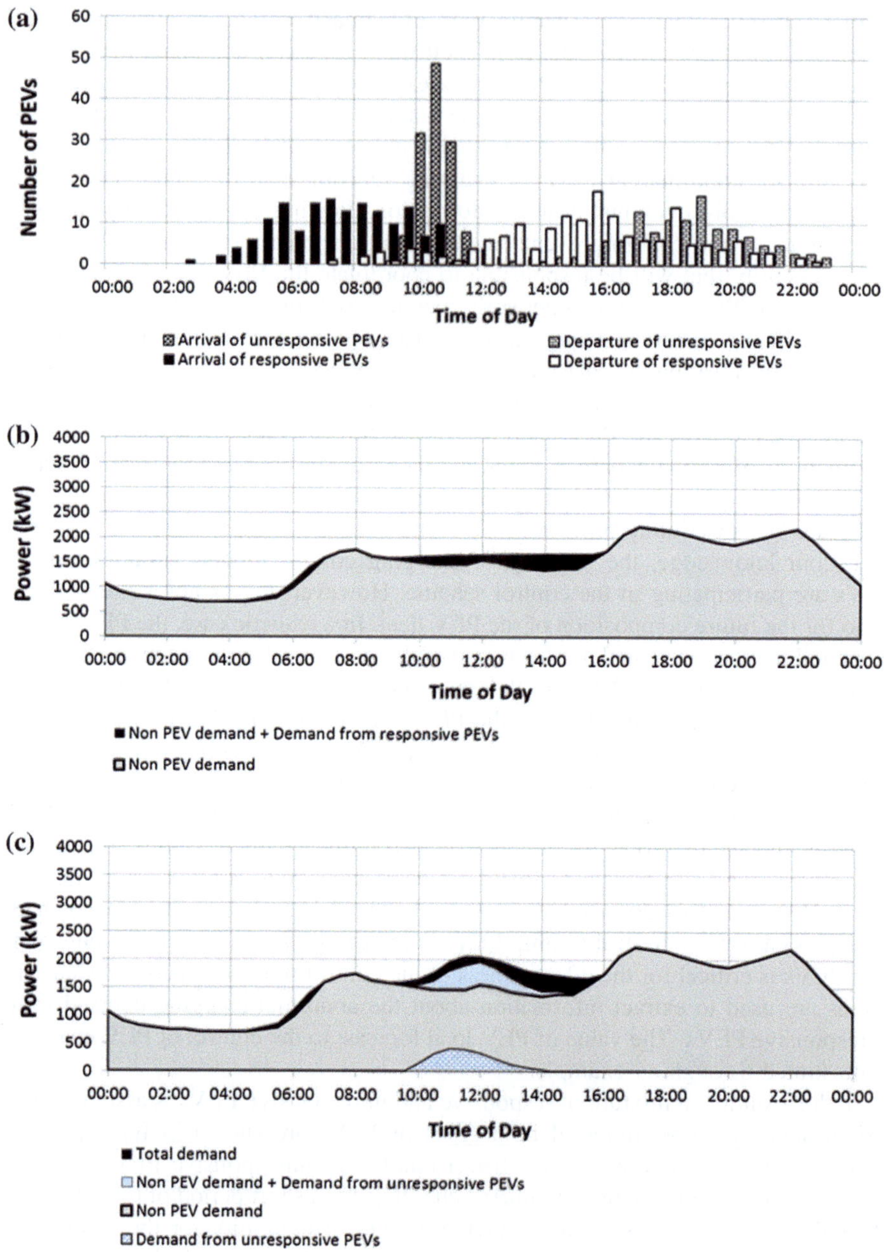

Fig. 5.6 **a** Distributions of arrival/departure times of responsive/unresponsive PEVs **b** Demand without adding the demand from unresponsive PEVs, **c** Total demand

PEVs were responsive to the control signals and as a result they are coordinated to charge at times when the demand is low. However, in a mix scenario like this example, without forecasting the demand from the unresponsive PEVs, the final result of the coordination algorithm is not optimal. Figure 5.6c indicates this weakness of the majority of the control models proposed in the literature.

5.3.2 The Integrated Model for PEV Load Management

A decentralized algorithm was developed to manage the PEV charging schedules, enhanced by PEV load forecast. The aim of the control algorithm was to achieve a valley-filling effect on the demand curve, avoiding a potential increase of the peak demand. The structure of this model follows the architecture of a multi agent system where each entity is an agent. Therefore, in this model there are three agent classes, the PEV agent, the PEV aggregator agent and the Distribution System Operator (DSO) agent. Figure 5.7 shows the location of each entity in an example network. PEV agents are located at the low voltage level and each Low Voltage (LV) feeder

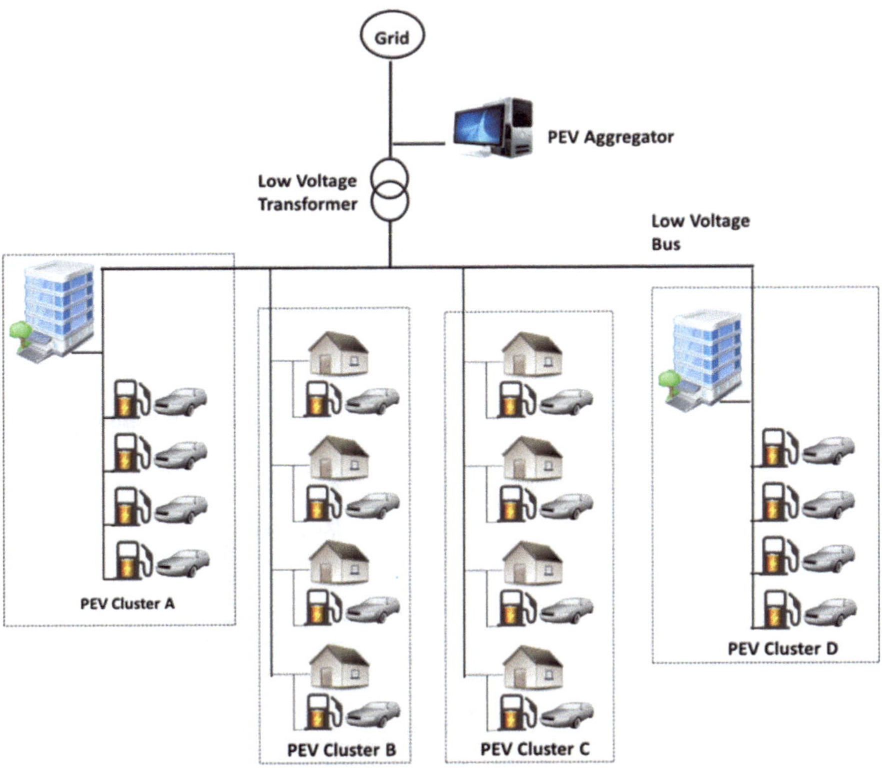

Fig. 5.7 Schematic example of a decentralized charging system

constitutes a PEV cluster. Each PEV cluster is a group of PEVs which are supplied with energy from the same LV feeder of which the technical constraints have to be respected. On the top level of the network, there is the DSO agent who is responsible to monitor the demand and voltage in the most significant parts of the network. PEV aggregator is an entity which is located in an intermediate level between PEVs and DSO. Based on the objectives of the control algorithm, the PEV aggregator can be located either in MV transformer or in a LV transformer.

The proposed control algorithm is designed based on the structure of a UK Generic Low Voltage Distribution Network obtained from [49], without affecting the generality of the model. The PEV aggregator is located on the Medium Voltage (MV) level, while the PEV agents are dispersed in the Low Voltage (LV) feeders. The PEV aggregator's role is to collect the historical charging data of the PEV fleet and apply machine learning algorithms to provide accurate forecasts of the future charging demand of unresponsive PEVs. The PEV Load Forecasting process uses Support Vector Machines, and is executed by the PEV Aggregator to improve the effectiveness of the algorithm. The PEVs are coordinated to achieve a local valley-filling effect in the demand curve of the LV feeder to which they are connected. In order to demonstrate the importance of the PEV load forecast algorithm in the proposed control scheme, different charging scenarios and composition of the PEV fleet were considered.

Figure 5.8 presents the basic operations of the PEVs and the PEV Aggregator. The DSO provides information to the PEV aggregator regarding the technical constraints of the network. This information is linked with the maximum power demand of the corresponding feeder, transformer loading and the thermal limits of

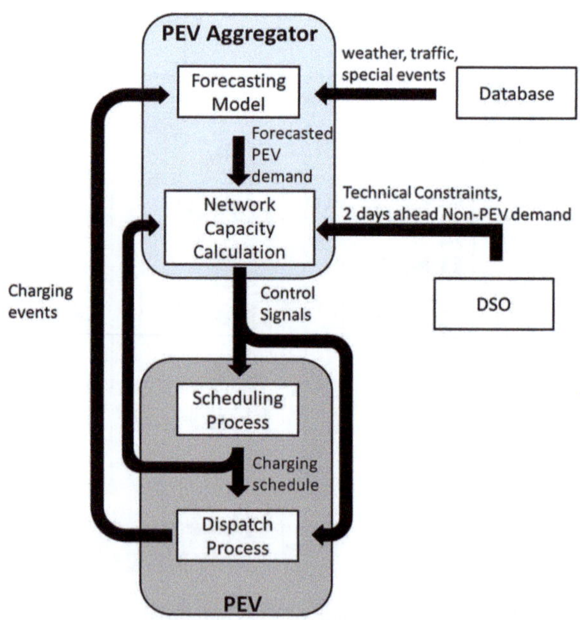

Fig. 5.8 Flow of information diagram

the network cables. In addition, PEV aggregator is receiving the forecasted non PEV demand of the next 2 days. In the proposed model, it is assumed that none PEVs owners leave their vehicles connected in a charging point for more than 24 h.

Initially, the PEV load forecast model is updated with charging data of the previous day. These data are processed based on the methodology presented in Sect. 5.2.2. Moreover, in case other sources of information like weather data or traffic measurements are accessible to the PEV aggregator, they are also included in the forecast model in order to increase the accuracy of the predictions. The forecasting model is updated with the latest data and provides the 2-days ahead forecasted charging demand from unresponsive PEVs. Once the output of the forecast model and relevant data from DSO are available to the PEV aggregator, the next stage includes the calculation of the control signals. The PEV aggregator calculates the network's capacity for PEV charging demand based on the PEV forecasted demand. The objective of the control model is a valley-filling effect on the demand of the LV feeders. Therefore, each PEV cluster is associated with a specific LV feeder and each group of PEVs receives the same control signals. These signals are related to the existing charging schedules and the predicted PEV demand for the corresponding part of the network. Based on those signals, each PEV defines its own charging schedule by selecting when to charge. The charging events are recorded, and used to update the forecasting procedure.

The timeframe resolution of the proposed model is measured in time steps (e.g. 10 min interval). The time steps affect the regularity of the control actions, for example a small time step indicates a more frequent delivery of control signal to the PEVs, and vice versa. However, the effectiveness of the proposed model is not affected by this interval. In this control model, 6 min time step duration is considered and thus every day is consisted of 240 time steps.

In the proposed control scheme, there are four main procedures which are repeated sequentially in a daily or a time step basis (see Fig. 5.9). At the beginning of each day, forecasting actions are taking place in order to estimate the future demand from the unresponsive PEVs.

The selected data mining method used in the PEV load forecast model is Support Vector Machines due to its high performance and its ability to extract information behind difficult patterns. Figure 5.10 presents the flowchart of the PEV load forecast model.

The PEV load forecast model is updated at the end of each day with all recorded charging events. The charging data include information about the connection times,

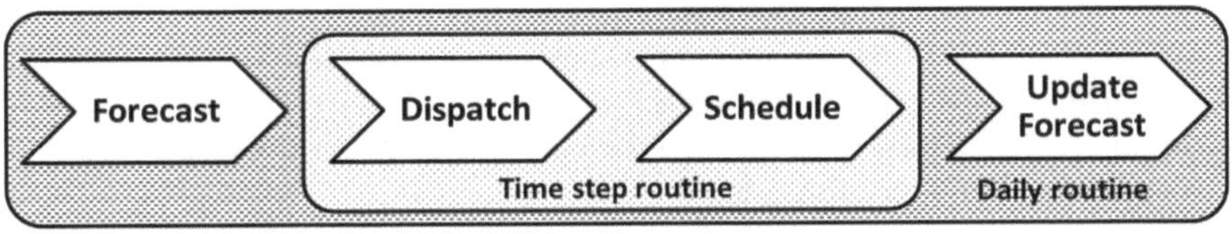

Fig. 5.9 Daily and time step routine in the control model

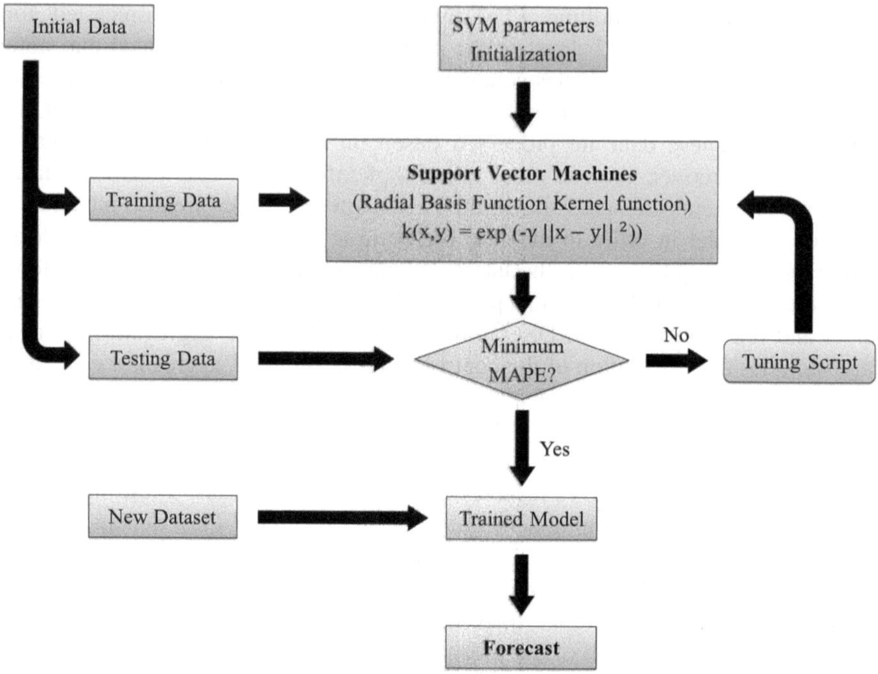

Fig. 5.10 PEV load forecast model for unresponsive PEVs

disconnection times and the energy requirement of the PEV fleet. In addition, each PEV has an ID and this is used to identify the charging pattern of a PEV owner. In addition, information about the ID of the PEV and its responsiveness to control signals is provided to the forecasting model. Based on the available information, the attributes used for the training and testing procedures are listed below:

1. *Two-day Unresponsive Load*: The aggregated charging demand from unresponsive PEVs of the previous 2 days for each time step.
2. *Two-day Responsive Load*: The aggregated charging demand from responsive PEVs of the previous 2 days for each time step.
3. *Two-day Unresponsive Load of previous week*: The unresponsive PEV charging demand of the same days of previous week for each time step.
4. *Two-day Responsive Load of previous week*: The responsive PEV charging demand of the same days of previous week for each time step.
5. *Day*: Number of the day (1–7) starting with Monday.
6. *Month*: Number of the month (1–12) starting with January.
7. 6 min *time step*: 1–240 parts of each day.
8. *Number of Unresponsive PEV Connections*: The number of unresponsive PEV connections for every time step.
9. *Number of Responsive PEV Connections*: The number of responsive PEV connections for every time step.

10. *Number of Unresponsive PEV Disconnections*: The number of unresponsive PEV disconnections for every time step.
11. *Number of Responsive PEV Disconnections*: The number of responsive PEV disconnections for every time step.

The target for the PEV forecast model is to forecast the PEV demand from unresponsive PEVs for each time step of the next 2 days. Once the training data are properly formed, the SVM and the RBF kernel parameters are initialized randomly. A script is executed in order to divide the sample in two separate datasets, one for the training and the other for the evaluation of the model. The testing dataset includes the values of the last 2 days of the initial datasets while the rest constitute the training dataset. Once the model is trained with the initial SVM parameters, it is evaluated through the testing dataset. The MAPE is calculated based on Eq. 5.2. The model is using a second script for updating the SVM parameters. The parameter C takes all integer values between the minimum and the maximum target value [30]. Additionally, parameter γ is updated within a range of $[0.85/n, 1.15/n]$ with a step of $(0.1/n)$, where n is the number of the attributes. The parameter ε is considered constant 0.001 (default value). All possible combinations of C and γ within the specified range are checked and the ones which result in the minimum MAPE are selected. Once this process is completed, the model is tested on the new dataset (which contains the attributes of the next 2 days) in order to provide a forecast of the charging demand from unresponsive PEVs for the next 2 days. The accuracy of this process is significant to the effectiveness of the control model.

In every time step, two main procedures are taking place namely "Dispatch" and "Schedule". Every process involves different tasks from both the PEV aggregator and the corresponding PEV. The "Dispatch" procedure involves the execution of the existing PEVs charging schedules. In this procedure, the PEV aggregator first runs power flows for the specific part of the distribution network according to the PEV's scheduled charging demand. According to this scheduled demand, the total non PEV demand and the demand from the unresponsive PEVs, the real time network constraints are calculated by the aggregator and sent to the corresponding PEV. After receiving these constraints, the PEV checks for a possible violation of the network constraints. In case the limits are violated, the PEV is rescheduling this charging demand in future time steps. This procedure is repeated until the scheduled demand of every existing PEV is either supplied or rescheduled. The "Dispatch" procedure is presented in Fig. 5.11.

In case new PEVs are connected (or the existing charging schedule violates the technical constraints of the network) the "Schedule" phase is activated (see Fig. 5.12). During this phase, each PEV will solve the scheduling problem to satisfy its charging requirements. Internal information such as the battery SOC and the charging station power rate, as well as information coming from the PEV aggregator (external) like the network's capacity and the forecasted PEV demand are used in the scheduling process.

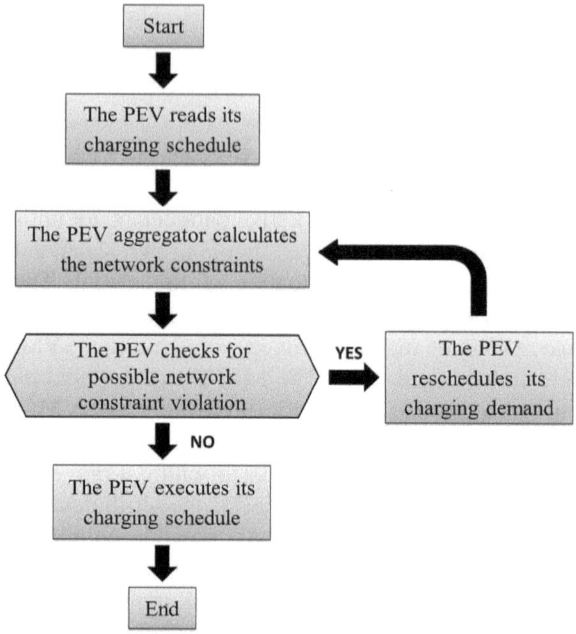

Fig. 5.11 Flowchart of the dispatch process

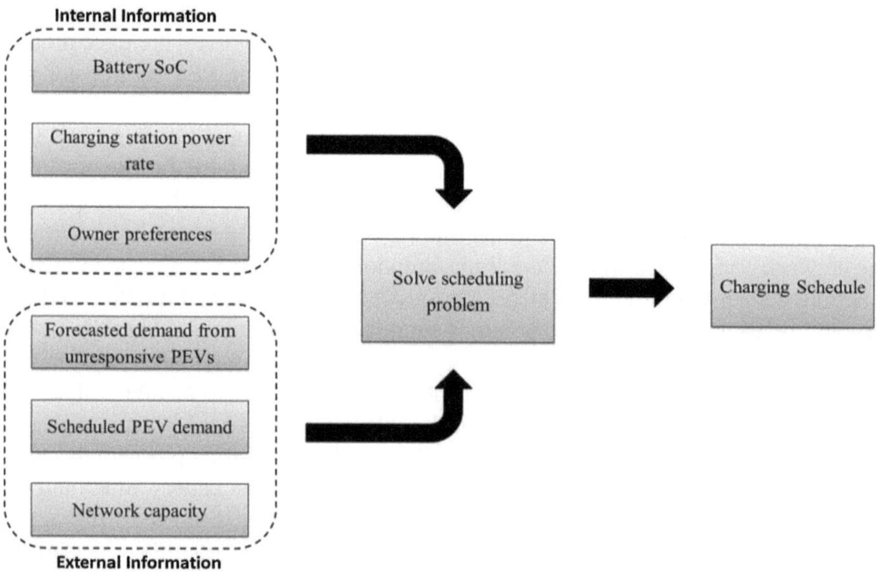

Fig. 5.12 Flowchart of the schedule process

The scheduling problem for PEV-n is formulated as follows:

$$\min \sum_{t_n}^{t_n+d_n} P_n(t) \cdot [V_n(t)] \tag{5.5}$$

where t_n is the connection time of PEV-n, d_n is the charging duration of PEV-n, $P_n(t)$ is the instantaneous charging power demand of PEV-n and $V_n(t)$ is the virtual cost value of a time step.

Every PEV tries to minimize a virtual cost function given in Eq. 5.5. The virtual cost values $V_n(t)$ are calculated by the PEV Aggregator according to the forecasted demand from unresponsive PEVs, the existing PEV charging schedules and the non PEV demand. The PEV aggregator sends to every PEV a vector $\boldsymbol{V_n}$. This vector contains the order sequence of the time steps with the lowest to highest demand for the period $[t_n, t_n + d_n]$. For example, the virtual cost value for the time step with the lowest demand is 1, while the one with the highest demand is d_n, and the intermediate time steps are taking values between 1 and d_n.

Minimizing Eq. 5.5 will result in an adaptive PEV behavior based on the local network's condition. Each PEV has knowledge of the future local aggregated demand and adjusts its charging schedule accordingly. The scheduling problem is subject to the following constraints:

$$\int_{t_n}^{t_n+d_n} P_n(t)dt = \left(SOC_{final_n} - SOC_{in_n}\right)\frac{C_{bat_n}}{\delta_{eff_n}} \tag{5.6}$$

$$P_n(t) \le P_{ch.nom_n} \tag{5.7}$$

$\forall n \in (1 \ldots N), \forall t$

where SOC_{final_n} is the desired SOC of PEV-n, SOC_{in_n} is the initial SOC of PEV-n, C_{bat_n} is the battery capacity of PEV-n, δ_{eff_n} is the efficiency of the charging station and $P_{ch.nom_n}$ is the nominal power rate of the charging station.

Equation 5.6 expresses the energy requirements of PEV-n. These requirements are satisfied during the connection period of the particular PEV $[t_n, t_n + d_n]$. The instantaneous charging power $P_n(t)$ must not exceed the power rating of the charging station ($P_{ch.nom_n}$) for every t, as described in Eq. 5.7. The next two constraints are related to the network topology and characteristics. Let us denote f as the LV feeder that a PEV is connected. Every such feeder has a group A_f that is consisted of all PEVs charging on LV feeder f at time t. Based on the network topology, the size of this group ($|A|$) has an upper boundary C_1 (maximum number of PEVs on feeder f). Additionally, denoting l as the MV/LV transformer that LV feeder f is attached, there is a group B_l containing all the corresponding feeders. C_2 expresses the number of feeders on a transformer. Equations 5.8 and 5.9 are used to keep the power demand of feeder f and the transformer l between the limits.

$$P_n(t) \leq P_{fd.nom_f} - \sum_m P_m(t) \tag{5.8}$$

$\forall f \exists A_f = \{n| \text{ connected on } f = \text{true}\}, \quad |A| \leq C_1$
$\forall m \in A_f - \{n\}$
$\forall n \in (1 \ldots N), \forall t$

where $P_{fd.nom_f}$ is the nominal power of feeder *f* and $P_m(t)$ is the power demand of PEV-*m* in feeder *f*. Equation 5.8 expresses that the charging power for PEV-*n* should not exceed the corresponding nominal feeder limit considering also all the other PEVs which are charging in the same feeder.

$$P_n(t) \leq P_{tr.nom_l} - \sum_f \sum_m P_m(t) \tag{5.9}$$

$\forall l \exists B_l = \{f| \text{ connected on } l = \text{ true}\}, \quad |\mathbf{B}| = C_2$
$\forall f \in B_l$
$\forall m \in A_f - \{n\}$
$\forall n \in (1 \ldots N), \forall t$

where $P_{tr.nom_l}$ is the nominal power limit of the transformer. Equation 5.9 expresses that the charging power for PEV-*n* should not exceed the corresponding nominal transformer limit considering also all the other PEVs which are charging in the same transformer.

5.4 Simulation Results

The following sections present the decentralized control for PEV charging. In order to demonstrate the importance of the PEV load forecast in the proposed control scheme, different charging scenarios and different composition of the PEV fleet were considered. A specific distribution network was used to test the performance of the control model. Different percentage of unresponsive PEVs were considered and the effectiveness of the control model is evaluated through case studies. In addition, the effect of the charging rate on the valley filling effect on the local demand curve is also presented.

5.4.1 Network Topology

The typical 33/11/0.4 kV UK generic distribution network model is based on [49]. The system is comprised of a 33 kV three-phase source, two 33/11.5 kV 15 MVA transformers with on-line-tap-changer and an 11 kV substation with five 11 kV

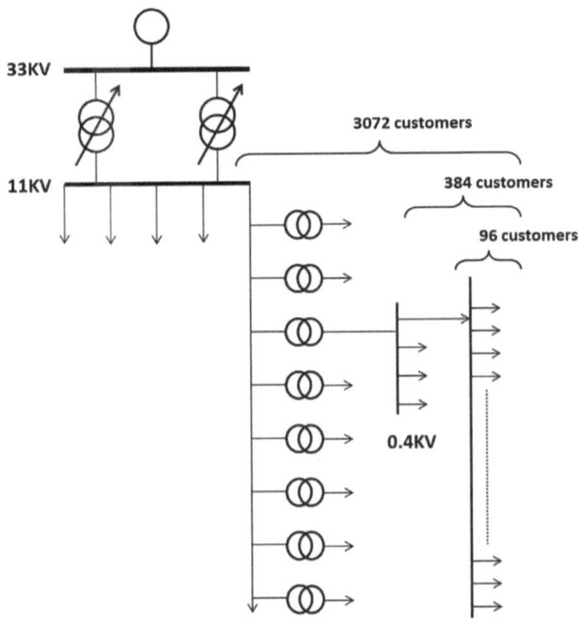

Fig. 5.13 Typical 33/11/0.4 UK generic distribution network

outgoing MV feeders. Each 11 kV feeder supplies eight 11/0.433 kV 500 kVA distributed transformers. Each MV/LV transformer has 4 LV feeders, and each LV feeder provides energy to 96 customers. The topology is presented in Fig. 5.13.

The PEVs are connected at the LV level, while the PEV aggregator is located on a MV feeder, and is responsible for 3,072 customers. In order to evaluate the control model, a realistic PEV fleet with the following characteristics is created, as shown in Table 5.4.

According to [50], the PEV uptake level of 20 % is considered as the Business as Usual Scenario. This uptake level was used for different case studies to investigate the impact of PEV load forecast on the effectiveness of the proposed control model. Representative charging rates of 3.6, 11 and 22 kW for the charging stations are considered to study their effect on the flexibility of a responsive PEV fleet. Non

Table 5.4 PEV fleet characteristics

PEV fleet variables	Mean value (μ)	Standard deviation (σ)
Battery capacity (kWh)	30	2
Initial SOC (%)	40	5
Final SOC (%)	90	10
Arrival time of responsive PEVs (h)	09:00	1
Departure time of responsive PEVs (h)	17:00	1
Arrival time of unresponsive PEVs (h)	10:30	0.5
Departure time of unresponsive PEVs (h)	13:30	0.5

PEV demand curves are obtained from [51] for a typical spring weekday. Different ratios of unresponsive/responsive PEVs are used to analyze the influence of this ratio to the effectiveness of the proposed model.

5.4.2 Case Study 1

In this case study a number of 614 PEVs were assumed, equivalent with 20 % uptake, having the characteristics presented in Table 5.4. Different PEV fleet synthesis with responsive and unresponsive PEVs is considered, charging at 11 kW charging stations. Two control options are presented, one without activating the forecasting modules of the model, and the second one that uses the forecasting model. Figure 5.14 shows the demand on the MV level for both options (without and with PEV load forecast) for different responsive/unresponsive PEVs ratio.

The results show that when PEV load forecast option is activated, PEVs are modifying their charging schedules in order to reduce the impact of unresponsive PEV charging on the demand curve. The charging demand of the responsive PEVs is adapted to the unresponsive PEV charging demand so that their aggregation results in a valley filling effect on the Non PEV demand curve. In most cases, this adaptive behavior of responsive PEVs leads to a reduction of the aggregated charging demand peak. For low levels of unresponsive PEVs (until 20 %), the control model is able to completely absorb the unresponsive PEV demand. On the other hand, high levels of unresponsive PEVs lead to inflexible demand, thus the capability of the proposed control model to reduce the peak charging demand is limited. Obviously, without having responsive PEVs in our system, the integrated model with PEV load forecast is not affecting the final charging demand.

5.4.3 Case Study 2

This case study investigates the effect of the charging stations' power rate on the effectiveness of the control model. The charging rates of 3, 11 and 22 kW and an uptake of 20 % PEVs are used for this analysis. The Peak-to-Average ratio (PAR) and peak reduction criteria are used to evaluate the performance of the model. Peak-to-Average Ratio is calculated according to Eq. 5.10. This index indicates the valley filling effect on the demand curve.

$$PAR = \frac{P_{\max}}{P_{average}}, \tag{5.10}$$

where P_{max} is the peak power demand of a day and $P_{average}$ is the mean power demand for the specific day.

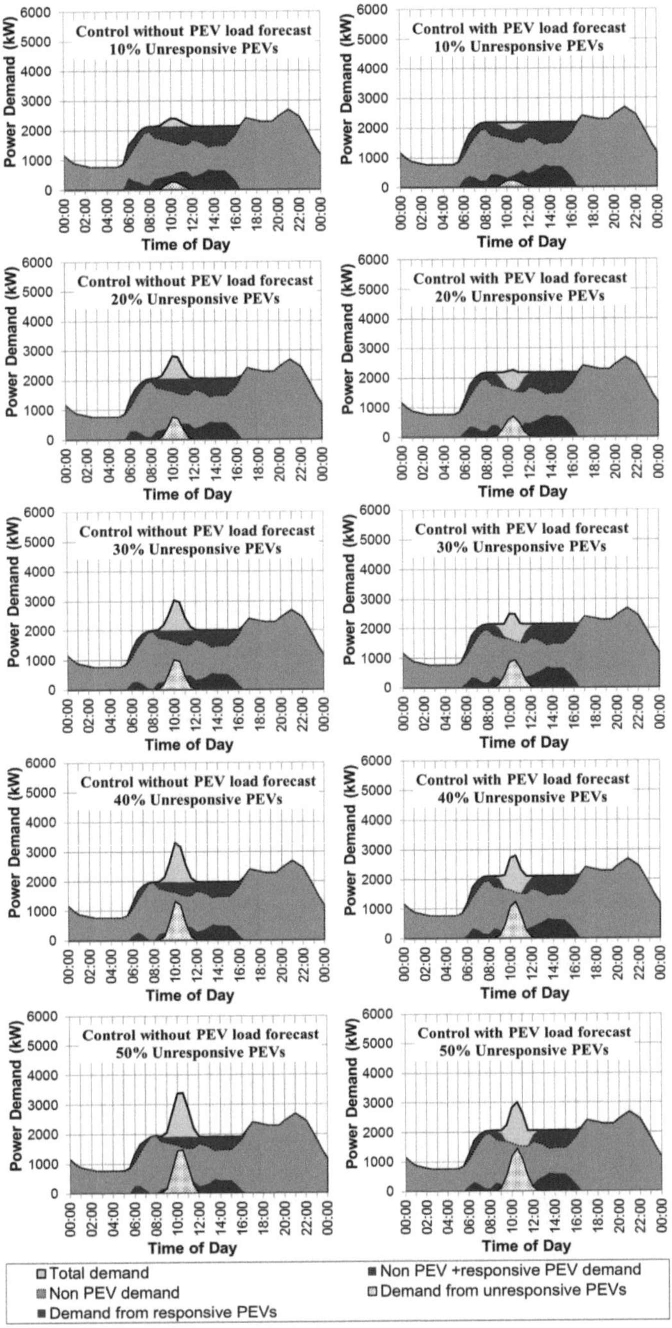

Fig. 5.14 Charging demand for different levels of unresponsive PEVs after two control strategies

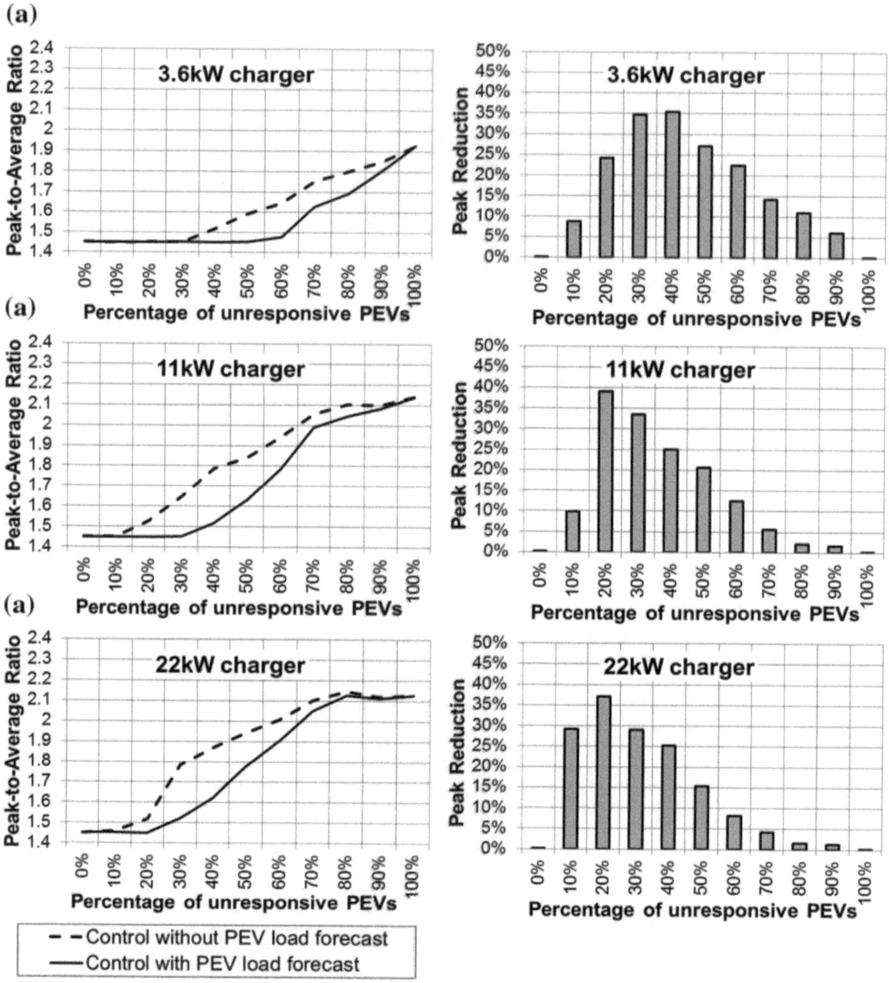

Fig. 5.15 Peak-to-average ratio and peak reduction for **a** 3.6 kW, **b** 11 kW and **c** 22 kW charger

As seen from Fig. 5.15, different charging rates have a different effect on PAR-index. At low charging rates (3.6 kW) the control model with PEV load forecast is capable to delay the increase of this index, even until a 50/50 ratio of responsive/unresponsive PEVs is achieved. For higher charging rates, unresponsive PEVs have a significant impact on PAR, even at low penetration levels. Despite this, the control model with PEV load forecast improves the results. At 0 and 100 % levels of unresponsive PEVs the results are identical, and both control options lead to the same demand curve. In every combination of responsive/unresponsive PEVs (except of course the extreme values of 0 and 100 %) there is a peak reduction due to the contribution of the forecasting model. For every charging rate, this reduction can reach up to 35 %. However, this reduction occurs at different percentage of unresponsive PEVs for each charging rate. For low charging rates a significant peak reduction is observed at a wide range of unresponsive PEVs percentages. On the contrary when the charging rate is increased, this range is narrower and the maximum peak reduction is found on lower unresponsive PEVs percentages.

5.5 Conclusions

The flexibility of the electricity sector in managing changes will have a significant influence to the success of the electric vehicle deployment. The development of the charging infrastructure is often seen as an essential investment to offer PEVs drivers the psychological support to overcome the range anxiety, one of the most inhibiting factors in electric vehicles adoption. In order to manage the electric vehicles charging in distribution networks, DNOs will have to upgrade their infrastructure or implement smart control techniques in parallel with the development of regulative measures to serve these new customers.

The "aggregator" is a new player which will control multiple PEVs. This research is proposing a decentralized control framework for a mixture of responsive and unresponsive PEVs enhanced by PEV load forecasting. The main aim of the control algorithm is to achieve a valley-filling effect on the demand curve. The effectiveness of the control algorithm was tested in a UK generic distribution network considering a geographical area with 3,072 customers. Two case studies were presented. The first case study considered a PEV fleet charging at 11 kW charging stations comprising of responsive and unresponsive PEVs. It was demonstrated that when the PEV load forecast option is activated the PEVs are adapting their charging schedule to reduce the impact of the unresponsive PEVs on the demand curve. The second case study investigated the effect of the charging station's power rate on the effectiveness of the decentralized control model. It was shown that when the forecasting module is activated there is a demand peak reduction for every combination of responsive/unresponsive PEVs considering charging rates of 3, 11 and 22 kW.

Smart management of PEVs charging based on aggregation enhanced by PEV load forecasting could be seen as a win-win strategy for both the DNO and the vehicle owner.

References

1. Department for Transport (2011) Transport energy and environment statistics. https://www.gov.uk/government/uploads/system/uploads/attachment_data/file/8947/energy-2011.pdf. Accessed 12 July 2014
2. Cipcigan LM, Wells P, Nieuwenhuis P, Davies H, Whitmarsh L, Papadopoulos P (2012) Electricity as a transportation fuel: bridging the gaps in the electric vehicles value chain. Paper presented at the EVVC-2012 European electric vehicle congress, Brussels, 19–22 November 2012

3. Papadopoulos P, Skarvelis-Kazakos S, Grau I, Cipcigan LM, Jenkins N (2010) Predicting electric vehicle impacts on residential distribution networks with distributed generation. Paper presented at the IEEE vehicle power and propulsion conference, Lille, France, 1–3 September 2010
4. Papadopoulos P, Akizu O, Cipcigan LM, Jenkins N, Zabala E (2011) Electricity demand with electric cars in 2030: comparing Great Britain and Spain. Proc Inst Mech Eng Part A: J Power Energy (PIA) 225(A):551–566
5. Amin SM, Wollenberg BF (2005) Toward a smart grid: power delivery for the 21st century. IEEE Power Energy Mag 3(5):34–41
6. Morrow K, Karner D, Francfort J (2008) Plug-in hybrid electric vehicle charging infrastructure review. US Department of Energy. http://avt.inl.gov/pdf/phev/phevInfrastructureReport08.pdf. Accessed 15 July 2014
7. Huang Y, Liu J, Shen X, Dai T (2013) The interaction between the large-scale EVs and the power grid. Smart Grid Renew Energy 4(2):137–143. doi:10.4236/sgre.2013.42017
8. Eichinger F, Pathmaperuma D, Vogt H, Muller E (2013) Data analysis challenges in the future energy domain. In: Yu T, Chawla N, Simoff S (eds) Computational intelligent data analysis for sustainable development. Chapman and Hall/CRC, Boca Raton, pp 182–233
9. Tugberk Isyapar M (2013) Classification of electricity customers based on real consumption values using data mining and machine learning techniques and its corresponding applications. Ph.D. dissertation, Middle East Technical University
10. Schuller A, Rieger F (2013) Assessing the economic potential of electric vehicles to provide ancillary services: the case of Germany. Zeitschrift für Energiewirtschaft 37(3):177–194
11. Tomic J, Kempton W (2007) Using fleets of electric-drive vehicles for grid support. J Power Sources 168(2):459–468
12. Lopes JAP, Soares FJ, Almeida PMR (2011) Integration of electric vehicles in the electric power system. Proc IEEE 99(1):168–183
13. Kempton W, Letendre SE (1997) Electric vehicles as a new power source for electric utilities. Transp Res Part D: Transp Environ 2(3):157–175
14. Karfopoulos E, Marmaras C, Hatziargyriou N (2012) Charging control model for EV supplier aggregator. Paper presented at the 3rd IEEE PES international conference and exhibition on innovative smart grid technologies (ISGT Europe), Berlin, 14–17 October 2012
15. Rodrigue JP, Comtois C, Slack B (2009) The geography of transport systems. Routledge, London and New York
16. DuBois & King, Inc. Vermont Energy Investment Corporation (2013) V trans electric vehicle fueling infrastructure plan. http://www.veic.org/docs/Transportation/201307_VTrans_EV_Charging_Plan_Final_Report_web.pdf. Accessed 21 July 2014
17. Pitt BD (2000) Applications of data mining techniques to electric load profiling. Ph.D. dissertation, Institute of Science and Technology, University of Manchester
18. Dutta S (2012) Data mining and graph theory focused solutions to smart grid challenges. Ph.D. dissertation, University of Illinois
19. Selbas R, Sencan A, Kucuksille EU (2011) Data mining method for energy system applications. In: Funatsu K (ed) Knowledge-oriented applications in data mining. InTech, Vienna, p 147–166
20. Han J, Kamber M, Pei J (2012) Data mining: concepts and techniques. Morgan Kaufmann Publishers, USA
21. Xydas E, Marmaras C, Cipcigan LM, Hassan AS, Jenkins N (2013) Forecasting Electric Vehicles Charging Demand Using Support Vector Machines. Paper presented at the 48th universities' power engineering conference (UPEC), Dublin, 2–5 September 2013
22. Xydas E, Marmaras C, Cipcigan LM, Hassan AS, Jenkins N (2013) Electric vehicle load forecasting using data mining methods. Paper presented at the 4th hybrid and electric vehicle IET conference (HEVC), London, 6–7 November 2013
23. Boser BE, Guyon IM, Vapnik VN (1992) A training algorithm for optimal margin classifiers. In: Haussler D (ed) Proceedings of the 5th annual workshop on computational learning theory (COLT'92), Pittsburgh, USA, 1992

24. Scholkopf B, Sung K, Burges CJC, Girosi F, Niyogi P, Poggio T, Vapnik V (1997) Comparing support vector machines with Gaussian kernels to radial basis function classifiers. IEEE Trans Signal Process 45(11):2758–2765
25. Osuna E, Freund R, Girosi F (1996) Support vector machines: training and applications. http://dspace.mit.edu/bitstream/handle/1721.1/7290/AIM-1602.pdf?sequence=2. Accessed 12 July 2014
26. Smola A, Scholkopf B (2001) A tutorial on support vector regression. Stat Comput 14 (3):199–222
27. Elattar EE, Goulermas J, Wu QH (2010) Electric load forecasting based on locally weighted support vector regression. IEEE Trans Syst Man Cybern Part C Appl Rev 40(4):438–447
28. Varewyck M, Martens JP (2011) A practical approach to model selection for support vector machines with a Gaussian kernel. IEEE Trans Syst Man Cybern Part B Cybern 41(2):330–340
29. Ikeda K (2006) Effects of kernel function on Nu support vector machines in extreme cases. IEEE Trans Neural Netw 17(1):1–9
30. Cherkassky V, Ma Y (2004) Practical selection of SVM parameters and noise estimation for SVM regression. Neural Netw 17(1):113–126
31. Muller KR, Mika S, Ratsch G, Tsuda K, Scholkopf B (2001) An introduction to kernel-based learning algorithm. IEEE Trans Neural Netw 12(2):181–201
32. Wu KP, Wang SD (2009) Choosing the kernel parameters for support vector machines by the inter-cluster distance in the feature space. Pattern Recogn 42(5):710–717
33. Hsu CW, Chang CC, Lin CJ (2003) A practical guide to support vector classification. http://www.csie.ntu.edu.tw/~cjlin/papers/guide/guide.pdf. Accessed 12 July 2014
34. Ben-Hur A, Weston J (2010) A user's guide to support vector machines. In: Carugo O, Eisenhaber F (eds) Data mining techniques for the life sciences. Humana Press, New York, pp 223–239
35. Kecman V (2001) Learning and soft computing: support vector machines, neural networks and fuzzy logic models. MIT Press, Cambridge
36. Suykens J, Horvath G, Basu S, Micchello C, Vandewalle J (eds) (2003) Advances in learning theory: methods, models and applications. IOS Press, Amsterdam
37. Shao S, Pipattanasomporn M, Rahman S (2012) Grid integration of electric vehicles and demand response with customer choice. IEEE Trans Smart Grid 3(1):543–550
38. Jin C, Tang J, Ghosh P (2013) Optimizing electric vehicle charging: a customer's perspective. IEEE Trans Veh Technol 62(7):2919–2927
39. Deilami S, Masoum AS, Moses PS, Masoum MAS (2011) Real-time coordination of plug-in electric vehicle charging in smart grids to minimize power losses and improve voltage profile. IEEE Trans Smart Grid 2(3):456–467
40. Fan Z (2012) A distributed demand response algorithm and its application to PHEV charging in smart grids. IEEE Trans Smart Grid 3(3):1280–1290
41. Ota Y, Taniguchi H, Nakajima T, Liyanage KM, Baba J, Yokoyama A (2012) Autonomous distributed V2G (Vehicle-to-Grid) satisfying scheduled charging. IEEE Trans Smart Grid 3 (1):559–564
42. Ma Z, Callaway DS, Hiskens IA (2013) Decentralized charging control of large populations of plug-in electric vehicles. IEEE Trans Control Syst Technol 21(1):67–78
43. Qi W, Xu Z, Max Shen ZJ, Hu Z, Song Y (2014) Hierarchical coordinated control of plug-in electric vehicles charging in multifamily dwellings. IEEE Trans Smart Grid 5(3):1465–1474
44. Cao Y, Tang S, Li C, Zhang P, Tan Y, Zhang Z, Li J (2012) An optimized EV charging model considering TOU price and SOC curve. IEEE Trans Smart Grid 3(1):388–393
45. Xi X, Sioshansi R (2014) Using price-based signals to control plug-in electric vehicle fleet charging. IEEE Trans Smart Grid 5(3):1451–1464
46. Karfopoulos E, Hatziargyriou N (2013) A multi-agent system for controlled charging of a large population of electric vehicles. IEEE Trans Power Syst 28(2):1196–1204
47. Gan L, Topcu U, Low SH (2013) Optimal decentralized protocol for electric vehicle charging. IEEE Trans Power Syst 28(2):940–951

48. Soares J, Ramos S, Vale Z, Morais H, Faria P (2012) Data mining techniques contributions to support electrical vehicle demand response. Paper presented at the 2012 IEEE PES transmission and distribution conference and exposition, Orlando, USA, 7–10 May 2012
49. Ingram S, Probert S, Jackson K (2003) The impact of small scale embedded generation on the operating parameters of distribution networks. http://webarchive.nationalarchives.gov.uk/20100919181607/http://www.ensg.gov.uk/assets/22_01_2004_phase1b_report_v10b_web_site_final.pdf. Accessed 12 July 2014
50. Element Energy Limited (2013) Pathways to high penetration of electric vehicles. http://www.theccc.org.uk/wp-content/uploads/2013/12/CCC-EV-pathways_FINAL-REPORT_17-12-13-Final.pdf. Accessed 12 July 2014
51. Electricity User Load Profiles by Profile Class (1997) UK Energy Research Centre, Energy Data Centre. http://data.ukedc.rl.ac.uk/browse/edc/Electricity/LoadProfile/data. Accessed 14 July 2014

Chapter 6
Optimal Charging Strategies of Plug-in Electric Vehicles for Minimizing Load Variance Within Smart Grids

Linni Jian, Guoqing Xu and C.C. Chan

Abstract With serious concerns on global warming and energy crisis, there are plenty of motivations for developing and commercializing plug-in electric vehicles (PEVs). It is believed that substitution of PEVs for conventional fuel vehicles can help reduce the greenhouse gases emission, increase the energy efficiency, enhance the integration of renewable energy, and so forth. At the same time, the impact of PEVs as an emerging electrical load for power grid has drawn increasing attention most recently. The possible challenge for power grids lies in that the penetration of large number of PEVs may trigger extreme surges in demand at rush hours, and therefore, harm the stability and security of the existing power grids. Nevertheless, there are also potential opportunities for power grids. An optimal scenario is to dig the potential of PEVs as moveable energy storage devices, which means PEVs withdrawing electricity from grid at off-peak hours and then feeding back energy deposited in the onboard batteries to grid at peak hours. This concept is also termed as vehicle-to-grid (V2G) technology. The key to the implementation of V2G is how to effectively integrate information into energy conversion, transmission and distribution. V2G should be carried out within the framework of smart grid, so that the status information of power grid can be perceived. In addition, the demand information of PEV owners should also be taken into account, so that the function of PEVs as transportation tools can be guaranteed. In this chapter, the possible scenarios of V2G implementation within both the household smart micro-grid and

L. Jian (✉)
Department of Electrical and Electronic Engineering,
South University of Science and Technology of China, Shenzhen, China
e-mail: jian.ln@sustc.edu.cn

G. Xu
Department of Mechanical and Automation Engineering,
The Chinese University of Hong Kong, Sha Tin, Hong Kong
e-mail: gqxu@mae.cuhku.edu.hk

C.C. Chan
Department of Electrical and Electronic Engineering,
The University of Hong Kong, Pokfulam, Hong Kong
e-mail: ccchan@eee.hku.hk

S. Rajakaruna et al. (eds.), *Plug In Electric Vehicles in Smart Grids*,
Power Systems, DOI 10.1007/978-981-287-317-0_6

the smart regional grid will be discussed. The related mathematical formulation will also be analyzed. It is essentially an optimization problem, and the objective is to minimize the overall load variance of power grids. Case studies will be conducted. The results demonstrate that V2G operation can definitely help flatten the overall power load curves and it enables power grid to contain newly added PEV loads to some extent without boosting its capacity.

Keywords Smart micro-grid · Smart regional grid · Vehicle-to-grid · Load variance minimization

6.1 Introduction

Most recently, a general consensus has been reached extensively, namely, energy crisis and global warming are becoming two critical issues which may threaten the sustainable development of our human beings. On one hand, the exploitable reverses of fossil fuels, such as coal, oil, and natural gas, may be exhausted in the near future, due to the rapid growth of global energy consumption. On the other hand, statistics indicate that the average global temperature has increased by about 0.8 °C since 1880, in which two-thirds of the warming has occurred since 1975 [1]. The greenhouse gases arising from burning fossil fuels are believed to be the major contributor to the climate change. Generally, energy use in transportation sector accounts for a significant proportion in the overall energy consumption. According to the latest *International Energy Outlook* issued by U.S. Energy Information Administration, almost 30 % of the world's delivered energy is used for transportation in 2013. Moreover, it is predicted that transportation sector accounts for the largest share (63 %) of the total growth in world consumption of petroleum and other liquid fuels from 2010 to 2040 [2]. Therefore, the transport electrification has been deemed as a promising solution to relieve energy crisis and global warming. It can greatly reduce the dependence on fossil fuels through diversifying the energy sources of transportation tools, and improve the efficiency of energy conversion. Consequently, the emission of greenhouse gases can be significantly reduced.

As far as transport electrification is concerned, plug-in electric vehicles (PEVs) definitely lies within the most popular topics. As for electric vehicles, electric motors are wholly or partially substituted for internal combustion engines (ICEs) to drive the wheels. The electricity needed can be either produced via onboard generation systems, such as hybrid electric vehicles and fuel cell electric vehicles, or recharged by any external electricity sources, for example, battery electric vehicles and plug-in hybrid vehicles. In the latter subcategory, large-capacity batteries are often equipped. In addition, these batteries can be charged by plugging into the power grid very conveniently. Therefore, this subcategory of electric vehicles is usually termed as PEVs or gridable EVs. Compared with conventional ICE vehicles, PEVs can offer very high efficiency of around 80 %, and extremely low operating cost. For example, in December 2011, the operating costs of Nissan Leaf and Chevrolet Volt are 3.5 cents per mile and 3.8 cents per mile, respectively, while for ICE vehicles, this number can reach as high as 12.5 cents per mile [3]. Nevertheless, the initial prices of PEVs are significantly more expensive than conventional ICE vehicles due to the additional cost of their battery packs. In summary, the characteristics of PEVs include:

- PEVs are equipped with large-capacity batteries, and can offer full-electrical operation mode. Therefore, PEVs are able to help reduce the greenhouse gases emission and the dependence on fossil fuels.
- Onboard batteries of PEVs are rechargeable by connecting into power grid. Thus PEVs are novel emerging electric loads for power grids.
- PEVs can also feed back electricity deposited in their onboard batteries to power grid, if necessary. Thus they may also play their potential role as movable and distributed energy storage systems for power grids.

The traditional method to charge the PEVs are either fast charging or slow charging [4]. The key difference between them lies in the charging power rating, thus the charging time. Generally, slow charging takes about 6–8 h to bring the battery to a full charge with a charging power of around 3 kW. Therefore, slow charging is typically associated with the overnight charging. Comparatively, fast charging is much quicker and with much more complex definitions. In a word, any scheme other than slow charging can be deemed as fast charging. Sometimes, it can be further specified as fast charging, rapid charging and quick charging as shown in Table 6.1. Both slow charging and fast charging operate regardless of the status of power grids. Hence, with the increase of penetration of PEVs, these uncertain intermittent power loads will definitely trigger extreme surges in demand side at rush hours, and therefore, threaten the stability and security of the power grid. Consequently, smart charging which carries information into power has been tentatively conducted to solve these problems [5]. In smart charging, both the real-time status of power grid and the demand of the PEV owners are taken into consideration. The charging power is adjusted on line under the smart charging strategies so as to avoid the conventional peak loads. Apparently, one-way power flows from grid to vehicles are adopted in both traditional charging and smart charging. Another step forward is the so called vehicle-to-grid (V2G) operation [6, 7], in which bi-directional power flow can be carried out. This means the PEVs can be charged at off-peak hours and then feedback energy deposited in onboard batteries to grid at peak hours. Therefore, by digging the potential of PEVs as moveable and

Table 6.1 Fast charging, rapid charging and quick charging for PEVs

Charging type	Charging power rating		
	Heavy duty (kW)	SUV (kW)	Small Sedan (kW)
Fast charging, 10 min, 100 % Soc	500	250	125
Rapid charging, 15 min, 60 % Soc	250	125	60
Quick charging, 60 min, 70 % Soc	75	35	20

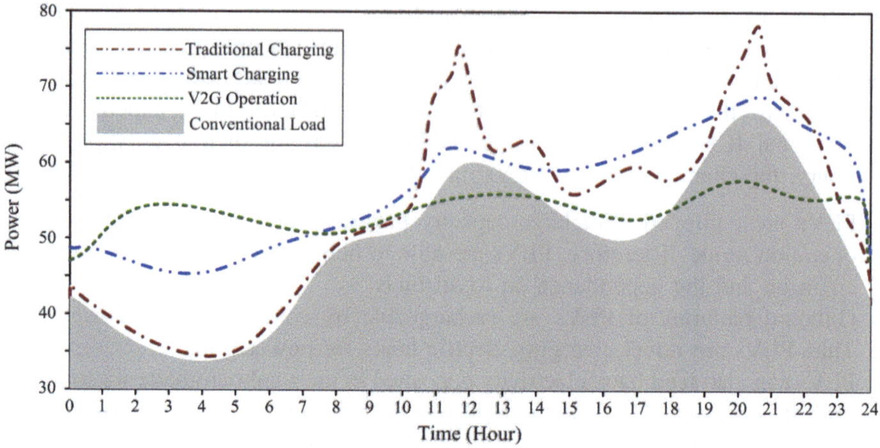

Fig. 6.1 Overall power load curves of regional power grid with different charging strategies

distributed energy storage devices, V2G operation can aid to flatten the power load curves [8–11], improve the stability and efficiency, as well as reduce the overall operating cost of power grid [12–19]. Figure 6.1 depicts the subtle differences among power grid's overall load curves in one-day period with traditional charging, smart charging, and V2G operation, respectively. Another remarkable benefit arising from conducting smart charging or V2G is to promote the integration of renewable energy sources, such as wind power, solar power, wave power, and so forth. Due to their severe dependence on climate conditions, renewable energy sources are significantly different from the traditional dispatchable power sources. Hence, how to integrate such unpredictable and unstable electricity into conventional power grids becomes a tough issue [20]. In the scheme of smart charging or V2G operation, onboard batteries of PEVs could help match the fluctuations between renewable energy production and power load consumption seamlessly [21, 22].

The key to the implementation of V2G is to what extent information can be effectively integrated into the energy conversion, transmission and distribution. There are several actual challenges ahead of us: Firstly, what are the crucial information related to the effective energy exchange between power grid and PEVs; Secondly, how does these information affect the overall energy efficiency and exhaust emission; Thirdly, how to acquire, transmit and process these information; Finally, how to effectively and correctly use these information to turn the complicated and chaotic process into somewhat ordered and effective process, so that the maximum benefits towards a win-win ecosystem can be achieved. These problems could be considered from three levels. The top level question is what the basic business model should be. What are the key players? Who invests? Who benefits? This is a question involving many aspects, such as people's lifestyles, geographical environment, degree of urbanization, condition of infrastructures and existing power grid, government policies, and so forth. The middle level question is

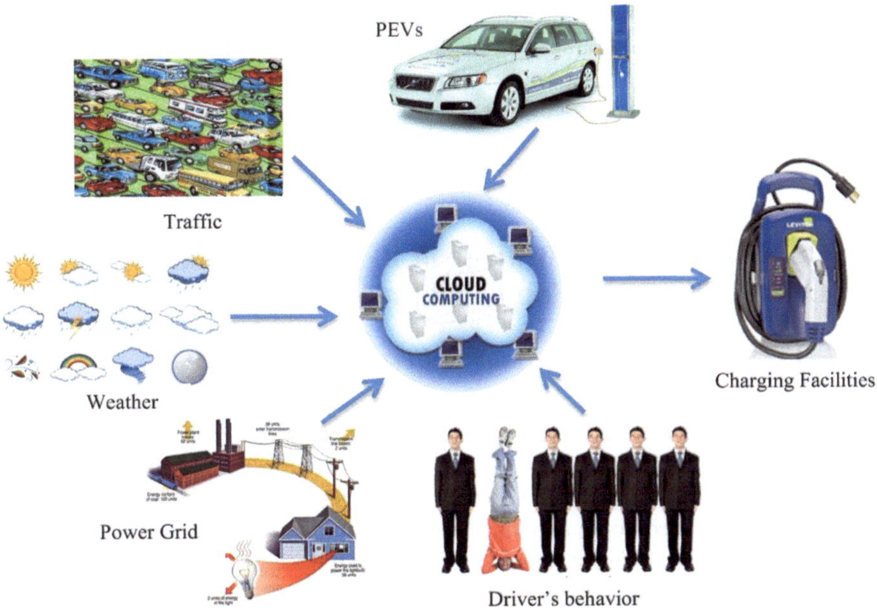

Fig. 6.2 Crucial information related to V2G implementation of PEVs

how to model and formulate the problem. What are the control objectives? What are the key variables? What are the constraints? How to solve the problem in real time so that the algorithm can be operated practically? The bottom level question is how to realize the designed business model and optimal charging strategy. This includes how to layout the charging posts or charging stations, how to build up the sensor networks and how to configure the computing resources. Therefore, V2G should be carried out within the framework of smart grid [23–25], so that the status information of power grid can be perceived. Another prerequisite is the massive data processing capability. As illustrated in Fig. 6.2, there are so much information should be taken into account, which includes traffic condition, weather condition, power grid condition, vehicle's condition, and conditions of charging facilities. The demand information of PEV owners should also be taken into account, so that the function of PEVs as transportation tools can be guaranteed.

This chapter will be focus on discussing the aforementioned problem formulation for V2G operation. Mathematically modeling concerning all kinds of crucial information can lead to optimal charging strategies of PEVs, which are aimed at maximizing the benefits of V2G operation. Generally, there are three kinds of optimization targets: minimizing the power losses of power grid, maximizing the load factors of power grid and minimizing the load variance of power grid. It has been theoretically demonstrated that, for practical systems, minimizing load variances will minimize power losses approximately. Additionally, maximizing load factor is almost equivalent to minimizing the load variance [17]. Herein, the optimization target of minimizing load variance is preferred since it will not involve any details on structural connection of the power grid, and has better universality for understanding, predicting and evaluating the performance of V2G operation. Two scenarios will be discussed, one is for the household smart micro-grid and the other is for smart regional grid.

6.2 Optimal Charging Strategy of PEVs for Minimizing Load Variance in Household Smart Micro-grid

6.2.1 V2G in Household Smart Micro-grid

Figure 6.3 depicts the scheme of a typical household smart micro-grid. Up till now, most of the home appliances are AC-powered, such as TV set, air conditioner, refrigerator, washing machine, computer, microwave oven, and so on. Still, more and more DC-powered home appliances are emerging nowadays, for example, LED lightings, fire alarms, exhaust fans, and so forth. So that, AC-DC power converters should be adopted to connect these equipments to the AC power line. PEVs can be plugged into the micro-grid through charging facilities when they are parked at the garage. These charging facilities can ensure bi-directional power flows between the PEVs and the micro-grid. The property of 'smart' refers to the sensor network and the power management unit (PMU) involved. The sensor network consists of smart meters, sensors for AC-powered devices, sensors for DC-powered devices and sensory components of the charging facilities for PEVs. They are employed to measure and acquire the comprehensive information regarding electricity usage within the household micro-grid and that of the outside grid delivered from the upper level operators. The PMU takes charge of coordinating the power flow patterns within the micro-grid according to the information acquired by the sensor network. Generally, micro-grids can work in either grid-connected mode or islanded mode [26], and intelligent control algorithms are performed by the PMU to automatically switch between different working modes, so as to satisfy the demands of power loads, reduce electric shock to the upper level grid, and achieve highly efficient operation of the whole system.

The micro-grid shown in Fig. 6.3 is essentially an AC-bus micro-grid. The red solid line and the green solid line are the AC power line and the DC power line, respectively, while the blue dash line representing the signal line. Generally speaking, DC-bus micro-grid may take advantages when renewable energy generation systems, such as on-roof photovoltaic power unit and yard-standing wind power unit, and energy storage systems, such as batteries and ultra capacitors, are involved. For DC-bus household micro-grid, a power isolation unit which takes charge of AC-DC power conversion and power factor correction is needed to connect the household micro-grid to the up level AC power grid.

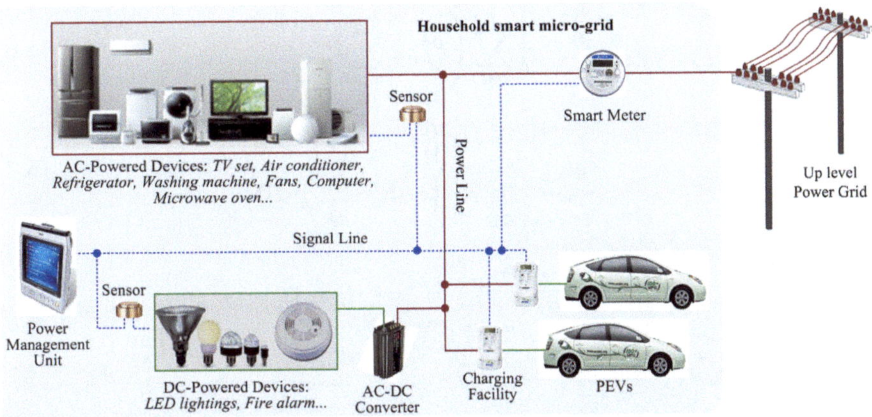

Fig. 6.3 Scheme of V2G operation in a typical household smart micro-grid

6.2.2 Optimal Problem Formulation for Minimizing Load Variance in Household Smart Micro-grid

Naturally, the one-day cycle from 00h00 to 24h00 is chosen for discussing. For convenience, the 24-h period is divided into T time-slots, and the length of each time-slot is given by Δt. In addition, each PEV is assumed to start charging at the beginning of some time-slot and complete charging at the end of another time-slot. Consequently, the problem can be formulated as:

$$\min \sum_{t=1}^{T} \left[\frac{1}{T} \left(P_t^A - \mu_T^H + \sum_{n=1}^{N} P_{t,n}^C \right)^2 \right] \tag{6.1}$$

Subject to:

$$P_t^A + \sum_{n=1}^{N} P_{t,n}^C \le P_{t,\max}^H, \; (t = 1 \sim T) \tag{6.2}$$

$$-P_{n,\max}^C \le P_{t,n}^C \le P_{n,\max}^C, \; \left(t = \Gamma_n^s \sim \Gamma_n^e; n = 1 \sim N\right) \tag{6.3}$$

$$soc_{\min,n} \le soc_{t,n} \le soc_{\max,n}, \; \left(t = \Gamma_n^s \sim \Gamma_n^e; n = 1 \sim N\right) \tag{6.4}$$

$$\sum_{t=\Gamma_n^s}^{\Gamma_n^e} \left[P_{t,n}^C \left(\eta_n^C\right)^{\mathrm{sgn}_{t,n}} \Delta t \right] = \Delta W_n = W_n^e - W_n^s, \; (n = 1 \sim N) \tag{6.5}$$

$$\mu_T^H = \frac{1}{T}\sum_{t=1}^{T}\left(P_t^A + \sum_{n=1}^{N} P_{t,n}^C\right) \tag{6.6}$$

$$\underbrace{\mathrm{sgn}_{t,n}}_{t=\Gamma_n^s \sim \Gamma_n^e,\ n=1\sim N} = \begin{cases} 1, & \left(P_{t,n}^C \geq 0\right) \\ -1, & \left(P_{t,n}^C < 0\right) \end{cases} \tag{6.7}$$

$$\underbrace{soc_{t,n}}_{n=1\sim N} = \begin{cases} \frac{W_n^s + P_{t,n}^C\left(\eta_n^C\right)^{\mathrm{sgn}_{t,n}}\Delta t}{Q_n}, & \left(t = \Gamma_n^s\right) \\ soc_{t-1,n} + \frac{P_{t,n}^C\left(\eta_n^C\right)^{\mathrm{sgn}_{t,n}}\Delta t}{Q_n}, & \left(t = \Gamma_n^s + 1,\ \ldots\ \Gamma_n^e\right) \end{cases} \tag{6.8}$$

where

Δt is the length of the time-slot (min);

T is the number of the time-slots in one-day cycle;

N is the number of the PEVs available for the micro-grid;

Γ_n^s is the serial number of the time-slot when the n-th PEV being connected to the micro-grid;

Γ_n^e is the serial number of the time-slot when the n-th PEV being disconnected from the micro-grid;

P_t^A is the power of the home appliance loads at the t-th time-slot (kW);

$P_{t,n}^C$ is the operating power of the n-th charger at the t-th time-slot (kW);

$\mathrm{sgn}_{t,n}$ is the sign function of $P_{t,n}^C$ [equal to either 1 or −1 defined by (6.7)];

$P_{t,\max}^H$ is the maximum operating power of the whole micro-grid at the t-th time-slot (kW);

$P_{n,\max}^C$ is the maximum operating power of the n-th charger at the t-th time- slot (kW);

η_n^C is the efficiency of the n-th charger (value lies in between 0 and 1);

μ_T^H is the average power of the whole micro-grid in one-day period (kW);

ΔW is the net charging quantity of the n-th PEV required by its owner (kWh);

W_n^s is the initial charge of its battery when the n-th PEV being connected to the micro-grid (kWh);

W_n^e is the finial charge of its battery when the n-th PEV being disconnected from the micro-grid (kWh);

$soc_{t,n}$ is the state of charge (Soc) of the n-th PEV at the t-th time-slot (value lies in between 0 and 1);

$soc_{\min,n}$ is the allowed minimum value of the Soc of the n-th PEV (value lies in between 0 and 1);

$soc_{\max,n}$ is the allowed maximum value of the Soc of the n-th PEV (value lies in between 0 and 1);

Q_n is the capacity of the battery equipped in the n-th PEV (kWh).

It is worth noting that the chargers' efficiencies vary with both the operating power and the work voltage. It has been demonstrated that they don't change fiercely in magnitude [27], thus herein they are set as constant for simplicity. Another simplification involved in the modeling is that the onboard batteries are deemed as ideal energy storage systems, namely, their transient performance during charging/discharging process are not taken into account.

In what follows, the operating power instructions for each charger during each time-slot which can satisfy the problem (6.1)–(6.8) the best will be determined. Nevertheless, this is a nonconvex problem due to the constraints (6.5) and (6.8). It is rather difficult to find the global optimal solution directly, especially for cases with large number of variables. Consequently, some tricks are engaged and the problem is solved as per following procedures:

Step 1: The original problem is slacked into a quadratic programming (QP) problem by ignoring the nonlinear factor $\left(\eta_n^C\right)^{\mathrm{sgn}_{t,n}}$ existing in constraints (6.5) and (6.8)

Step 2: Make the indicator $j = 1$. Find the global optimal solution of the QP problem obtained in step 1, and denote them by $P_{t,n}^C(j)$

Step 3: Determine the values of $\mathrm{sgn}_{t,n}$ according to $P_{t,n}^C(j)$, and substitute them into constraints (6.5) and (6.8). Then, the problem (6.1)–(6.8) becomes a QP problem

Step 4: Solve the QP problem obtained in step 3, and denote the global optimal solution by $P_{t,n}^C(j+1)$

Step 5: If the signs of $P_{t,n}^C(j+1)$ are identical with that of $P_{t,n}^C(j)$, cease the procedure, and $P_{t,n}^C(j+1)$ is the final solution, otherwise, go back to step 3.

It is well known that the global optimal solution of QP problem can be efficiently solved in polynomial time [28]. Thus the resulted optimization problem is of polynomial time-space complexity.

6.2.3 Case Study

In case 1, it is assumed that there are two PEVs available for the smart household micro-grid, viz., PEV1 and PEV2. The parameters involved are listed in Table 6.2. For a typical weekday, PEV1 is disconnected from the micro-grid at 07:00 in the morning and connected into grid at 18:00 in the evening, while PEV2 is disconnected from the micro-grid at 14:00 in the afternoon and connected into grid at 17:00 in the evening. The initial charge W_n^s of the onboard battery of PEV1 and PEV2 are set as 2.8 and 2.2 kWh, respectively. The initial Soc values are both 0.2 for the two PEVs. Correspondingly, the net charging quantities ΔW_n are set as 8 and 6 kWh, respectively. Herein, ΔW_n is defined as the difference between the final charge and the initial charge of the onboard battery, and it represents the

Table 6.2 Parameters involved in case 1

	Γ_n^s	Γ_n^e	$soc_{\min,n}$	$soc_{\max,n}$	Q_n (kWh)	W_n^s (kWh)	ΔW_n (kWh)	$P_{n,\max}^C$ (kW)	η_n^C	Δt (min)	$P_{t,\max}^H$ (kW)
PEV1 ($n = 1$)	72	28	0.2	0.8	14	2.8	8	1	0.92	15	6
PEV2 ($n = 2$)	68	56	0.2	0.8	11	2.2	6	1	0.92	15	6

demanded energy consumed by the PEVs on their way for the next itinerary. The length of the time-slot Δt is set as 15 min, thus there are 96 time slots in the one-day cycle. Actually, a tradeoff should be made when setting the length of the time slot. On one hand, if the time slot is too long, a coarse optimization procedure can not ensure satisfactory effect on the reduction of the load variance. One the other hand, if the time slot is too short, the frequent adjustment on the charging power instruction is no good for both the power electronic circuits and the batteries. In addition, very short time slot will result in considerable large number of variables, and make the problem too difficult to solve.

Figure 6.4 gives a typical load profile of household micro-grid in one-day cycle when PEV loads are excluded. With the development of smart grid technology, the 24-h-ahead conventional power load (excluding PEV loads) forecasting based on meteorological information and historic data [29] is believed to be with acceptable accuracy in the future. The mean and the variance of the conventional power load profile given in Fig. 6.4 are 4.15 kWh and 0.57632, respectively. The optimal charging profiles of the two PEVs for minimizing the load variance can be obtained as illustrated in Fig. 6.5. It can be observed that both PEVs absorb electrical energy from the micro-grid with different power during each time slot. Sometimes, they may even feed back the deposited energy to the micro-grid in order to compensate the energy demands by other loads.

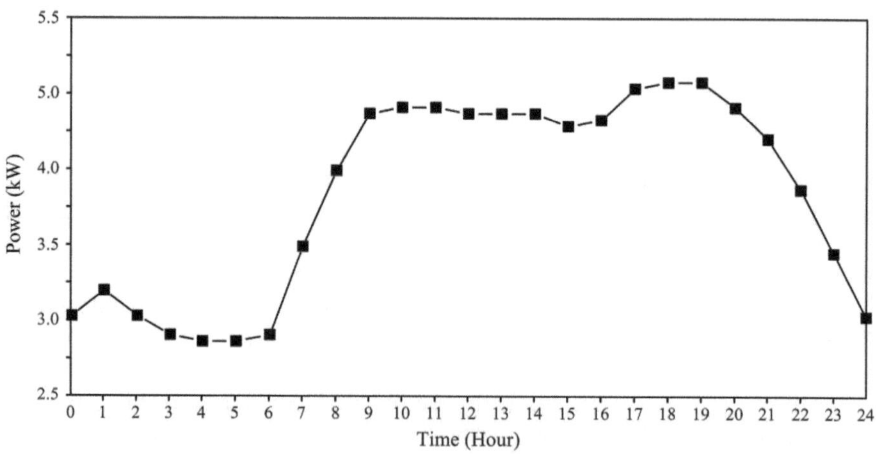

Fig. 6.4 Typical load profile of household micro-grid (excluding PEVs) in 24-h period

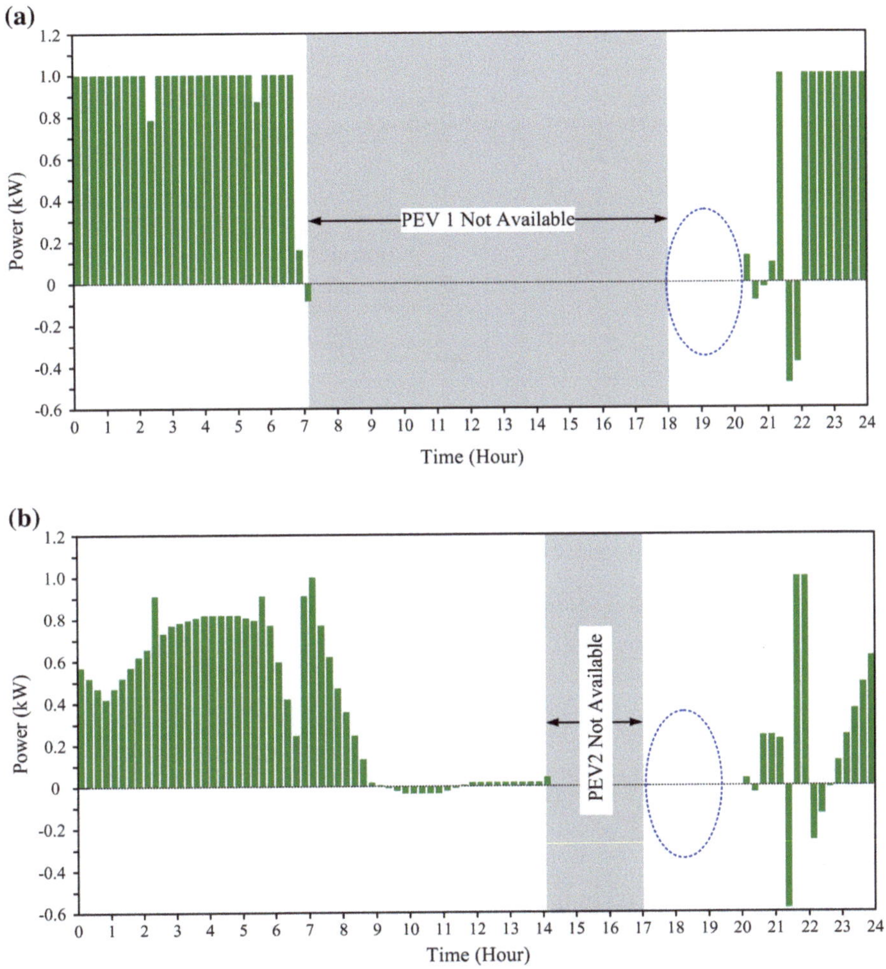

Fig. 6.5 Calculated optimal charging profiles in case 1: **a** for PEV1, **b** for PEV2

Figure 6.6 gives calculated V2G operation results in this case. The comparisons of the conventional load profile, the load profile with optimal charging and the average power load of the micro-grid are illustrated in Fig. 6.6a. It can be observed that the overall power load demand of the micro-grid can be dramatically smoothed out via V2G operation. The resulted mean and variance with optimal charging are 4.7884 kW and 0.00547, respectively. Figure 6.6b gives the Soc curves of the two PEVs. The Soc values when disconnected from the grid are 0.7714 and 0.7818 for PEV1 and PEV2, respectively.

In case 1, the initial charges of the two PEVs are given directly based on assumptions. Nevertheless, it is easy to understand that these initial statuses do have profound impact on the overall performance of the V2G operation. In order to find

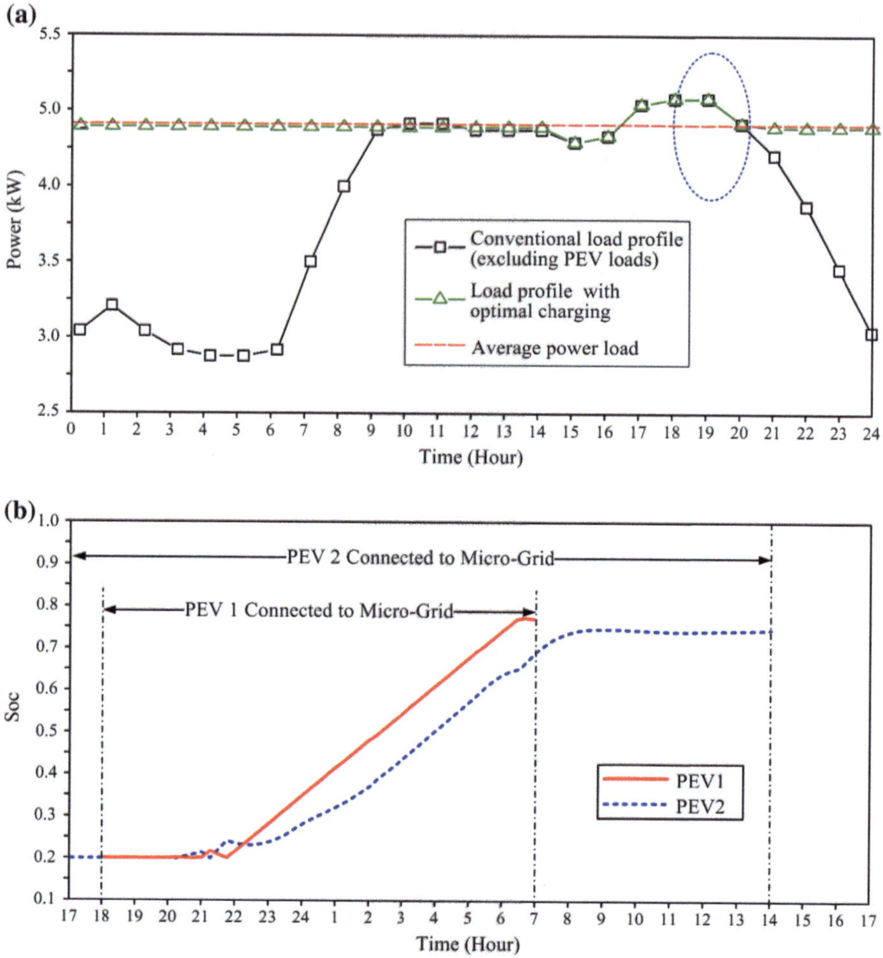

Fig. 6.6 Calculated V2G operation results in case 1: **a** comparison of load-power profiles, **b** Soc *curves* of two PEVs

out the greatest capability to subdue load variance by using optimal charging, the case with optimized initial charges of batteries should be investigated. Therefore, in case 2, the initial charges of the onboard batteries of the two PEVs are also set as variables. Then, they are solved along with the other variables aiming at satisfying the aforementioned optimal charging problem the best. All the other parameters are kept exactly the same with those in the case 1.

Figure 6.7 illustrates the resulted optimal charging profiles of the two PEVs with optimal initial charges of the batteries (case 2). It can be observed that there are subtle differences between the resulted charging patterns of case 1 and 2, especially in the parts enclosed by the ellipses in dash line. In case 2, the PEVs feedback more

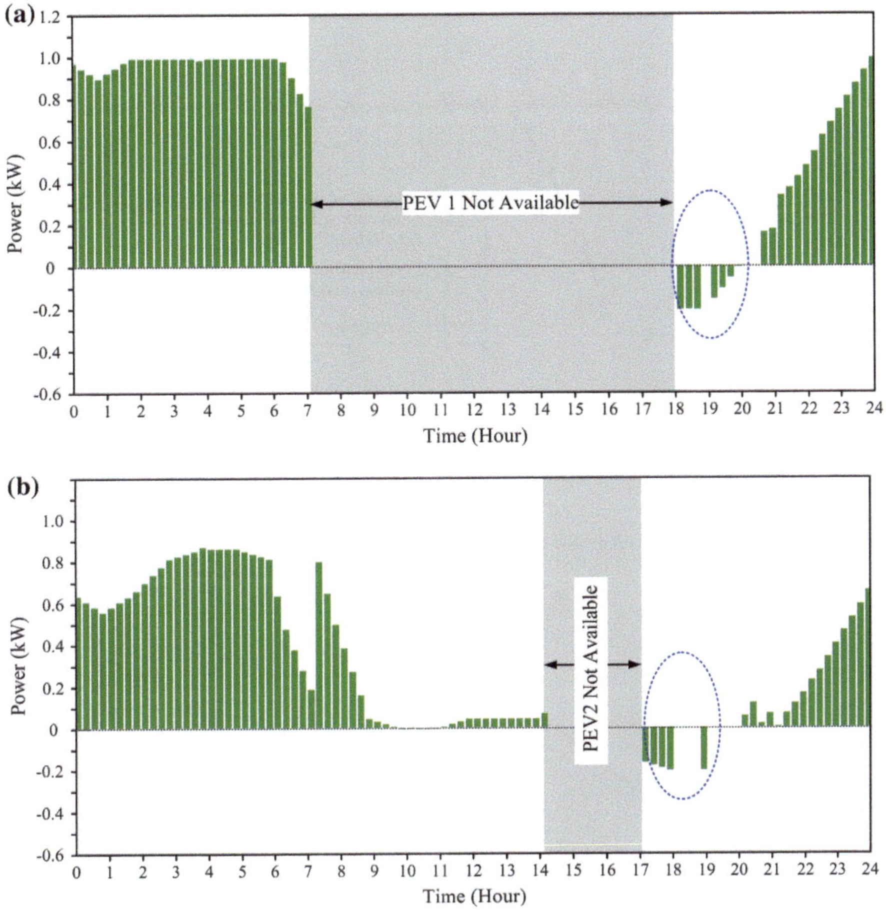

Fig. 6.7 Calculated optimal charging profiles in case 2: **a** for PEV1, **b** for PEV2

electricity into grid than they did in case 1. This means the initial statuses of the onboard battery really can affect the energy exchange between the PEVs and the micro-grid. What is more, the comparisons of the conventional load profile, the load profile with optimal charging and the average power load of the micro-grid are illustrated in Fig. 6.8a. Since the PEVs are able to feed back power immediately when connected to the micro-grid, the resulted variance of the load power is 0.00132, much less than that resulted in case 1. The mean of the load power equals 4.7873 kW in case 2. Figure 6.8b gives the Soc curves of the two PEVs. The values of Soc when disconnected from the grid are 0.7943 and 0.7800 for PEV1 and PEV2, respectively.

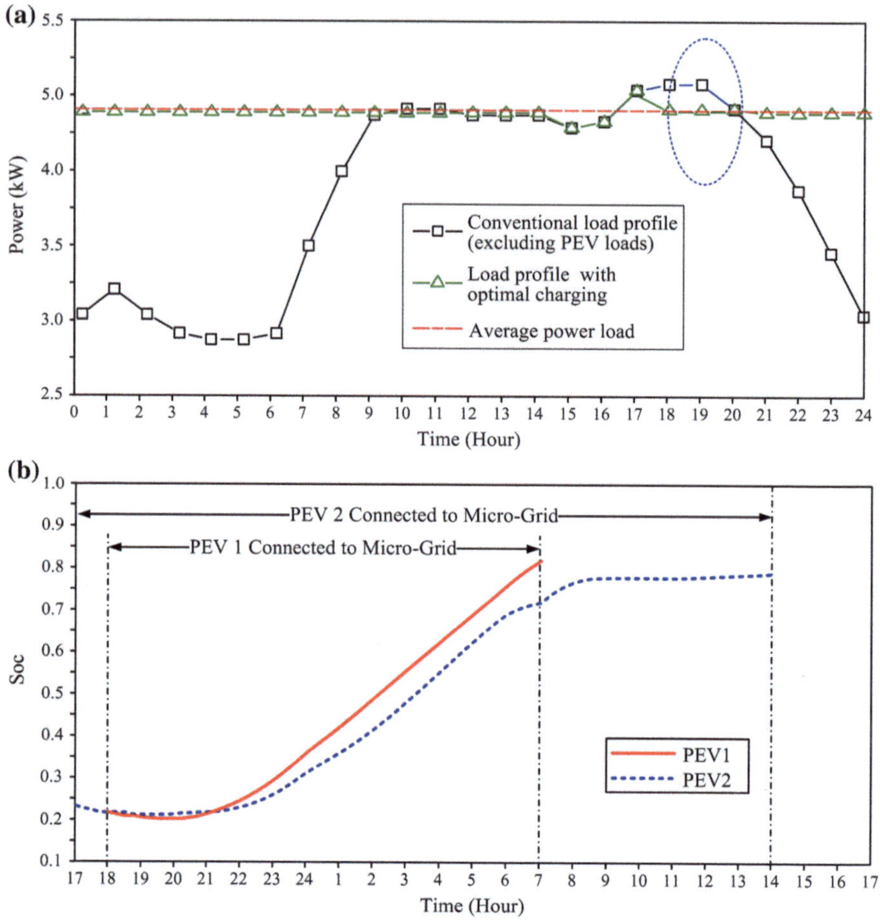

Fig. 6.8 Calculated V2G operation results in case 2: **a** comparison of load-power profiles, **b** Soc *curves* of the two PEVs

Through optimizing the initial charges of onboard batteries, both the variance and the mean of the load power profile can be further reduced. Moreover, the mean of the load power curves are different in the two cases, which implies that there are extra energy losses caused by the regulated charging process. As we all know, energy losses will occur when it flows through the power electronic circuits. The total energy losses E_{LT} during the one-day cycle can be given by:

$$E_{LT} = \sum_{n=1}^{N} \sum_{t=\Gamma_n^s}^{\Gamma_n^e} \text{sgn}_{t,n} P_{t,n}^C (1-\eta)\eta^{\frac{\text{sgn}_{t,n}-1}{2}} \Delta t \tag{6.9}$$

Apart from the energy losses E_{LW} which is responsible for withdrawing energy from the micro-grid by the PEVs, the energy losses E_{LF} aroused by feeding back energy to the micro-grid from the PEVs can be obtained by:

$$E_{LF} = E_{LT} - E_{LW} = E_{LT} - \sum_{n=1}^{N} (1 - \eta)\left(W_n^e - W_n^s\right)/\eta \tag{6.10}$$

Let's consider a new case, viz., the case 3, in which the discharging of the PEV batteries is not allowed by amending the constraint (6.3) into:

$$0 \leq P_{t,n}^C \leq P_{n,\max}^C, \left(t = \Gamma_n^s \sim \Gamma_n^e; n = 1 \sim N\right) \tag{6.11}$$

Then, the optimal charging pattern of case 3 can be solved. It should be noted that the initial charge of the onboard batteries have no impact on the overall performance in this case, since the energy can not be fed back to the micro-grid. The overall performances including mean and variance of the power load profile with optimal charging and the power losses involved of these three cases are listed in Table 6.3 for comparison. It can be found that by feeding back power to the micro-grid when necessary, it can further reduce the load variance by 77.04 %, from 0.00575 to 0.00132, with consuming 0.0782 kWh more electricity. Although it is a trifle cost, there is no reason to require customers to pay for the extra bill for their good deeds. Subsidy mechanism could be developed by power companies under the guidance of governments since they can really benefit from promoting V2G operation of PEVs. Recently, the floating electricity pricing has become a worldwide topic, and in some regions, it has been implemented. With no doubt, it can benefit costumers by depositing extra electricity during low-price time periods, and then compensate the demands during high-price time periods. However, there may still be two problems: (1) it is not easy to design suitable pricing mechanism which can ensure the optimal overall performance; (2) by conducting floating pricing, the optimization target of the regulated charging discussed herein may be shifted into maximizing the profits, and it may contradict the minimization of load variance. Thus, we believe that a better way to develop the subsidy mechanism may lie in investigating the comprehensive benefits arising form reducing the load variance, including the increase of the grid's efficiency, stability, and the reduction of the greenhouse gases emission. Then, governments and power companies should share

Table 6.3 Comparison of performance in different cases

Performance index	Case studied		
	Case 1	Case 2	Case 3
Mean	4.7884 kW	4.7873 kW	4.7841 kW
Variance	0.00547	0.00132	0.00575
E_{LW}	1.2174 kWh	1.2174 kWh	1.2174 kWh
E_{LF}	0.1053 kWh	0.0782 kWh	0

these benefits with the customers by refunding them, so as to reach the win-win situation.

Next, the impacts of the capacity of onboard batteries and the requested net charging quantities on the performance of V2G operation concerning minimizing the power load variance are investigated. It is worth noting that the case with optimized initial statuses can eliminate the impact of the initial charge of the onboard batteries. Thus, in what follows, all the discussion and analysis will be based on such kind, so as to ensure the fairness of the comparisons.

Figure 6.9 gives the calculated results with different requested net charging quantities ΔW of the two PEVs. It can be observed that the requested net charging

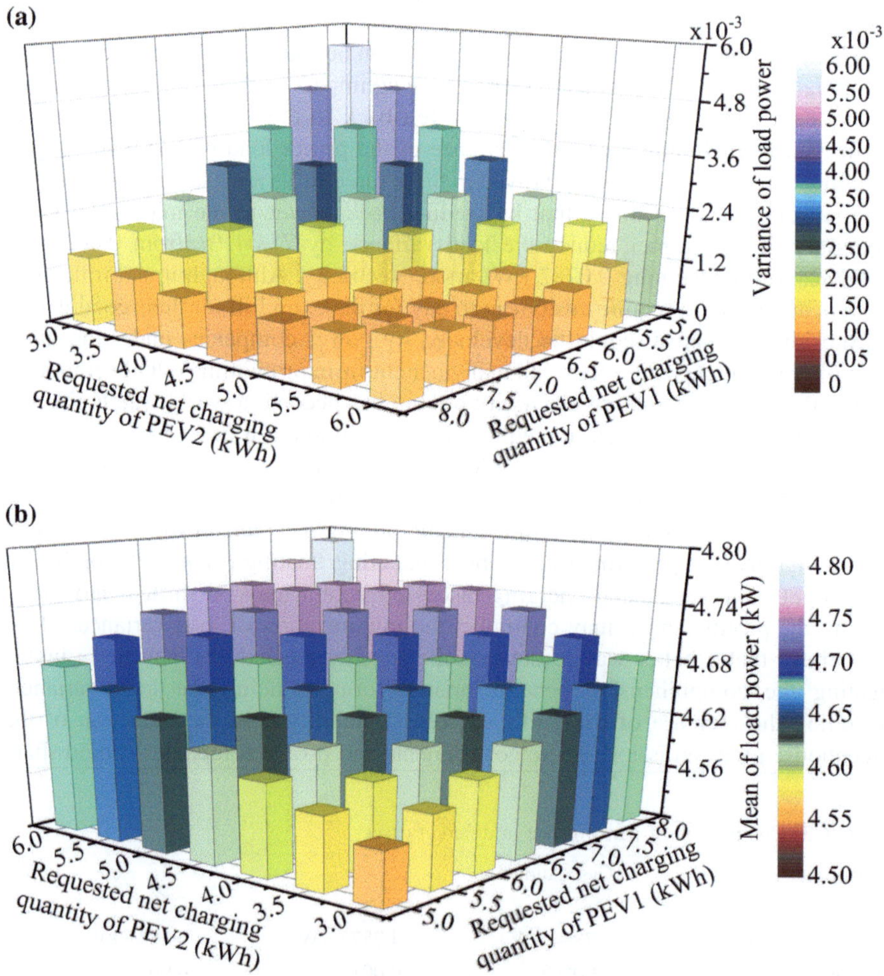

Fig. 6.9 Calculated optimal charging results with different requested net charging quantities of PEVs. **a** Variance of load power. **b** Mean of load power

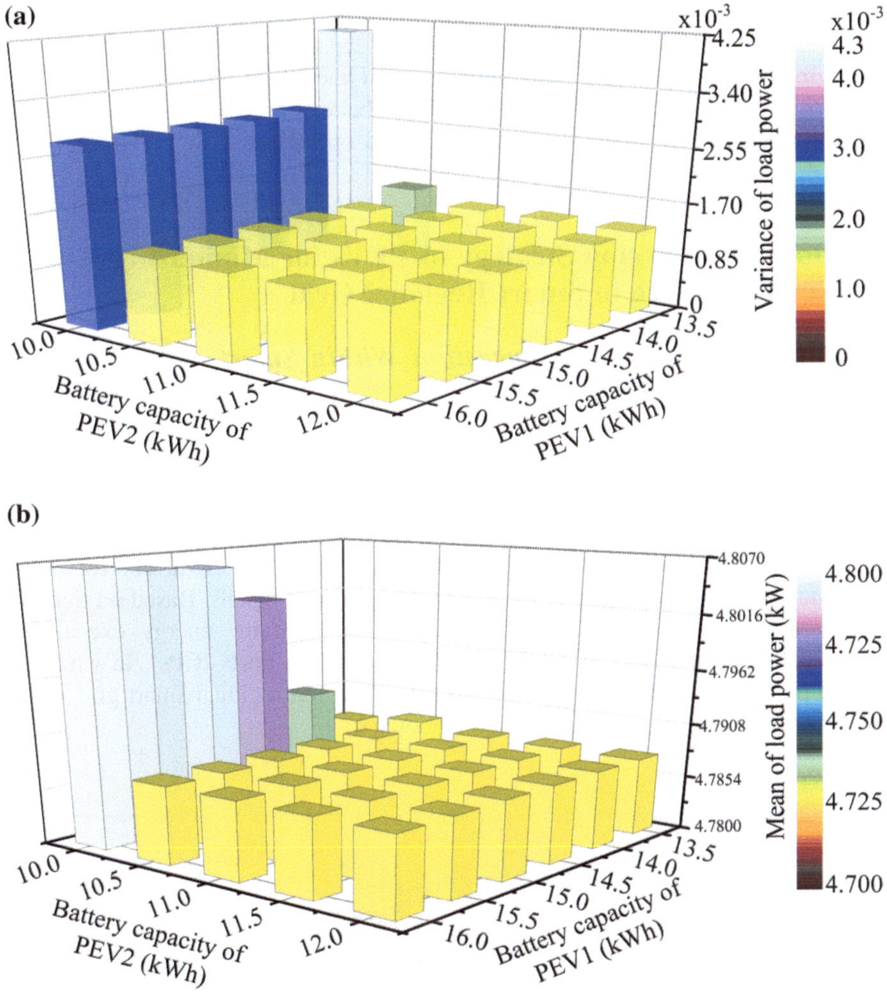

Fig. 6.10 Calculated optimal results with different battery capacities of the PEVs. **a** Variance of load power. **b** Mean of load power

quantity has an effect on the resulted variance and mean of the load power. For this case, the bigger the requested net charging quantity is, the smaller the variance of the load power will be. However, this does not fit for all other cases. Actually, the impact of ΔW is also influenced by the time slots when the PEVs are connected into the micro-grid and the pattern of the load curve of the home appliances.

Figure 6.10 gives the calculated results with different battery capacities Q of the two PEVs. Customers may take it for grant that by using batteries with larger capacities better performance of the V2G operation could be achieved. Nevertheless, it is quite interesting to observe that once the battery capacities of PEV1 and

PEV2 are larger than 14.0 and 11.0 kWh, respectively, there is no significant improvement by using batteries with even larger capacities. Since the battery is one of the main reasons for the high initial cost of the PEVs, customers ought not to choose the PEVs with unnecessarily large battery capacities, just out of the consideration for expanding the benefits of V2G operation.

6.3 Optimal Charging Strategy of PEVs for Minimizing Load Variance in Smart Regional Grid

6.3.1 A Scenario of V2G Operation Within Smart Regional Grid

V2G implementation in regional grid is a rather more sophisticated issue than that in micro-grid. It involves new technique patterns, innovative business models and even novel industrial rules. Figure 6.11 illustrates the information flow and the energy flow concerning V2G in regional smart grid. The central control center (CCC) takes charge of acquiring and processing the key signals. Based on that, the CCC generates optimal charging instructions to guide the energy exchanging between PEVs and power grid bridged by the charging posts (CPs). In what follows, a possible scenario for implementing V2G within regional smart grid which involves a booking mechanism will be elaborated.

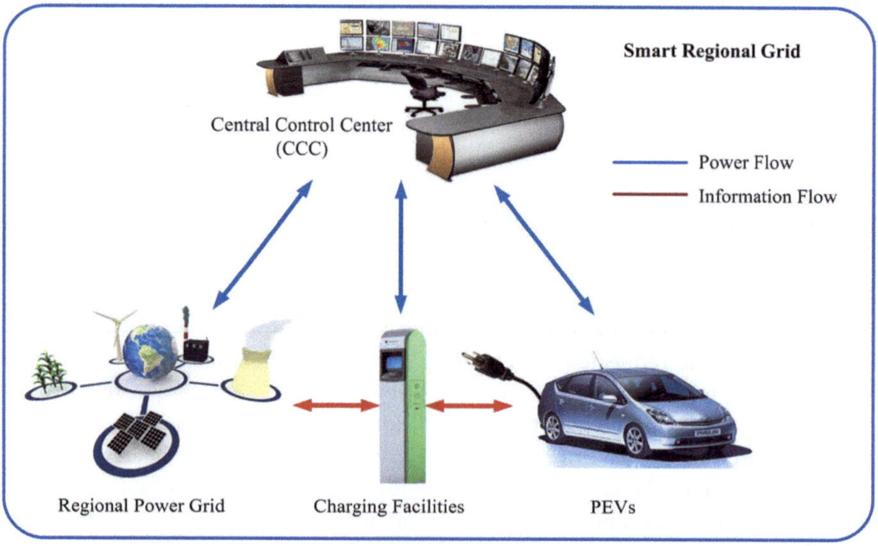

Fig. 6.11 Scheme of V2G operation in smart regional grid

Firstly, the CCC should be aware which CPs are available for the V2G operation, and what their key charging-planning-related parameters are. All the CPs, no matter whether they are public or private, should be reported to the CCC. Consequently, the CCC can build data tuples on its server to record every CP involved:

$$CP = [CP_{ID}, CP_{Loc}, P_{CP}^{\max}, Flag] \tag{6.12}$$

where the entities are:

CP_{ID} the identity number of the CP;
CP_{Loc} the location of the CP;
$P_{CP}^{\max}$ the allowed maximum charging power of the CP;
$Flag$ to indicate whether the CP is available for public use. $Flag = 0$ means this is a public CP, while, $Flag = 1$ represents this is a private CP.

Secondly, the CCC should also understand the necessary information on PEVs involved. Vehicle owners who are willing to associate their PEVs with V2G in the smart regional grid should register their vehicles in advance by, for example, submitting the registration form online to the CCC. Then, CCC will allot a unique identity number for each PEV that are successfully registered. Consequently, CCC builds data tuples on its server to record every possible PEV involved:

$$EV = [EV_{ID}, EV_{Mod}, BAT_{Typ}, BAT_{Cap}, Soc_{upper}, Soc_{lower}] \tag{6.13}$$

where the entities represent:

EV_{ID} the identity number of the PEV;
EV_{Mod} the model of the PEV;
BAT_{Typ} the type of the battery equipped;
BAT_{Cap} the capacity of the battery equipped;
Soc_{upper} the allowed upper limit for the Soc value of battery;
Soc_{lower} the allowed lower limit for the Soc value of battery.

The battery Soc is an equivalent quantity of a fuel gauge for battery packs. It indicates the amount of electric energy left in a battery compared with the energy it has when it is fully charged [30, 31]. In order to extend the lifetime of batteries, upper limit and lower limit for the Soc value should be set to avoid over charging and deep discharging, which both can harm the physical constitution of batteries.

Thirdly, PEV owners propose request for joining V2G Operation. Normally, the CCC plans the charging schedule for every 24 h (one-day cycle), for example, from 06:00 to 05:59 a.m. (next day). Thus, each vehicle owner who intends to join V2G operation for the coming one-day cycle should propose request to the CCC before by the due time (06:00 a.m.), and tell when and where their PEVs will be connected to grid. This can be conducted by logging into the designed online booking system. For our mathematical modeling, some key information should be reported to the CCC: (1) The estimated latest moment when this PEV can be connected into grid;

(2) The estimated earliest moment when this PEV will be detached from grid; (3) Which charging post this PEV will be connected to; (4) The estimated battery Soc value when this PEV connected into grid; (5) The required battery Soc value when this PEV detached from grid. For point 3, if the vehicle owner plans to connect his/her PEV into his/her private charging post, he/she reports the ID Number of the charging post directly. Otherwise, he/she reports the location or area where the PEV will stop, and request the CCC to allocate a public charging post for the PEV. For point 4, the battery Soc value when the PEV connected into grid is affected by many factors, such as the energy efficiency of the PEV, the initial Soc when the last charging is completed, the travel way of the PEV before being connected in, the on board passengers and loads, and so on. We believe that with better understanding on the people's lifestyle and the social eco-environment in the region considered, an 'Estimator' program can be developed to help the vehicle owner exactly estimate the battery Soc when his/her PEV connected into grid.

Fourthly, PEV owners' request is confirmed and the CCC starts data preparation for system modeling. After the vehicle owner submits his/her request, the CCC attempts to include the proposed PEV into the V2G operation by allocating an available charging post for it. On the server of the CCC, there is a database to record the mappings between the location where PEV stops and all the charging posts installed nearby. The charging posts are available on the principle of first-proposed first-served. The allocation may be failed for two reasons: (1) There is not any charging post located within the area where the PEV will be parked; (2) The charging posts nearby have all been allocated to other PEVs. The CCC feedbacks its allocation results to the vehicle owners and asks for their confirmation by the online booking system. If the plan is confirmed, this case will be included into the V2G operation. After that, the CCC conduct data preparation for mathematical formulation according to the confirmed allocation plans. For the first step, it generates a set of the active charging posts as:

$$S_{ACP} = [A_{CP}^{1}, A_{CP}^{2}, A_{CP}^{3}, \ldots, A_{CP}^{N}] \tag{6.14}$$

where A_{CP}^{n}, $n = 1, 2, 3, \ldots, N$, represents the n-th active charging post that is assigned to offer charging services in the next 24-h, and N is the number of the active charging posts.

For the second step, the CCP builds up a 2D data tuple for each charging post A_{CP}^{n} as given by:

$$A_{CP}^{n} = \begin{bmatrix} \Gamma_{ACP-n}^{s-1}, & \Gamma_{ACP-n}^{e-1}, & EV_{ACP-n}^{ID-1}, & BAT_{ACP-n}^{Cap-1}, & Soc_{ACP-n}^{upper-1}, & Soc_{ACP-n}^{lower-1}, & Soc_{ACP-n}^{s-1}, & Soc_{ACP-n}^{e-1} \\ \Gamma_{ACP-n}^{s-2}, & \Gamma_{ACP-n}^{e-2}, & EV_{ACP-n}^{ID-2}, & BAT_{ACP-n}^{Cap-2}, & Soc_{ACP-n}^{upper-2}, & Soc_{ACP-n}^{lower-2}, & Soc_{ACP-n}^{s-2}, & Soc_{ACP-n}^{e-2} \\ \vdots & \vdots & \vdots & \vdots & \vdots & \vdots & \vdots & \vdots \\ \Gamma_{ACP-n}^{s-k}, & \Gamma_{ACP-n}^{e-k}, & EV_{ACP-n}^{ID-k}, & BAT_{ACP-n}^{Cap-k}, & Soc_{ACP-n}^{upper-k}, & Soc_{ACP-n}^{lower-k}, & Soc_{ACP-n}^{s-k}, & Soc_{ACP-n}^{e-k} \\ \vdots & \vdots & \vdots & \vdots & \vdots & \vdots & \vdots & \vdots \\ \Gamma_{ACP-n}^{s-K(n)}, & \Gamma_{ACP-n}^{e-K(n)}, & EV_{ACP-n}^{ID-K(n)}, & BAT_{ACP-n}^{Cap-K(n)}, & Soc_{ACP-n}^{upper-K(n)}, & Soc_{ACP-n}^{lower-K(n)}, & Soc_{ACP-n}^{s-K(n)}, & Soc_{ACP-n}^{e-K(n)} \end{bmatrix} \tag{6.15}$$

where the quantities are:

Γ_{ACP-n}^{s-k} the k-th charging service offered by A_{CP}^{n} will start at the beginning of the Γ_{ACP-n}^{s-k}-th time-slot;

Γ_{ACP-n}^{e-k} the k-th charging service offered by A_{CP}^{n} will finish at the ending of the Γ_{ACP-n}^{e-k}-th time-slot;

EV_{ACP-n}^{ID-k} the ID Number of the PEV connected to A_{CP}^{n} during the k-th charging service offered by A_{CP}^{n};

BAT_{ACP-n}^{Cap-k} the battery capacity of the PEV connected to A_{CP}^{n} during the k-th charging service offered by A_{CP}^{n};

$Soc_{ACP-n}^{upper-k}$ the allowed battery Soc upper limit of the PEV connected to A_{CP}^{n} during the k-th charging service offered by A_{CP}^{n};

$Soc_{ACP-n}^{lower-k}$ the allowed battery Soc lower limit of the PEV connected to A_{CP}^{n} during the k-th charging service offered by A_{CP}^{n};

Soc_{ACP-n}^{s-k} the estimated battery Soc value of PEV when the k-th charging service offered by A_{CP}^{n} starts;

Soc_{ACP-n}^{e-k} the required battery Soc value of PEV when the k-th charging service offered by A_{CP}^{n} ends;

$K(n)$ Charging post A_{CP}^{n} is assigned to offer $K(n)$ times charging service in the coming one-day cycle.

Up till now, the CCC gets enough information on the PEVs, the charging posts, and the vehicle owners' expectations. Some other key points should be gotten known before the CCC can carry out optimal charging planning for every charging service, and this will be elaborated in the following sections. We must make it clear that the scenario of V2G implementation presented herein is come out by focusing on the essential functions V2G operation, and it does not concern any considerations on economic incentives, business models or government policy makings.

6.3.2 Optimal Problem Formulation for Minimizing Load Variance in Smart Regional Grid

The optimal charging planning is to determine the charging power at each time slot for every charging post when it is offering charging services for PEVs. For each charging post, its charging power in the same time slot is kept unchanged. The objective is to minimize the overall load variance of the regional grid during the coming one-day cycle. Hence, the problem can be formulated as:

$$\min \sum_{t=1}^{T} \left[\frac{1}{T} \left(P_{Con}^{t} - P_{Avg} + \sum_{n=1}^{N} P_{ACP-n}^{t} \right)^{2} \right] \tag{6.16}$$

Subject to:

$$P^t_{Con} + \sum_{n=1}^{N} P^t_{ACP-n} \le P^t_{\max},\ t \in [1, T] \tag{6.17}$$

$$P_{Avg} = \sum_{t=1}^{T} \left(P^t_{Con} + \sum_{n=1}^{N} P^t_{ACP-n} \right) \Big/ T \tag{6.18}$$

$$-P^{\max}_{ACP-n} \le P^t_{ACP-n} \le P^{\max}_{ACP-n}, t \in \cup_{k=1}^{k=K(n)} \left[\Gamma^{s-k}_{ACP-n}, \Gamma^{e-k}_{ACP-n}\right],\ n \in [1, N] \tag{6.19}$$

$$P^t_{ACP-n} = 0, t \in [1, T] - \cup_{k=1}^{k=K(n)} \left[\Gamma^{s-k}_{ACP-n}, \Gamma^{e-k}_{ACP-n}\right],\ n \in [1, N] \tag{6.20}$$

$$\sum_{t=\Gamma^{s-k}_{ACP-n}}^{\Gamma^{e-k}_{ACP-n}} \left[\Delta t \cdot P^t_{ACP-n}\right] = \left(Soc^{e-k}_{ACP-n} - Soc^{s-k}_{ACP-n}\right) \cdot BAT^{Cap-k}_{ACP-n}, \quad k \in [1, K(n)],\ n \in [1, N] \tag{6.21}$$

$$Soc^{lower-k}_{ACP-n} \le Soc^k_{ACP-n}(j) \le Soc^{upper-k}_{ACP-n}\ j \in \left[\Gamma^{s-k}_{ACP-n}, \Gamma^{e-k}_{ACP-n}\right], \quad k \in [1, K(n)],\ n \in [1, N] \tag{6.22}$$

$$Soc^k_{ACP-n}(j) = Soc^{s-k}_{ACP-n} + \sum_{t=\Gamma^{s-k}_{ACP-n}}^{j} \left[\Delta t \cdot P^t_{ACP-n} \Big/ BAT^{Cap-k}_{ACP-n}\right] \tag{6.23}$$

where

P^t_{ACP-n}	is the charging power of A^n_{CP} in the t-th time-slot;
P^t_{Con}	is the estimated conventional power in the t-th time-slot;
P_{Avg}	is the estimated average power of the regional grid during the coming one-day cycle;
$P^t_{\max}$	is the maximum total power that can be supplied by the regional grid in the t-th time-slot;
$P^{\max}_{ACP-n}$	is the allowed maximum working power of A^n_{CP};
$Soc^k_{ACP-n}(j)$	is the battery Soc value at the end of the j-th time-slot when A^n_{CP} offers its k-th charging service.

As mentioned in Sect. 6.2, it is assumed that with the development of smart grid, the 24-h-ahead load forecasting for conventional power loads (excluding PEV loads) can be achieved with high accuracy.

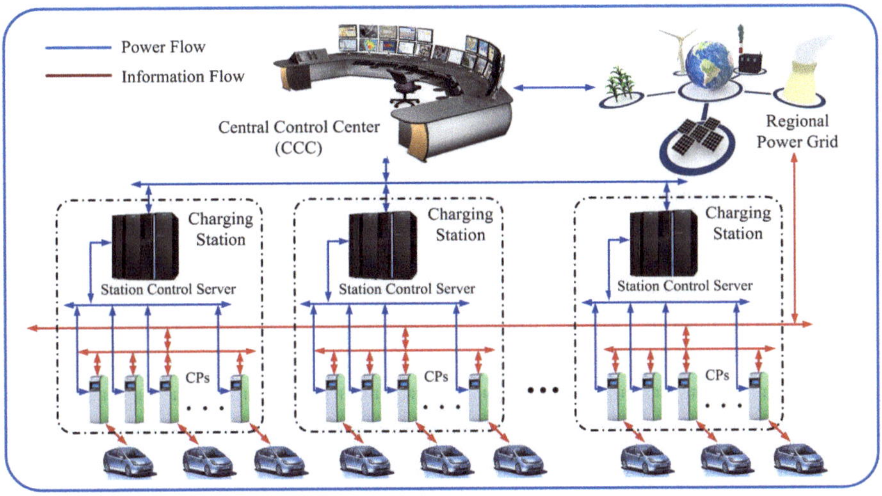

Fig. 6.12 Scheme of proposed double-layer optimal charging strategy

6.3.3 Double-Layer Optimal Charging Strategy

It can be observed from the above problem formulation that the number of variables increase linearly with the number of charging posts involved, and the number of restraints are related to both the number of charging posts and the number of charging services they plan to offer which is partly decided by the number of PEVs involved. This means, with the large scale of penetration of PEVs and charging posts, the computational complexity will become tremendously high. Therefore, a double-layer optimal charging (DLOC) strategy is proposed for dealing with this challenge. Figure 6.12 illustrates the scheme of the proposed DLOC strategy, in which both the energy flow and the information flow are indicated. The basic idea is to categorize all the charging posts in the regional grid under the administration of several charging stations. The charging posts located in the same area, or connected to the same node transformer can be classified into the same charging station, such as those installed on the same streets, in the same parking lot, or in the same residential community. In the first layer optimization (FLO), the CCC plans the optimal operating power schedule for each charging station as a whole aiming to minimize the overall load variance. Then in the second layer optimization (SLO), the station control server plans the charging power for each charging post under its governance, aiming to meet the instructions ordered by CCC which has been generated in the first layer optimization.

Let's discuss the FLO problem first. The data tuple given by (6.12) should be updated due to the introduction of charging stations:

$$CP = [CP_{ID}, CP_{Loc}, P_{CP}^{\max}, CS_{ID}, Flag] \tag{6.24}$$

where the added entity CS_{ID} denotes the identity number of charging station into which this charging post is classified. Then, the set given by (6.14) is changed to:

$$S_{ACP}^{CS-h} = [A_{CP}^{1-h}, A_{CP}^{2-h}, A_{CP}^{3-h}, \ldots, A_{CP}^{N(h)-h}] \tag{6.25}$$

where S_{ACP}^{CS-h} is the set of the active charging posts in the h-th charging station, $h \in [1, H]$, and H is the number of the active charging station in the V2G operation for the coming one-day cycle. A_{CP}^{n-h} denotes the n-th active charging post in the h-th charging station, $n \in [1, N(h)]$, $N(h)$ is the number of the active charging post in the h-th charging station. A_{CP}^{n-h} is also a 2D data tuple with the same structure as illustrated in (6.15), but the entities are updated into:

Γ_{ACP-n}^{s-k-h} the k-th charging service offered by A_{CP}^{n-h} will start at the beginning of the Γ_{ACP-n}^{s-k-h} -th time-slot;

Γ_{ACP-n}^{e-k-h} the k-th charging service offered by A_{CP}^{n-h} will finish at the ending of the Γ_{ACP-n}^{e-k-h} -th time-slot;

EV_{ACP-n}^{ID-k-h} the ID Number of the PEV connected to A_{CP}^{n-h} during the k-th charging service offered by A_{CP}^{n-h};

$BAT_{ACP-n}^{Cap-k-h}$ the battery capacity of the PEV connected to A_{CP}^{n-h} during the k-th charging service offered by A_{CP}^{n-h};

$Soc_{ACP-n}^{upper-k-h}$ the allowed battery Soc upper limit of the PEV connected to A_{CP}^{n-h} during the k-th charging service offered by A_{CP}^{n-h};

$Soc_{ACP-n}^{lower-k-h}$ the allowed battery Soc lower limit of the PEV connected to A_{CP}^{n-h} during the k-th charging service offered by A_{CP}^{n-h};

Soc_{ACP-n}^{s-k-h} the estimated battery Soc value of PEV when the k-th charging service offered by A_{CP}^{n-h} starts;

Soc_{ACP-n}^{e-k-h} the required battery Soc value of PEV when the k-th charging service offered by A_{CP}^{n-h} ends.

The target variables become the operating power of charging stations at every time slot. Thus, the objective function given in (6.16) becomes:

$$\min \sum_{t=1}^{T} \left[\frac{1}{T} \left(P_{Con}^{t} - P_{Avg} + \sum_{h=1}^{H} P_{CS-h}^{t} \right)^{2} \right] \tag{6.26}$$

where P^t_{CS-h} is the operating power of the h-th charging station at the t-th time slot. The restraints (6.17) and (6.18) become:

$$P^t_{Con} + \sum_{h=1}^{H} P^t_{CS-h} \leq P^t_{\max}, t \in [1, T] \tag{6.27}$$

$$P_{Avg} = \sum_{t=1}^{T} \left(P^t_{Con} + \sum_{h=1}^{H} P^t_{CS-h} \right) \Big/ T \tag{6.28}$$

Moreover, restraints (6.19) and (6.20) become:

$$-\sum_{n=1}^{N(h)} \left(\xi^h_n(t) P^{\max-h}_{ACP-n} \right) \leq P^t_{CS-h} \leq \sum_{n=1}^{N(h)} \left(\xi^h_n(t) P^{\max-h}_{ACP-n} \right) \tag{6.29}$$

$$\xi^h_n(t) = \begin{cases} 0, & \text{if} : t \in [1, T] - \cup_{k=1}^{k=K(n)} \left[\Gamma^{s-k}_{ACP-n}, \Gamma^{e-k}_{ACP-n} \right] \\ 1, & \text{if} : t \in \cup_{k=1}^{k=K(n)} \left[\Gamma^{s-k}_{ACP-n}, \Gamma^{e-k}_{ACP-n} \right] \end{cases} \tag{6.30}$$

where $P^{\max-h}_{ACP-n}$ is the allowed maximum working power of the charging post A^{n-h}_{CP}.

The restraints (6.21)–(6.23) are set to guarantee the demanded charging quantities of each charging service, and to make sure that the batteries are neither over charged nor deeply discharged. Considering the k-th charging service that going to be offered by the charging post A^{n-h}_{CP}, the charging process can be illustrated by the change of battery Soc versus time as shown in Fig. 6.13. As long as the battery Soc value is located in the shadow area, the restraints (6.21)–(6.23) can be satisfied.

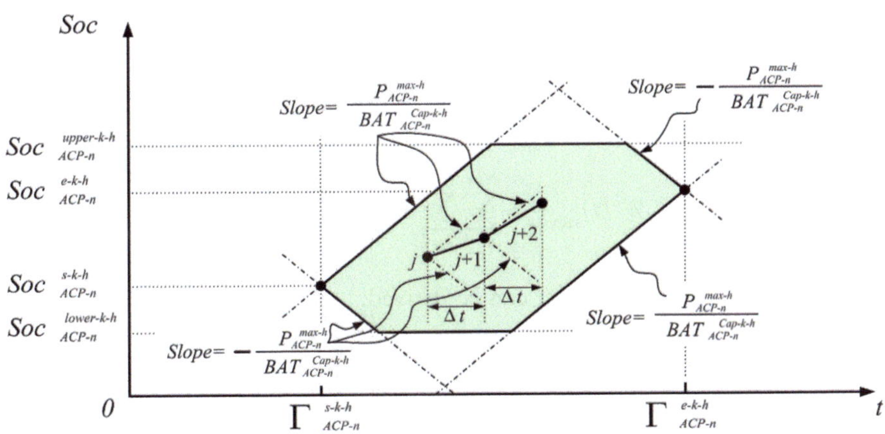

Fig. 6.13 Variation of Soc during a charging process

Denoted by $W_{ACP-n}^{k-h}(t)$ the accumulated charging quantity from the first time slot to the t-th time slot offered by the charging post A_{CP}^{n-h} in its k-th charging service, it can be known from Fig. 6.13 that the lower boundary and upper boundary of $W_{ACP-n}^{k-h}(t)$ are given by:

$$W_{ACP-n}^{k-h}(t)\big|_{lower}=\begin{cases}0 & \text{, if}: t<\Gamma_{ACP-n}^{s-k-h} \\ \left(Soc_{ACP-n}^{k-h}(t)\big|_{lower}-Soc_{ACP-n}^{s-k-h}\right)BAT_{ACP-n}^{Cap-k-h} & \text{, if}: \Gamma_{ACP-n}^{s-k-h}\le t\le\Gamma_{ACP-n}^{e-k-h} \\ \left(Soc_{ACP-n}^{e-k-h}-Soc_{ACP-n}^{s-k-h}\right)BAT_{ACP-n}^{Cap-k-h} & \text{, if}: t>\Gamma_{ACP-n}^{e-k-h}\end{cases} \tag{6.31}$$

$$W_{ACP-n}^{k-h}(t)\big|_{upper}=\begin{cases}0 & \text{, if}: t<\Gamma_{ACP-n}^{s-k-h} \\ \left(Soc_{ACP-n}^{k-h}(t)\big|_{upper}-Soc_{ACP-n}^{s-k-h}\right)BAT_{ACP-n}^{Cap-k-h} & \text{, if}: \Gamma_{ACP-n}^{s-k-h}\le t\le\Gamma_{ACP-n}^{e-k-h} \\ \left(Soc_{ACP-n}^{e-k-h}-Soc_{ACP-n}^{s-k-h}\right)BAT_{ACP-n}^{Cap-k-h} & \text{, if}: t>\Gamma_{ACP-n}^{e-k-h}\end{cases} \tag{6.32}$$

where

$$Soc_{ACP-n}^{k-h}(t)\big|_{lower}=\max\left\{Soc_{ACP-n}^{lower-k-h},\left[Soc_{ACP-n}^{s-k-h}-\frac{P_{ACP-n}^{\max-h}\left(t-\Gamma_{ACP-n}^{s-k-h}+1\right)\Delta t}{BAT_{ACP-n}^{Cap-k-h}}\right],\left[Soc_{ACP-n}^{e-k-h}+\frac{P_{ACP-n}^{\max-h}\left(t-\Gamma_{ACP-n}^{e-k-h}\right)\Delta t}{BAT_{ACP-n}^{Cap-k-h}}\right]\right\} \tag{6.33}$$

$$Soc_{ACP-n}^{k-h}(t)\big|_{upper}=\min\left\{Soc_{ACP-n}^{upper-k-h},\left[Soc_{ACP-n}^{s-k-h}+\frac{P_{ACP-n}^{\max-h}\left(t-\Gamma_{ACP-n}^{s-k-h}+1\right)\Delta t}{BAT_{ACP-n}^{Cap-k-h}}\right],\left[Soc_{ACP-n}^{e-k-h}-\frac{P_{ACP-n}^{\max-h}\left(t-\Gamma_{ACP-n}^{e-k-h}\right)\Delta t}{BAT_{ACP-n}^{Cap-k-h}}\right]\right\} \tag{6.34}$$

Thus, the accumulated charging quantity from the first time slot to the t-th time slot offered by the h-th charging station as a whole has the lower boundary and the upper boundary, given by:

$$W^h(t)\big|_{lower}=\sum_{n=1}^{N(h)}\sum_{k=1}^{K(n)}W_{ACP-n}^{k-h}(t)\big|_{lower} \tag{6.35}$$

$$W^h(t)\big|_{upper}=\sum_{n=1}^{N(h)}\sum_{k=1}^{K(n)}W_{ACP-n}^{k-h}(t)\big|_{upper} \tag{6.36}$$

Hence, the following restraint can be derived:

$$W^h(t)\big|_{lower}\le\sum_{t=1}^{m}\left(\Delta t\cdot P_{CS-h}^{t}\right)\le W^h(t)\big|_{upper}, m\in[1,T] \tag{6.37}$$

Next, the SLO problem is discussed. In the SLO, the station control server plans the charging power for each charging post, aiming to meet the instructions ordered by CCC which has been generated in the FLO. Therefore, The problem can be formulated as:

$$\min \sum_{t=1}^{T} \left[\frac{1}{T} \left(\sum_{n=1}^{N(h)} P_{ACP-n}^{t-h} - P_{CS-h}^{t} \right)^2 \right], \; h \in [1, H] \tag{6.38}$$

Subject to:

$$-P_{ACP-n}^{\max-h} \le P_{ACP-n}^{t-h} \le P_{ACP-n}^{\max-h}, t \in \cup_{k=1}^{k=K(n)} \left[\Gamma_{ACP-n}^{s-k-h}, \; \Gamma_{ACP-n}^{e-k-h}\right], n \in [1, N(h)] \tag{6.39}$$

$$P_{ACP-n}^{t-h} = 0, t \in [1, T] - \cup_{k=1}^{k=K(n)} \left[\Gamma_{ACP-n}^{s-k-h}, \Gamma_{ACP-n}^{e-k-h}\right], \; n \in [1, N(h)] \tag{6.40}$$

$$\begin{gathered} \sum_{t=\Gamma_{ACP-n}^{s-k-h}}^{\Gamma_{ACP-n}^{e-k-h}} \left[\Delta t \cdot P_{ACP-n}^{t-h}\right] = \left(Soc_{ACP-n}^{e-k-h} - Soc_{ACP-n}^{s-k-h}\right) \cdot BAT_{ACP-n}^{Cap-k-h}, \\ k \in [1, K(n)], \; n \in [1, N(h)] \end{gathered} \tag{6.41}$$

$$\begin{gathered} Soc_{ACP-n}^{lower-k-h} \le Soc_{ACP-n}^{k-h}(j) \le Soc_{ACP-n}^{upper-k-h}, \; j \in \left[\Gamma_{ACP-n}^{s-k-h}, \Gamma_{ACP-n}^{e-k-h}\right], \\ k \in [1, K(n)], \; n \in [1, N(h)] \end{gathered} \tag{6.42}$$

$$Soc_{ACP-n}^{k-h}(j) = Soc_{ACP-n}^{s-k-h} + \sum_{t=\Gamma_{ACP-n}^{s-k-h}}^{j} \left[\Delta t \cdot P_{ACP-n}^{t-h} \Big/ BAT_{ACP-n}^{Cap-k-h}\right] \tag{6.43}$$

The proposed DLOC strategy can effectively reduce the computational complexity. For the first layer, the number of variables depends on the number of charging stations, which is dramatically shrunk compared to the number of all charging posts in the regional grid. For the second layer, the optimization program can be executed at the same time for all the charging stations. The results and performance will be presented in the following section.

6.3.4 Case Study

Several cases are studied to assess the performance of the proposed V2G and the DLOC strategy. A set of programs are designed to randomly generate the data needed for simulation studies. The one-day cycle starts at 06:00 a.m. and ends at

Table 6.4 Specifications and computing time consumed

	Case 1	Case 2	Case 3	Case 4
Number of PEVs	100	500	1,000	2,000
Number of charging posts	100	500	1,000	2,000
Number of charging stations	10	22	31	44
Number of services	336	1,738	3,433	6,919
Computing time (model 1)	1.38 s	12.72 s	32.60 s	81.67 s
Computing time (model 2)	0.64 (layer 1) + 0.38 (layer 2) = 1.02 s	1.51 (layer 1) + 1.17 (layer 2) = 2.68 s	2.32 (layer 1) + 2.06 (layer 2) = 4.38 s	3.67 (layer 1) + 3.60 (layer 2) = 7.27 s

05:59 a.m. (next day). Also, the time slot is set to be 15 min. Some practical situations are taken into account when designing the random data generation programs, for example, the conventional load is likely to reach peak values at noon and in the evening, the PEVs are likely to be connected to grid at night and at noon, and so forth.

Totally four cases with different problem scales as listed in Table 6.4 are simulated. For each case, two models are employed. One is the single-layer model (model 1) introduced in Sect. 6.3.2 and the other is the double-layer model (model 2) proposed in Sect. 6.3.3. Figure 6.14 gives the performances of V2G operation in selected cases 1, 3, and 4. It can be observed that the overall load curves are successfully flattened with the V2G operation of the PEV loads. What is more, it can also be found that the peak value of the total load is slightly lower than that of the conventional load, attributed to the energy feedback of the PEVs. This demonstrates that power grid is able to contain newly added PEV loads to some extend without boosting its capacity, if the V2G operation can be effectively carried out.

Calculation results demonstrate that the PEV load curves obtained by using single-layer model (model 1) and the other is the double-layer model (model 2) are almost the same. This implies that the proposed DLOC strategy agrees very well with the design objectives. However, this does not mean that model 2 is exactly equivalent to model 1. Figure 6.15 gives the optimal charging schedules of the 10 charging stations in case 1. It can be found that there are tiny differences between the results obtained by using these two models. Fortunately, these tiny differences will not cause obvious degradation in the performance of our proposed DLOC strategy.

In addition, Fig. 6.16 gives the optimal charging schedules of two selected charging post (CP #8 and 69) in case 1 obtained by using two models. Both charging posts are assigned to offer five charging services (I-IV) by the CCC in the coming 24-h. The upper part of each plot gives the resulted optimal charging power provided by the corresponding charging post. With these regulated charging

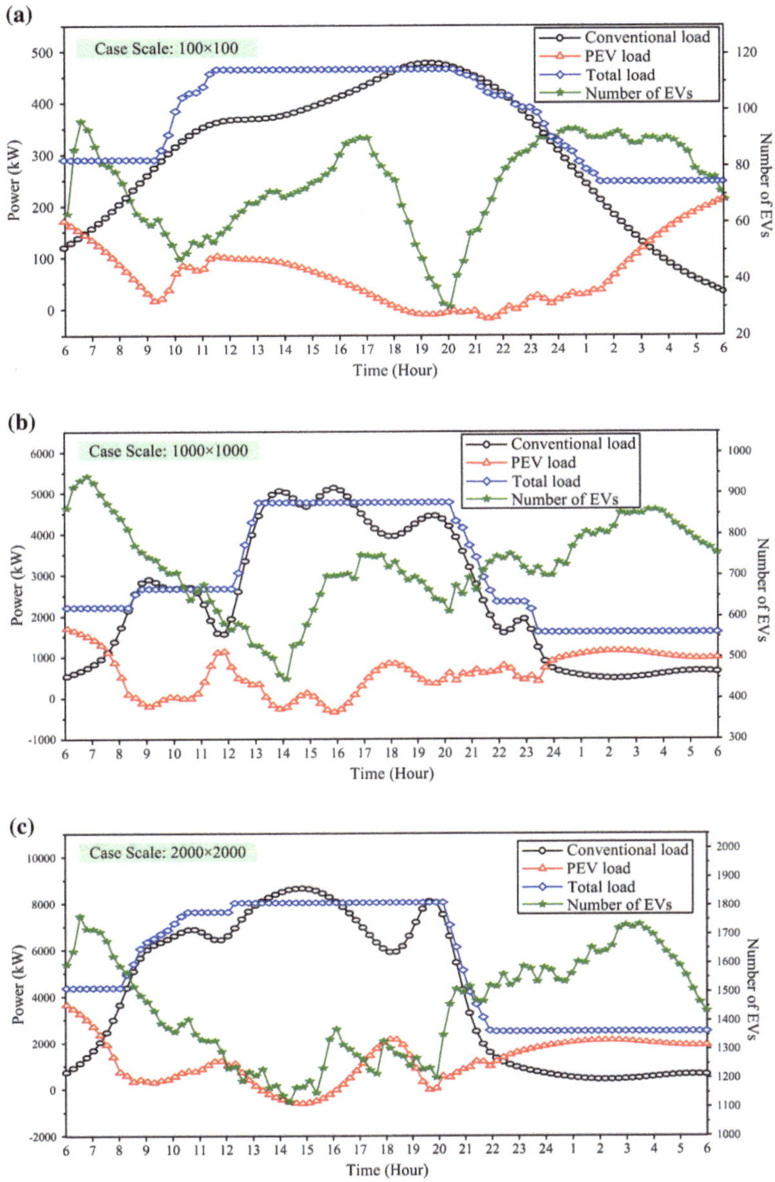

Fig. 6.14 Calculated power load profiles with V2G operation: **a** case 1, **b** case 3, **c** case 4

profiles, the electricity charging demand of the PEV connected can be guaranteed, moreover, the minimized overall load variance can be achieved. The lower part of each plot illustrates the battery Soc curve of the PEV connected to the

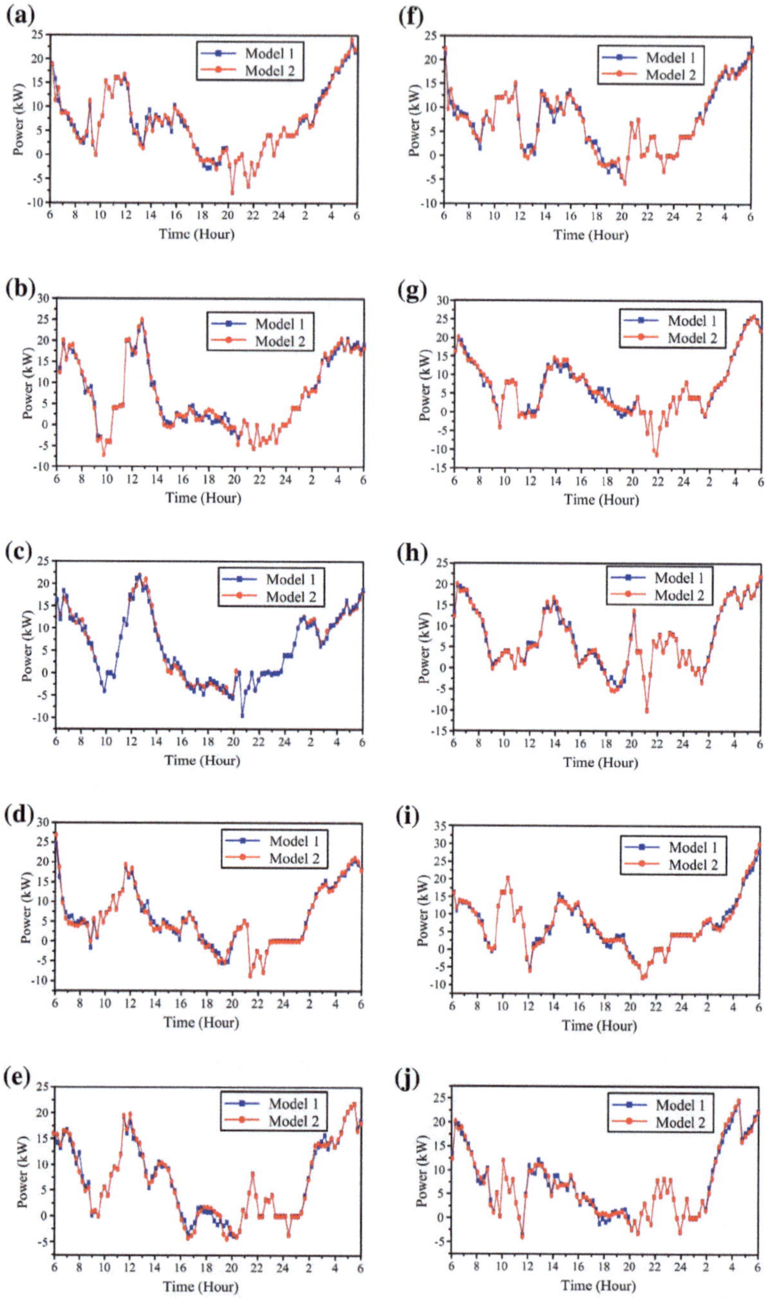

Fig. 6.15 Calculated optimal charging schedules in case 1: **a** charging station # 1, **b** charging station # 2, **c** charging station # 3, **d** charging station # 4, **e** charging station # 5, **f** charging station # 6, **g** charging station # 7, **h** charging station # 8, **i** charging station # 9, **j** charging station # 10

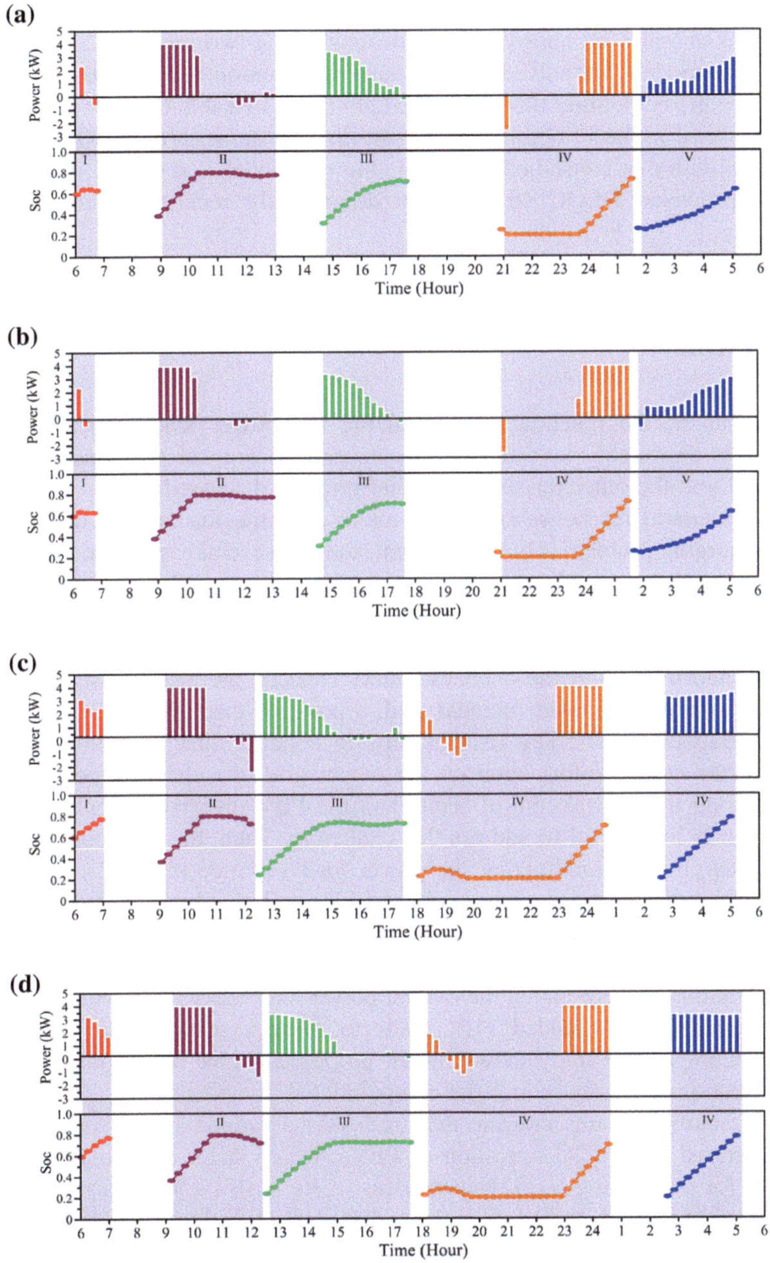

Fig. 6.16 Calculated optimal charging services in case 1: **a** CP #8 with model 1, **b** CP #8 with model 2, **c** CP #69 with model 1, **d** CP #69 with model 2

corresponding charging post by engaging the optimal charging pattern. Also, subtle distinctions can also be observed in the results of the two models.

The two models are solved on the same workstation (CPU 3.20 GHz, RAM 6 GB), and the computing time consumed are listed in Table 6.4. For model 2, the time consumed in the FLO plus the longest time consume in the SLO is given for comparing with that consumed by calculating the model 1. It clearly demonstrates that the proposed DLOC strategy can dramatically reduce the computational complexity.

6.4 Conclusions

In this chapter, the potential in minimizing the power load variance of V2G operation is thoroughly investigated. Two scenarios one for the household smart micro-grid and the other for smart regional grid are discussed.

In the scenario for household smart micro-grid, the mathematic model of the optimal charging problem is built up. Then, study of several cases is conducted, and it demonstrated that by planning the charging profiles of PEVs involved the variance of the load power in the micro-grid can be dramatically reduced. Several related issues, such as energy losses, subsidy mechanism, impacts of the charging demands, and impacts of the onboard battery capacity are also investigated.

In the scenario for smart regional grid, a possible method of V2G implementation is elaborated. The key issue is how to conduct effective communications regarding the crucial information on the power grid, the charging posts, and the PEVs, among the central control center and the PEV owners. An online booking mechanism is introduced to address this challenge. Then, the problem concerning optimal charging for minimizing the power load variance is mathematically formulated. With the increase of the scale of PEVs and charging posts involved, the computational complexity may become tremendously high. Therefore, a double-layer optimal charging strategy is proposed. Case studies also demonstrate that the V2G operation can help flatten the overall power load curves and it enables power grid to contain newly added PEV loads to some extent without boosting its capacity. Comparative study shows that the proposed double-layer optimal charging strategy can dramatically reduce the computational complexity.

The simulation results indicate that tremendous economic and social interests can be derived from V2G operation of PEVs, which states the reasonability and necessity for developing V2G. Nevertheless, there is still a long way ahead of us before it can be implemented in practice. Two major prerequisites are the development of smart grid and the improvement of massive data processing capability. Some other techniques, such as conventional power load forecasting, floating electricity prices, long life-cycle batteries, and so on, should be further developed in order to make V2G a more pragmatic issue.

References

1. NASA Goddard Institute for Space Studies (2010) GISS surface temperature analysis. December 2010
2. International Energy Outlook (2013) http://www.eia.gov/forecasts/ieo/pdf/0484(2013).pdf
3. Eric Evarts (2011) Leaf, Volt tests show electric cars cost less per mile to operate. Consumer reports. Accessed 10 Dec 2011
4. Botsford C, Szczepanek A (2009) Fast charging vs. slow charging: pros and cons for the new age of electric vehicles. In: Proceeding of EVS24, Stavanger, Norway, May 2009
5. Guildhary F (2012) IBM, Honda, and PG&E enable smarter charging for electric vehicles. http://www-03.ibm.com/press/us/en/pressrelease/37398.wss. Accessed 12 Apr 2012
6. Guille C, Gross G (2009) A conceptual framework for the vehicle-to-grid implementation. Energy Policy 37(11):4379–4390
7. Kempton W, Tomic J (2005) Vehicle-to-grid power implementation: from stabilizing the grid to supporting large-scale renewable energy. J Power Source 144:280–294
8. Amoroso F, Cappuccino G (2012) Advantages of efficiency-aware smart charging strategies for PEVs. Energy Convers Manage 54(1):1–6
9. Tomic J, Kempton W (2007) Using fleets of electric-drive vehicles for grid support. J Power Source 168:459–468
10. Clement-Nyns K, Haesen E, Driesen J (2011) The impact of vehicle-to-grid on the distribution grid. Electr Power Syst Res 81:185–192
11. Jian J, Xue H et al (2013) Regulated charging of plug-in hybrid electric vehicles for minimizing load variance in household smart micro-grid. IEEE Trans Industr Electron 60 (8):3218–3226
12. Judd S, Overbye T (2008) An evaluation of PHEV contributions to power system disturbances and economics. In: The 40th North American power symposium
13. Masoum A, Deilami S et al (2011) Smart load management of plug-in electric vehicles in distribution and residential networks with charging stations for peak shaving and loss minimization considering voltage regulation. IET Proc Gener Transm Distrib 5(8):877–888
14. Deilami S, Masoum A, Moses P, Masoum M (2011) Real-time coordination of plug-in electric vehicle charging in smart grids to minimize power losses and improve voltage profile. IEEE Trans Smart Grid 2(3):456–467
15. Mullen S (2009) Plug-in hybrid electric vehicles as a source of distributed frequency regulation. Doctoral dissertation, The University of Minnesota
16. Clement-Nyns K, Haesen E, Driesen J (2010) The impact of charging plug-in hybrid electric vehicles on a residential distribution grid. IEEE Trans Power Syst 25(1):371–380
17. Sortomme E, Hindi M, Macpherson S, Venkata S (2011) Coordinated charging of plug-in hybrid electric vehicles to minimize distribution system losses. IEEE Trans Smart Grid 2 (1):198–205
18. Singh M, Kar I, Kumar P (2010) Influence of EV on grid power quality and optimizing the charging schedule to mitigate voltage imbalance and reduce power loss. In: The 14th international power electronics and motion control conference
19. Acha S, Green T, Shah N (2010) Effects of optimized plug-in hybrid vehicle charging strategies on electric distribution network losses. In: Transmission and distribution conference and exposition
20. Wang L, Singh C, Kusiak A (2012) Guest editorial: special issue on integration of intermittent renewable energy resources into power grid. IEEE Syst J 6(1):2–3
21. Saber A, Venayagamoorthy G (2011) Plug-in vehicles and renewable energy sources for cost and emission reductions. IEEE Trans Industr Electron 58(4):1229–1238
22. Jian J, Xu G et al (2011) Intelligent multi-mode energy-refreshing station for electric vehicles within the framework of smart grid. In: IEEE international conference on information and automation, pp 534–539

23. Steer K, Wirth A, Halgamuge S (2012) Decision tree ensembles for online operation of large smart grids. Energy Convers Manage 59(1):9–18
24. Silva M, Morais H, Vale Z (2012) An integrated approach for distributed energy resource short-term scheduling in smart grids considering realistic power system simulation. Energy Convers Manage 64:273–288
25. Slootweg H (2009) Smart grids—the future or fantasy. IET conference: smart metering—making it happen, Feb 2009
26. Mehrizi-Sani A, Iravani R (2010) Potential-function based control of a microgrid in island and grid-connected modes. IEEE Trans Power Syst 25(4):1883–1891
27. Musavi F, Edington M, Eberle W, Dunford W (2011) Energy efficiency in plug-in hybrid electric vehicle chargers: evaluation and comparison of front end AC-DC topologies. In: IEEE energy conversion congress and exposition, pp 273–280
28. Sun W, Yuan Y (2006) Optimization theory and methods: nonlinear programming. Springer, Berlin
29. Kanchev H, Colas F, Lazarov V, Francois B (2011) Energy management and operational planning of a microgrid with a PV-based active generator for smart grid applications. IEEE Trans Industr Electron 58(10):4583–4592
30. Agarwal V, Uthaichana K, DeCarlo R, Tsoukalas L (2010) Development and validation of a battery model useful for discharging and charging power control and lifetime estimation. IEEE Trans Energy Convers 25(3):821–835
31. Dogger J, Roossien B, Nieuwenhout F (2011) Characterization of li-ion batteries for intelligent management of distributed grid-connected storage. IEEE Trans Energy Convers 26(1):256–263

Chapter 7
A Model Predictive Control-Based Approach for Plug-in Electric Vehicles Charging: Power Tracking, Renewable Energy Sources Integration and Driver Preferences Satisfaction

Alessandro Di Giorgio and Francesco Liberati

Abstract This chapter presents a model predictive control (MPC) framework for controlling in real-time the charging processes of a set of plug-in electric vehicles (PEVs) located in a load area (LA), namely a distribution system operator (DSO)-defined portion of the grid under a secondary substation. The LA considered in the reference scenario hosts remotely controlled, IEC 61851-compliant electric vehicle supply equipment (EVSE), where the PEVs are plugged to recharge the batteries, and a share of generation from renewable energy sources (RES). The proposed framework works regardless of the EVSE technology and power level (direct current, alternating current, single phase or three phases). The controller, named load area controller (LAC), works under the requirements of: (i) seeking costs minimization while respecting *drivers' preferences* on the amount of energy to recharge (or desired final state of charge) and the time flexibility for recharging specified by the driver; (ii) tracking of a LA-level power reference established by the DSO on a day-ahead basis and possibly updated intraday; (iii) integrating RES by, e.g., maximizing the share of photovoltaic power absorbed by the LA, thus ensuring economic saving while avoiding the injection into the grid of possibly intermittent power profiles. The design of the controller is based on the analysis of a possible future charging scenario in an unbundled electricity system, but is general enough to be tailored to a large number of possible regulatory frameworks and business models. Starting from the available equipment and the role of actors possibly involved, use cases are presented and controller functional requirements and technical specifications identified; based on that, the reasons for using MPC methodology are explained and the discrete time optimal control problem at its

A. Di Giorgio (✉) · F. Liberati
Department of Computer, Control and Management Engineering "Antonio Ruberti", Sapienza University of Rome, Via Ariosto 25, 00185 Rome, Italy
e-mail: digiorgio@dis.uniroma1.it

F. Liberati
e-mail: liberati@dis.uniroma1.it

S. Rajakaruna et al. (eds.), *Plug In Electric Vehicles in Smart Grids*,
Power Systems, DOI 10.1007/978-981-287-317-0_7

basis is presented. The issue of estimating the battery state of charge is discussed, which constitutes a delicate point for the deployment of the control system in a real environment. A set of incremental simulations is presented in order to validate the concept and show its effectiveness.

Keywords Plug-in electric vehicles charging control · IEC 61851 · Model predictive control · Demand side management · Renewable energy sources integration

7.1 Introduction

During the last decade regulatory and technological advancements have been quickly driving the renovation of legacy power systems towards the future smart grids. The unbundling process has created new grid players and some others are expected to arise in coming years while progressively emerging new technical solutions for grid efficiency and interdependences with other systems. This is the case of the mobility sector, where a significant shift from fossil fuels to electromobility is expected for the coming years, creating huge opportunities and challenges in the way distribution electricity grids will be operated.

On one hand relevant investments for network upgrade and the establishment of new business models are necessary [1], on the other massive penetration of plug-in electric vehicle (PEV) technology will have a significant technical impact, as highlighted in [1–4]. A first consequence will be a relevant change in the magnitude and shape of distribution lines loading, considering the significant difference between the traditional electricity demand and the current mechanical power on the road [5]. Further, strengthening the coupling between transportation and electrical systems will increase uncertainty and intermittency of load profiles, which are typical "side effects" associated with renewable energy sources (RES). As a result, grid operation will become more complex, in terms of load balancing, survivability of network elements and overall power quality [6], asking for charging strategies aimed at providing the new load with a more regular behavior.

Nevertheless, PEV technology also represents a valuable opportunity [7]. The rapid integration of RES, recognized as a priority for an eco-sustainable growth of industrialized countries [8], asks for the availability of negative and positive balancing power as a basic requirement for mitigating the effects of RES volatility on grid stability and reliability [9]. Depending on the size and placement of RES, the balancing task can be performed at different levels, according to the basic principle "the smaller the distance between renewables and consumption, the higher the benefit for the grid". In this sense early works [10] have recognized that a proper control of PEVs charging at fleet level can contribute to meet this requirement. Such result can be achieved by combining the control of fleet charging power [11] with the control of reverse energy flows from the PEVs to the grid (vehicle to grid—V2G) [5], then exploiting the flexibility of drivers through proper demand side management (DSM) programs and the huge potential of PEVs of acting as distributed storage systems. In the light of above a charging control system appears as the enabler of local matching between demand and supply, and the regulator of power flow exchanges between the charging area and the distribution grid, according to the needs of the distribution system operator (DSO) and the distributed energy resources (DER) operator owning RES in the charging area.

In this chapter, a model predictive control (MPC) framework for automatic control of a set of charging sessions running in a load area (LA) is discussed. A LA is a DSO-defined portion of the distribution grid under a secondary substation [12]. In the considered scenario, the LA is equipped with photovoltaic (PV) plants and hosts a set of remotely controlled, IEC 61851-compliant electric vehicle supply equipment (EVSE), also known as charging stations (CSs), which are the stations where the PEVs are brought and plugged in by the PEVs' users to have the batteries recharged. The proposed framework works regardless of the particular EVSE technology and charging power level (direct current, alternating current, single phase or three phases). The controller, named load area controller (LAC), works under the main requirements of: (i) pursuing the minimization of the charging costs to be sustained by the PEVs' users, while respecting the PEVs users' preferences regarding the maximum available time for charging and the amount of energy to be recharged (or the desired final state of charge (SoC)); (ii) tracking of a LA-level power reference defined by the DSO according to its own criteria (as clarified in the following) on a day-ahead basis and possibly updated intraday; (iii) favoring the integration of RES into the grid by, e.g., maximizing the share of PV power absorbed by the LA and flattering the overall LA load profile, thus avoiding the injection into the grid of possibly intermittent power profiles and hence resulting into greater economic benefits for both the DSO and the RES operators. As regards the costs minimization requirement, the LAC is designed to work under both designed and market indexed pricing models, and is able to react to DSM signals. Secondly, differently from other works in literature [4, 13], which integrate grid constraints directly in the charging control problem formulation (via, e.g., power flow constraints), this work, in line with the electromobility business chain, regards the LAC as a software module belonging to a specific electromobility business actor (e.g. the EVSE operator, i.e. the business actor managing the EVSEs) and able to provide smart charging services to the interested actors. In this way, in a DSO-oriented scheme for example, the DSO works out (via its SCADA/distribution management system (DMS) running dedicated power flow routines) a desired daily LA-level power reference for electromobility, to ensure safe and efficient operation of the grid. The LAC takes the DSO-generated power reference as an input and ensures, given the power tracking requirement, that the charging sessions are dynamically updated so that the aggregated charging power matches the reference. RES integration can be achieved either via the establishment, by the DSO, of a proper LA power reference taking into account RES profiles, or, as shown in this work, by ensuring the maximization of RES self-consumption by the LA (by properly controlling the charging processes); for this purpose the controller is able to update the control when receiving a new RES generation forecast.

The control problem is formulated as a MPC problem, based on mixed integer quadratic programming. An instance of the problem is built and solved by the LAC on a periodic basis considering all the meaningful events triggering the controller (such as, a new user request, a price signal, a volume DSM signal, notifications of availability of new RES forecasts, etc.). The objective function is given by a linear weighted combination of the (controlled) cost for satisfying the charging requests currently managed, of the L-2 norm of the error between the aggregated (controlled) charging power and the LA power reference, and a term for RES self-consumption maximization. The control variables are given by the PEV charging rates. They are either Boolean or semi-continuous in nature, depending on if on-off charging is chosen (as reasonably the case for slow charging) or the charging power is modulated (when different from zero) between a minimum positive value and a maximum positive value, in accordance with standard IEC 61851 (this latter choice being relevant in case of high-power charging processes); discharging (i.e. V2G) is also considered. Constraints are given by User Preferences (UP) (thus directly integrated at problem formulation level, and which can be updated at any time by the driver), by technical limits at EVSE-level imposed by IEC 61851, by technical limits of the PEV battery, by the overload power at LA level, etc. In particular, a battery control model is integrated in order, for the LAC, to be able to predict the future SoC of the PEVs based on the assigned charging load profile and the measure or estimation of the current SoC; considering the current unavailability of real time data about the battery SoC from car manufacturers, two practical strategies are reported for SoC estimation, based on the use of (i) meter readings and; (ii) a very detailed, highly non-linear model [14] of the batteries considered representative of a real battery pack. A set of simulations is proposed in order to validate the concept and show its effectiveness.

The remainder of the chapter is organized as follows. Section 7.2 discusses the state of the art in control approaches for PEVs charging. Section 7.3 presents the reference charging scenario, detailing the role of actors involved and components making part of the architecture. In Sect. 7.4 the considered use cases are presented, and consequently a set of functional requirements and technical specifications for the controller are listed. Based on that, the proposed control system flow of operations and the open loop optimal control problem at the basis of the closed loop MPC approach are presented in Sects. 7.5 and 7.6, respectively. Section 7.7 is dedicated to the delicate point of SoC estimation, which is necessary to achieve a closed loop control system. Simulation results are shown in Sect. 7.8 and finally the conclusions are drawn in Sect. 7.9. Starting from some key concepts proposed within the reference ADDRESS project [12] like the one of LA, aggregator and DSM signals, the solution reported in the chapter is the result of the investigation performed by the authors in the framework of the European Union SMART V2G [15] and MOBINCITY [16] projects, where the local charging control problem has been studied on an incremental basis, collecting use cases, requirements and specifications through the interaction with DSOs and other players of the electromobility sector.

7.2 State of the Art and Proposed Innovation

The emerging concept of smart grid is based on the deployment of a multi-level control architecture, with the aim of reaching a deeper integration between generation and demand. Demand response, DSM and active demand are terms which all refer to new central paradigms evoked for referring to a direct influence of demand on the technical and economical balance of the grid [17]. Load management problems have received an increasing attention from academics and industry during the last decade. Industry has been the driving sector for many years and it is also the first one for which pioneer DSM programs have been proposed [18–20]. Load shifting concept is being deeply investigated also in the residential sector, with the purpose of optimally controlling smart household appliances, storage devices and local renewable energy sources [21–23].

The concepts established in the aforementioned works in the field of load management find a natural application also to electromobility. An interesting approach for coordinating charging operation of multiple EVs in a smart grid system is presented in [13]. A maximum sensitivities selection (MSS) optimization problem is established, with the aim of minimizing cost of energy consumption and network losses. PEVs are divided into priority groups, depending on UP and sensitivity of system losses due to each PEV. Moreover, voltage constraints at each EVSE of the network and congestion constraints are considered. Grid variables are computed through simulation, via a standard Newton-based load flow routine. The main drawbacks of this work are: (i) charging control signals are continuous in nature but not IEC 61851 compliant; (ii) there is not a strict control over the time needed to provide the charging service and on the desired final state of charge of the batteries; (iii) backfeeding is not considered. These are rather common drawbacks in the relevant literature.

A similar approach is presented in [24], where the authors set up an optimization problem seeking to maximize the amount of energy available for charging operations, while considering constraints on voltage levels, charging rates changes, network congestion and thermal loading of network components. Voltage levels and thermal loadings are calculated based on load flow analysis. Interestingly, a weighted objective function is proposed, in order not to penalize charging points characterized by a high sensitivity (in radial networks, voltage level is generally more sensitive to addition of load far from the transformers). Among the drawbacks of the work there is the fact that charging control signals are continuous in nature but not IEC 61851 compliant; moreover, UP are poorly modelled (the overall energy available for charging operations is maximized, not taking into account the precise amount of energy demanded by each PEV, or the time preferences for charging operations set by the users). Also in this work the authors do not take into account the possibility of delivering *active demand* services to the grid.

An original control approach is presented in [25]. The charging process is controlled by using a distributed additive increase multiplicative decrease (AIMD) feedback control algorithm, known for its use in telecommunication resource management problems. The main advantage of the approach is related to its distributed nature, which keeps low the number of communications needed to achieve the objectives. The main drawbacks are also related to the AIMD concept. It requires that the PEVs have the ability to vary their charge rate in a continuous manner from zero to a maximum value, a very common assumption which, again, is not compliant with the standard IEC 61851. Moreover, the vehicle-to-grid concept is not considered in that work.

Another interesting contribution from the control methodology point of view is given in [26], where the authors apply sliding mode control principles to achieve stability and robustness with respect to system uncertainties. The authors derive a simple centralized control strategy in which a unique charging rate signal for all the PEVs is adjusted in order for the aggregated charging power profile to track the available power trajectory resulting from both renewable and traditional generation. The interesting achievements of this work are the stability and robustness to the collective effects of system uncertainties (in particular, drivers' arrival at the EVSE and power generation from RES). However, only the high level behavior of the system is investigated; driver preferences are not considered in the problem formalization and the applied control is the same for all the PEVs. So doing the benefits for the drivers are not differentiated in relation to their degree of flexibility.

Another work taking inspiration from communication engineering is [27], in which the author proposes a distributed framework for PEVs charging, based on the concept of congestion pricing in Internet traffic control. The work is based on concepts already well known and studied also in smart grid research: each PEV is modeled as an agent with an associated utility function. The objective of the agent is to maximize its individual surplus (the utility minus costs). The cost of energy is calculated by the agent based on the unit price of energy, which is centrally updated depending on the state of congestion of the network. Based on this information, agents update their charging rates, depending also on the so called "willingness to pay preference", a parameter which has a similar role as the quality of service class parameter in telecommunications. The work does not include network constraints and does not give detailed suggestions for the selection and tuning of utility functions (which is the delicate point of such an approach). Therefore, it is not clear how UP could be modeled via utility functions.

Concluding, works in literature mostly differentiate depending on: (i) the control architecture (centralized control schemes opposed to decentralized control schemes); (ii) the control methodology (optimization techniques, optimal control, nonlinear control, telecommunication algorithms, etc.); (iii) the nature of control variables (on/off signals or continuous charging rate signals); (iv) quality and effectiveness of UP modeling; (v) inclusion of backfeeding in the problem formulation. In this respect, the characterizing aspects of this work are:

- The charging rate is modelled as a Boolean variable or as a semi-continuous variable (in compliance with the standard IEC 61851), depending on the maximum allowed charging power
- Backfeeding is also modelled in a Boolean or semi-continuous manner. Nowadays there is not a commonly accepted and standardized vision on V2G power from the technical point of view. By reasonably extending the technical requirement of the charge mode to backfeeding, it is shown the relevance of such a concept for the fulfilment of grid and drivers' requirements
- The controller works on a time driven basis. It updates the control signals periodically, taking into account all the events triggering it during the sampling period, such as charging requests, user preferences updates, RES forecast updates and DSM signals, then adapting its behavior to the uncertainty of mobility dynamics and different grid players' needs
- The expected cost for the charging service is notified to the driver just after the charging request is made. A modification to the cost in reaction to possible DSM signals is taken into account in order to establish the minimum rebate for drivers' acceptance
- Each PEV is associated with its own control signal, which is built and updated according to the time of arrival, the UP and the user flexibility in terms of parking time. So doing, the controller is able to exploit the time varying nature of the energy price and the backfeeding capability to guarantee a higher economic benefit to the drivers with the higher level of flexibility
- The controller performs the tracking of an aggregated power reference for charging. By properly managing drivers' flexibilities, the effects of multiple charging sessions are mitigated so that large excursions in power withdrawal are avoided
- The controller acts so as to maximize the self-consumption from PV generation, then mitigating the intermittency of generation and allowing the penetration of RES into the electricity system
- Battery aging is taken into account through the inclusion of a depreciation term in the objective function which depends on the control. So doing, it is possible to achieve a balance between the benefit coming from multiple activations of the battery and the decrease in the battery life cycle
- The state of charge is used as feedback signal for control purposes.

It is to remark that, although grid constraints could be included in the problem formulation, the peaks shaving of charging power results in acceptable losses and voltage levels, as shown in [13]. In this sense, differently from most of the works appearing in the relevant literature, in which the aggregated power can freely evolve together with other physical variables within given thresholds, the power reference here considered for tracking purpose can be seen as a signal validated by the DSO for a reliable operation of the electrical infrastructure, in compliance with an unbundled scenario where, in principle, the owners of the charging infrastructure and the electricity grid are two different grid players, as detailed in Sect. 7.3.

7.3 Charging Scenario

7.3.1 Actors and Components

The charging scenario considered in this chapter (Fig. 7.1) is limited to a set of PEVs connected to EVSEs located in the same LA, the size and topology of which are established in advance by the DSO. The sample LA considered in the picture is represented by the portion of the distribution grid under a secondary substation; however, its extension can be further limited, as established within the ADDRESS project [12]. A PV generator is also connected in the LA and managed by a DER operator. The owner of the charging infrastructure is supposed to be an EVSE operator, which makes the EVSEs available to drivers having a retailing contract with qualified electric vehicle service providers (EVSPs); the charging sessions are managed in real time by the LAC, a software module hosted by the EVSE operator back-end, the basic functionality of which is to control the power withdrawals by dynamically solving a load shifting problem according to drivers' contracts and needs, PV generation forecasts received by the DER operator and a set of boundary conditions established by the DSO on a day-ahead and intra-day basis.

At current time, pioneer charging infrastructures in operation or used within demonstration projects are typically owned by the DSO itself, due to the need of

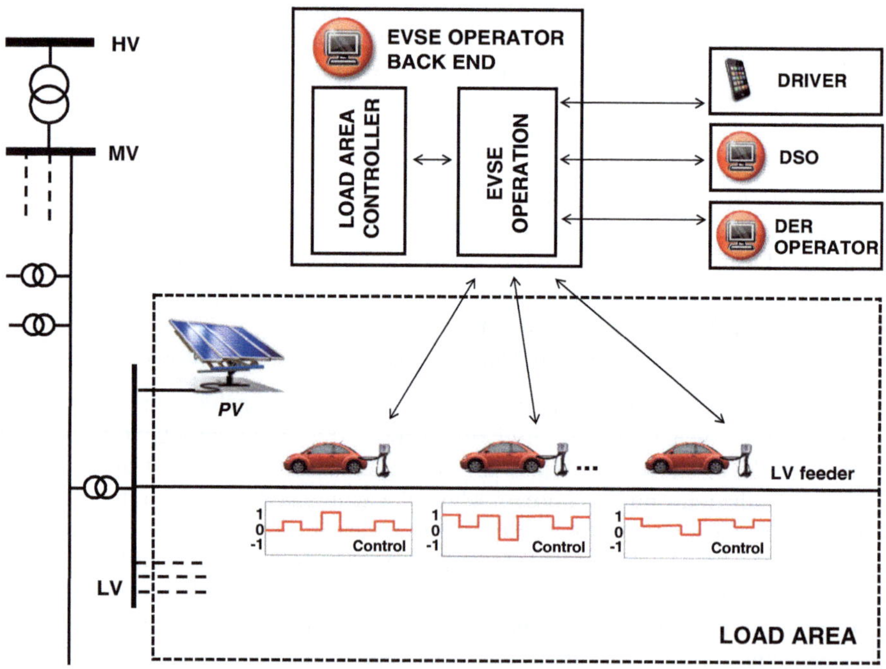

Fig. 7.1 Reference scenario

validating the system and understanding the best way to manage it while the penetration of PEVs increases. For the coming future, as a consequence of the unbundling process driving the renovation of electricity sector in a large number of industrialized countries, the idea of establishing EVSE operator companies as owners of local charging infrastructures appears reasonable. As regards this chapter, in absence of a commonly accepted business and regulatory framework, the control system presented in the following is sufficiently general to cover the proposed reference scenario and also a number of other possible situations, such as: (i) EVSPs competing in the LA maintaining the ownership of the EVSEs or not (which implies static or dynamical EVSEs assignment, respectively); (ii) a DSO directly controlling the service and making available the metering data to the EVSPs for billing purpose only; (iii) a municipality which guarantees the charging service based on a bilateral energy contract subscribed with a generation company. Finally, also the concept of LA is sufficiently general to be tailored to specific DSO needs.

The main actors involved in the reference scenario are the PEV drivers, the DSO, the EVSP, the EVSE operator and the DER operator [28], the role of which is more specifically detailed in the following.

- Driver. It is interested in obtaining the charging service at low price and in respect of his/her UP. The PEV driver subscribes a contract for charging service provisioning with an EVSP, receiving an radio-frequency identification (RFID) card for the authentication at the EVSE; depending on his/her flexibility in PEV charging, the contract establishes if the charging has to be uncontrolled (always at maximum power) or "smart" (power modulation); in the latter case some additional clauses can regard incentives for the acceptance of power modulation and the participation in DSM programs, and additional costs for the update of UP during the charging
- DSO. It is the owner of the electricity distribution infrastructure and is responsible for the safe operation of the network. It establishes maximum and reference power withdrawal at LA level on a day-ahead basis. In particular, the reference power curve for electromobility could be established by the DSO according to a range of criteria, with the possible objectives of, e.g., (i) ensuring safe operation of the network (e.g. by choosing flat or shaved profiles); (ii) supporting network through, e.g., balancing of peaks from RES; (iii) ensuring that the load profiles at primary substation level agreed on a day-ahead basis with the transmission system operator are met (i.e. by properly choosing the references for the LACs "under" the primary substations in question), etc. Furthermore, the presence of a tunable reference for the LAC controllers is a degree of flexibility which could be exploited in the future by broader control schemes for optimal balance of the energy resources in a macro load area, as explained in [29]. Also, depending on the grid status acquired in real time by the SCADA system and its possible evolution, the DSO is expected to trigger the EVSE operator with DSM signals, then calling for charging rescheduling, and provide it and flexible drivers with a remuneration which depends on the grid operation saving coming from the rescheduling action

- EVSP/Retailers. Business companies qualified to act in the electricity markets for the acquisition of the electric energy and to offer proper energy contracts to the drivers. As a result of the day-ahead market trading, each EVSP/Retailer is associated with daily energy and cost profiles, with hourly resolution. Based on that, EVSP/Retailer defines daily energy tariffs for the drivers, which can be the same every day or indexed to the market; also, the tariffs can include incentives for flexible drivers participating in DSM programs. Contract schemes including DSM are expected to be key elements allowing PEVs to be part of a future intra-day local balancing market, where managing short term requirements for the balance of demand and supply from traditional and renewable energy sources; as a matter of fact the EVSP can play the role of intermediary between grid players asking for power modulation and the drivers, taking a margin from the provided remunerations
- DER operator. It is the owner of PV generators installed in the LA. In order to support the maximization of the hosting capacity and then sustain its business, it provides the EVSE operator with generation forecasts at LA level, then enabling the local demand/supply matching. Also, the DER operator is expected in the future to trigger grid players with DSM signals for charging rescheduling
- EVSE operator. It is the owner of the charging infrastructure; it allows each charging session to take place only after an authentication process involving an EVSP, aimed at verifying the existence of an energy contract for the driver making the request. Charging processes are managed according to drivers' contract and UP, PV generation forecasts and boundary conditions established by the DSO at LA level. It is remunerated by the EVSP, the DSO and DER operator, as a consequence of its ability to provide the charging service to drivers, assure the respect of grid requirements and minimize the effect of fluctuating power generation.

The equipment making part of the reference architecture can be summarized as follows:

- Plug-in electric vehicles. Fully PEVs [30] are considered, characterized by the following technical parameters: (i) the capacity of the battery pack; (ii) the input/output battery performance coefficients; (iii) the maximum and minimum allowed charge levels and; (iv) the maximum and minimum charge/discharge rates
- EVSE. Depending on the circuit and on the current and voltage levels, different charging levels are today available [31]. This chapter deals with two different kinds of slow charging taking place at 230 V voltage level: (i) single-phase charging with 16 A maximum current (about 3.6 kW) and three-phase charging with 32 A maximum phase current (about 22 kW), being them quite common in practice; however the proposed control algorithm is also suitable for other charging levels. The power flow from the EVSE to the PEV cannot be varied continuously from zero to the maximum value: the standard IEC 61851 establishes that, beyond the standby mode (no power flow), the charging current has to be in the range from 6 to 48 A, being then a semi-continuous variable. This

range can be further limited by the EVSE manufacturers, as for the case here considered

- EVSE operator back-end. It is the charging monitoring and controlling platform; it allows drivers authentication and EVSE socket unlock, monitoring of EVSE meter readings and remote control of the charging current. It represents the heart of the infrastructure, managing in real time data of different players and able to react to different kind of events like charging requests, forecasts updates and DSM signals for charging rescheduling. It hosts two main subsystems:
  - The EVSE operation, responsible for driver authentication, socket unlock, events recording, trigger of LAC, power to phase current conversion and communication of load profiles to the EVSEs for actuation to the PEVs
  - The Load Area Controller, a control entity logically acting at LA level, responsible for real time computation of the charging power. The LAC calculates the control signals for each ongoing charging session, namely the power withdrawal over the time and the budgeted charging costs for drivers.

From the communication point of view, it is assumed that a data connection can be established between (i) the PEV and the EVSE (e.g.: via power line, wireless, GSM, etc.); (ii) the EVSE and the EVSE operator back-end (e.g. via the Internet). Reference documents on this topic are IEC 61851 and IEC 62196 and ISO 15118.

7.3.2 Use Cases

Four relevant use cases are considered:

- UC1: charging request. The driver arrives at the EVSE site, makes the authentication through its RFID card and plugs the PEV; he/she makes a charging request by using the dashboard of the EVSE (or a mobile Internet enabled device), specifying the PEV model, initial state of charge as read on the PEV dashboard and the following UP:
  - The desired final level of charge
  - The time at which the charging process can start (typically the current time)
  - The time within which the charging process has to be terminated.

  This is in the following called a "charging request (CR) event". The control system is expected to notify the driver of the optimal charging cost and to provide the EVSE with the optimal charging load profile
- UC2: user preferences update. During charging the driver realizes to need the PEV charged at the desired level of charge before the departure time declared when making the charging request, then he/she sends an update of the UP (specifically the departure time) to the EVSE operator back-end by using an Internet enabled device. In the following this is referred to as "user preferences update event," for which the control system is expected to react by updating the

load profiles for all the ongoing charging sessions and calculating a new budgeted cost for the driver. This event represents the breach of the charging conditions established at the moment of making the charging request; then there is no need of keeping the original budgeted cost as target for the whole charging session and a new one for the remaining part of the charging session has to be calculated
- UC3: forecast update. In this case the DER operator notifies the EVSE operator back-end of an update in PV generation forecast for the coming hours. In the following, this is referred to as "forecast event". The control system uses this new data as boundary condition for the future calculations, and is expected to update the load profiles for the ongoing charging sessions
- UC4: demand side management. In this case the involved actor is the DSO, which notifies the EVSE operator back-end of an intra-day change in the reference available power for a specific temporal slot (volume signal). The control system is expected to react to this event by updating the control signals for the EVSE and evaluating the related changes in the cost for flexible drivers, which gives rise to minimum rebates for them.

7.4 Controller Requirements and Specifications

7.4.1 Functional Requirements

The analysis of the proposed use cases results in requirements and specifications for all the involved equipment. In the following, a set of requirements and specifications are reported, to be intended as referred to the LAC component, the design of which represents the focus of this chapter. The functional requirements for the LAC can be broken down into categories as follows, depending on the grid player asking for them:

- Driver perspective:
  - The LAC has to be able to provide each individual EVSE with a cost-effective charging power profile which satisfies driver preferences on charging
  - The LAC has to be able to provide the budgeted cost for charging in response to a charging request and to guarantee a waiting time for the driver in line with a real time application
  - The LAC has to be able to guarantee that the real cost evaluated at the end of the charging session does not differ significantly from the budgeted one.
- DSO perspective:
  - The LAC has to be able to flatter the aggregated charging curve in the LA while managing the dynamical and asynchronous arrival of charging requests

  - The LAC has to be able to provide ancillary services for short term grid needs, in reaction to price/volume signals
  - The LAC has to be able to produce a schedule of the control signals both for the present and the coming hours, so that a lack in communications from the EVSE operator back-end to the EVSEs does not preclude the actuation of the controlled load profiles, even if suboptimal.

- DER operator perspective:

  - The LAC has to be able to maximize the self-consumption from PV while managing the dynamical and asynchronous arrival of charging requests, then minimizing the injection of fluctuating power profiles into the distribution grid.

7.4.2 Technical Specifications

The technical specifications for the LAC are as follows:

- The aggregated cost for charging has to be minimum
- The budgeted cost for charging has to be available in response to each charging request
- The difference between the real cost and the budgeted one must not exceed a given bound
- The minimum rebate for drivers has to be calculated in reaction to DSM signals
- The charging can take place only during the time period notified by the driver when asking for service
- The final level of charge has to be the one notified by the driver when asking for service
- Self-consumption from PV has to be maximized
- The net aggregated power withdrawal in the LA has to track the reference given by the DSO
- The tracking error of the net aggregated power reference has to be minimum
- The net aggregated power withdrawal must not exceed a given threshold
- The power flow for 22 kW three-phase charging has to be limited according to the standard IEC 61851
- The gradient of power flow over the time for 22 kW three-phase charging has to be limited
- The power flow for 3.6 kW single-phase charging has to be subject to on/off control
- During charging, the state of charge must not exceed a given upper bound
- During charging, the state of charge must not be lower than a given lower bound
- The number of charging sessions managed simultaneously has to be compatible with the size of a typical LA
- Feasible suboptimal control and cost have to be provided if a given allowed maximum computational time is overcome.

7.5 Control System Working Logic

The plant to be controlled is constituted by a time varying set of charging PEVs, while the controller to be designed has to work taking into account a set of boundary conditions including the UP established by the drivers, the power reference established by the DSO, the PV generation forecast provided by the DER operator and the energy tariff, all of them subject to possible updates during the operation. Considering the heterogeneity of PEVs to be charged, and the stringent objective to provide them with the desired level of charge at a given time, it is reasonable to let the controller relying on a PEVs SoC prediction model and the measure or an estimation of the PEVs SoC as feedback signal. Also, considering the use of digital systems and communications in a real implementation (e.g. the electronic meter hosted in the EVSE), it is reasonable to let the controller work in a discrete time framework. All these considerations suggest to design a MPC framework, by which the optimal control (the load profiles for the PEVs) over a specified control horizon is obtained at each sampling time by retrieving the PEVs SoC and solving an open loop optimal control problem. A principle scheme is reported in Fig. 7.2. The problem to be solved at each iteration can be based on a target function to be minimized taking into account the cost for charging and the

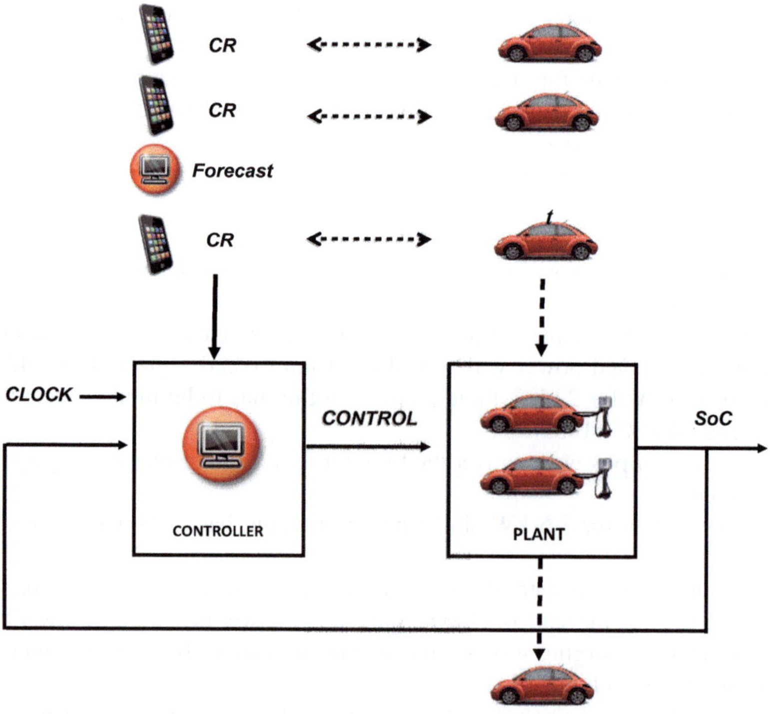

Fig. 7.2 Control system concept

tracking error, subject to a set of constraints modeling the technical specifications previously detailed, including the main one of guaranteeing the desired PEVs level of charge.

The optimal solution found at each iteration of the algorithm is intrinsically open-loop, loosing optimality over the time as a consequence of new events triggering the controller. This issue is solved by collecting the new boundary conditions and iterating optimization at the next sampling period; as customary in MPC system design, the new calculated control sequence replaces the portion of the previous control sequence that has not been actuated yet. Then the calculation of control is time-driven, while the update of boundary conditions is event-based. This approach allows to properly manage system model inaccuracies and react to the asynchronous dynamics of the environment, whatever the arrival frequency of new events will be.

In the light of above, each sampling period is characterized by the same base set of sequential steps and a number of possible events, as shown by the sequence diagram reported in Fig. 7.3. More in detail, the considered flow of operations is as follows:

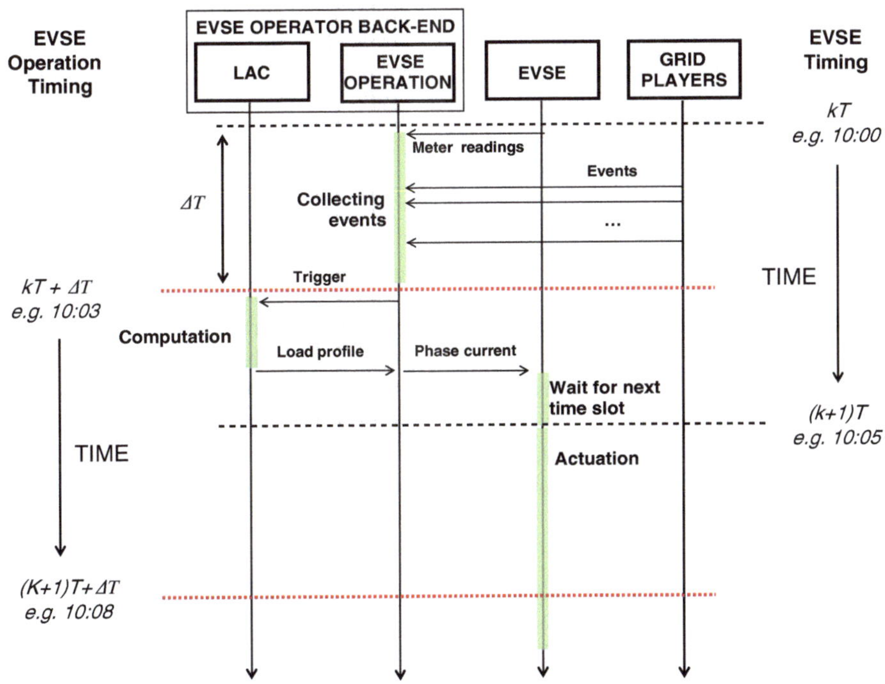

Fig. 7.3 Sequence diagram

- Each EVSE sends the metering data to the EVSE operation module, including the energy absorbed since the beginning of the charging session, the current power withdrawal and the current SoC, in case it can be retrieved and communicated by the PEV
- The EVSE operation collects all the events coming from drivers, DSO and DER operator
- The EVSE operation triggers the LAC providing it with all the new boundary conditions
- The LAC calculates the load profiles and sends them to the EVSE operation. Also the budgeted cost for charging in case of a new charging request is provided (not reported in the figure)
- The EVSE operation converts the load profiles to phase current profiles, depending on the three-phase or single-phase nature of the charging sessions, and sends them to the EVSEs
- At new sampling time, EVSEs actuate the control signals.

Considering the time needed by the LAC to solve the open loop optimal control problem, it is important to remark that, though the EVSE operation and EVSEs use the same time resolution T, the related sampling instants have to be shifted by a proper time period $\Delta t < T$. This precaution, together with the choice of a proper timeout for the solver, guarantees that the new control signals are actually available (i.e. they have been computed) at the moment in which the actuation command is given (i.e. when they have to be sent to the EVSEs for actuation). As far as concerns the meter readings from the EVSEs, this data can be used to build an estimation of the PEVs SoC in case a direct measure is not available; details on this point are given in Sect. 7.7.

7.6 Problem Formalization

The following subsections detail the discrete-time, open-loop optimal control problem that the LAC solves each time it is triggered by the EVSE operation and the PEVs' SoC feedback is estimated. The mathematical formulation of the problem is discussed starting from the objective function and then detailing the different set of constraints considered. Finally, in the last subsection, an equivalent mathematical formulation of the problem, suitable for implementation on the calculator, is given.

7.6.1 Target Function

Let T denote the discretization step of the optimal control problem and I the first time interval of problem definition (i.e. the time when the LAC is triggered for computation). The target function J can be written as

$$J = J^{cost} + J^{DER} + J^{reg} \tag{7.1}$$

where: (i) J^{cost} accounts for the cumulative cost associated to the ongoing controlled charging sessions; (ii) J^{DER} is a discount term taking into account DER exploitation and; (iii) J^{reg} is a term accounting for the remuneration associated to the tracking of a DSO-defined load profile. This function is such that its minimization leads the EVSE operator to find a trade-off between the optimal costs for PEV users, RES self-consumption and tracking error. Some notation is introduced to define the terms of J in details. Let M denote the set of charging sessions to be controlled at time I, U_{mk} the control signal associated to the mth charging session at time k ($m \in \mathrm{M}$ univocally identifies the charging session), C_k the tariff at time k. Furthermore, let $\Delta\mathrm{P}_m$ denote the maximum charging power applicable during the charging session (which is given by the minimum between the power rating of the PEV, of the charging cable and the EVSE) and E_m the last allowed end time for the charging session (set by the PEV user). The term J^{cost} can be therefore written as

$$J^{\mathrm{cost}} = \sum_{m \in \mathrm{M}} \sum_{k=I}^{\mathrm{E}_m} \Delta\mathrm{P}_m \mathrm{T} \mathrm{C}_k U_{mk} \tag{7.2}$$

in which the control variable U_{mk} specifies the charging rate and, therefore, $U_{mk}\Delta P_m$ the power actually flowing in the cable connecting the mth PEV to the corresponding EVSE. The term J^{cost} can be further expanded to model the cost of batteries' wear deriving from charging/discharging operations. For each PEV, a depreciation term can be added, which is proportional to the amount of energy exchanged with the grid during each sampling interval. The effect of the depreciation term is such that V2G is chosen only if the deriving economic benefit at least is greater than the cost of the associated components' wear. It should be noted that the inclusion of a depreciation term accounts for a phenomenon (wear) that is different from that of batteries' energy losses (modelled as well in the following), although the deriving effect which can be observed (decrease of V2G power flow) is similar. The term J^{cost} can be thus finally written as

$$J^{\mathrm{cost}} = \sum_{m \in \mathrm{M}} \sum_{k=I}^{\mathrm{E}_m} \Delta\mathrm{P}_m \mathrm{T} \left\{ \mathrm{C}_k U_{mk} + \mathrm{C}_m^{dep} |U_{mk}| \right\} \tag{7.3}$$

where the coefficient $\mathrm{C}_m^{\mathrm{dep}}$ can be computed based on the cost of the batteries and their expected life time.

Then, the term J^{DER} in the objective function introduces a discount proportional to the amount of self-consumed DER energy. It is written as

$$J^{DER} = -\sum_{k=\mathrm{I}}^{\mathrm{E}} \mathrm{P}_k^{\mathrm{DER}} \mathrm{T} \mathrm{C}^{\mathrm{DER}} \theta_k \tag{7.4}$$

where $\mathrm{P}_k^{\mathrm{DER}}$ is the last available forecast for DER power at time k, $\mathrm{C}^{\mathrm{DER}}$ is a discount parameter associated with self-consumption and θ_k a continuous variable indicating the share of DER power self-consumed at time k $(0 \leq \theta_k \leq 1 \forall k)$. Coefficient E denotes the last time interval of problem definition (which therefore spans from I to E), and is given by

$$\mathrm{E} = \max_{m \in \mathrm{M}} \{\mathrm{E}_m\} \tag{7.5}$$

The effect of J^{DER}, being it a discount term appearing with the minus sign in a minimization problem, is such that the controller tends to increase as much as possible at each k the value of variable θ_k, representing the share of self-consumed power from DER. By posing a constraint (see (7.12) below) stating that, at each k, the amount of self-consumed DER power $\theta_k \mathrm{P}_k^{\mathrm{DER}}$ shall be less or equal than the charging power at the same instant, it is made sure that the power $\theta_k \mathrm{P}_k^{\mathrm{DER}}$ accounted in the objective function actually "matches" a portion of the load curve from electromobility (eventually, that charging load is shifted under the curve of DER generation).

Finally, the last term J^{reg} appearing in the objective function is a regulation term, which allows the EVSE operator to shape the aggregated charging power according to a positive power reference $\mathrm{P}_k^{\mathrm{ref}}$, set by the DSO. J^{reg} can be written as

$$J^{reg} = \mu \sum_{k=I}^{\mathrm{E}} \lambda_k \left(P_k - \mathrm{P}_k^{\mathrm{ref}} - \mathrm{P}_k^{\mathrm{DER}}\right)^2 \tag{7.6}$$

where P_k is the aggregated controlled charging power in the LA at time k, which can be written as

$$P_k = \mathrm{P}_k^s + \sum_{m \in M_k} \Delta \mathrm{P}_m U_{mk} \tag{7.7}$$

being $M_k \subseteq \mathrm{M}$ a subset of M defined as

$$M_k = \{m \in \mathrm{M} \ : \mathrm{I} \ \leq k \leq \mathrm{E}_m\} \tag{7.8}$$

and representing the set of flexible PEVs involved in the smart charging operation at time interval k. $\mathrm{P}_k^{\mathrm{s}}$ denotes the aggregated power consumption of those PEVs whose charging profile cannot be rescheduled (i.e. PEVs which do not allow smart charging). $\lambda_k \in [0, 1]$ is a shaping factor which is included to differently weight the tracking requirement along the control horizon. Acting on λ_k it is possible to influence the way the charging power is allocated in the time window ahead of the current time I. A general rule of thumb for λ_k is that it has to go to zero as k goes to E (or, in any case, to values considerably smaller than the ones taken for k close to I), since the tracking requirement cannot be stringent for the time periods close to E, where there might be not enough demand for charging power to let the aggregated load profile follow the reference (as a matter of fact, charging requests arrive sequentially, and hence, at time I, the demand close to time E might be small). The parameter λ_k can be also chosen adaptively, in order to adjust the way power is allocated depending on, e.g., the congestion rate of the grid (e.g. the arrival rate of new charging requests). Finally, $\mu \in R$ is a positive weight which, together with λ_k is associated to the remuneration assured for the provisioning of the tracking service.

Concluding, (7.1) represents the objective function of the optimization problem to be solved in order to optimally schedule the charging sessions according to the requirements (posed in Sect. 7.4.1) of (i) minimizing charging costs, (ii) maximizing self-consumed DER power and, (iii) tracking of a DSO-defined, LA-level reference load curve for electromobility. Formula (7.1) also represents the function through which the exact cost minimization values associated to the found solutions can be determined. It is important to notice that the actual costs/revenues for the different actors involved in the PEV load management problem (i.e. the PEV users, the EVSE operator, the DSO, the retailer) are determined starting from metering data and based on agreed billing/revenue repartition policies, which are not dealt with in this chapter. For example, the overall charging costs for the PEV users could be determined by subtracting from the actual costs incurred for charging (accounted for by the term J^{cost}) proper remuneration from the DER operator and the DSO, in consideration of the fact that it is actually the flexibility provided by the PEV users taking part in smart charging the key factor which enables the DER operator and the DSO to extract value from the process of PEV charging control.

The next subsection starts the review of the problem's constraints.

7.6.2 Control Model

The *control model* is a tractable mathematical representation of the process under control, given here by the set of PEVs and the associated EVSEs to be controlled. The control model is employed to derive a relation between the control variables (the charging rates) and the controlled output of the plant (the PEVs' SoC). It has to be therefore simple enough to keep low the complexity of the resulting control problem, and yet accurate enough to capture the main dynamics of the physical process under control.

The following first-order control model is considered

$$\begin{cases} x_{mk} = x_{m,k-1} + \Delta P_m T(U_{mk} - \xi_m |U_{mk}|) \\ x_{m,I-1} = X_m^0 \end{cases} \quad \forall k \in [I, E_m], \quad m \in M \tag{7.9}$$

which allows to *predict* the future SoC of the PEVs and write it as a simple (linear) function of the initial SoC and the control sequence (notice that SoC here denotes the *absolute*, not percentage, value of the energy stored). The captured phenomena are the integral behaviour of the battery pack and the losses in the PEV converters and in the battery pack (a simple constant loss factor ξ_m is considered). X_m^0 denotes the initial SoC of the mth PEV (which is either communicated by the PEV user via, e.g., a smartphone or the PEV/EVSE dashboard, or is automatically exchanged between the PEV and the EVSE as soon as the charging session is authorized). Note that X_m^0 represents one of the feedback signals considered in the presented control scheme (see the explanation in Sect. 7.5): in case the current SoC can be read and automatically communicated by the PEV to the EVSE, the value of X_m^0 can be updated at each iteration of the problem, thus allowing to counteract disturbances and model uncertainties, and making sure that the controlled charging process actually ends with a final SoC that is in accordance with the UP.

Finally, a second, very detailed model of the EVSE/battery pack, referred to as *simulation model*, will be given in Sect. 7.7 for the purpose of validating the proposed strategy on a simulation basis. As explained in Sect. 7.7, the simulation model accurately replicates the non-linear dynamics of the battery pack, and is therefore used for the purposes of: (i) simulating the feedback of the SoC from the field and, (ii) evaluating the actual SoC evolution resulting from the implementation of the proposed strategy, then verifying that all the posed requirements (see Sect. 7.4) are met. In particular, it will be shown via the simulations how the combined effect of reoptimization and feedback from the field lets the system recover from the inaccuracies of the control model.

7.6.3 Control Constraints

The first set of control constraints is related to the nature of the control variables U_{mk} and θ_k. Standard IEC 61851 prescribes that the charging power shall be either zero (when recharging is paused), or limited between a minimum positive value and a maximum positive value (when charging is in progress). By reasonably extending the same specification to the discharge phase (which is not addressed by IEC 61851), the following set of constraints for U_{mk} arises

$$U_{mk} \in \begin{cases} cc \in [\alpha_m, 1] & \text{in case of charging} \\ 0 & \text{in case of standby} \\ dd \in [-1, -\alpha_m] & \text{in case of discharging} \end{cases} \qquad \forall k \in [\mathrm{I}, \mathrm{E}_m], \quad m \in \mathrm{M} \tag{7.10}$$

where α_m is the ratio between the minimum positive charging power and the maximum allowed charging power (without loss of generality we assume the same ratio holds for the discharging phase). For each charging session, α_m has to be determined depending on maximum/minimum allowed positive charging rates of the EVSE, of the charging cable and the PEV (α_m shall be computed at the beginning of the charging session, when these values are discovered). It is worth mentioning that more restrictive values for α_m can be also set when designing the control strategy. For example, it can be decided to allow charging power modulation only in case of high power charging sessions (e.g. when ΔP_m is equal or greater than 22 kW), while performing on-off charging in case of low power charging sessions (i.e. single phase charging, ΔP_m around 3.6 kW), for which it may not be worth to modulate the charging power. The latter can be simply achieved by setting $\alpha_m = 1$ for the charging sessions to be controlled in an on-off manner. Formula (7.10) translates into mathematical formulation the technical specifications posed in Sect. 7.4.2 about IEC 61851-compliant control of 22 kW three-phase charging processes and on/off control of single-phase 3.6 kW charging processes.

As regards θ_k, it is not a control variable input to a real process; it represents the share of RES power self-consumed by the PEV fleet at time k. The following simple constraint then holds

$$\theta_k \in [0, 1] \quad \forall k \in [\mathrm{I}, \mathrm{E}] \tag{7.11}$$

The second set of control constraints is related namely to the definition of variables θ_k, and reads as follows

$$\mathrm{P}_k^{\mathrm{DER}} \theta_k \leq \mathrm{P}_k^s + \sum_{m \in M_k} \Delta \mathrm{P}_m U_{mk} \quad \forall k \in [\mathrm{I}, \mathrm{E}] \tag{7.12}$$

which assures that the share of RES power self-consumed is properly computed (by definition it cannot exceed the allocated charging power). Notice that θ_k either saturates at 1 (in case the allocated charging power exceeds RES power), or it is limited by the above constraint (in case all the charging power is "matched" by DER power). Constraints (7.12), (7.11) and the term J^{DER} allow to translate into mathematical formulation the requirement posed in Sect. 7.4.2 about maximization of self-consumption of DER power.

The next set of control constraints aims at avoiding that the aggregated charging power exceeds the LA threshold P_k^* set by the DSO at each k. The difference between the threshold and the reference can be seen as the maximum displacement which is allowed without penalties. Moreover, the threshold could also be established by the DSO during the emergency operation of the distribution grid. Such set of overload constraints can be written as

$$\mathrm{P}_k^s + \sum_{m \in M_k} \Delta \mathrm{P}_m U_{mk} \leq \mathrm{P}_k^* + \mathrm{P}^{\mathrm{DER}} \quad \forall k \in [\mathrm{I}, \mathrm{E}] \tag{7.13}$$

The former constraints were related, respectively, to the RES and the DSO. The next set of constraints is instead explicative of those constraints imposed by PEV users. In particular, until this point, only cost minimization and technical constraints satisfaction related to the entire set of PEVs under control have been addressed. However, it may happen that a cost-efficient and technically feasible solution for the entire fleet does not equally distribute the cost (or the saving) among the PEVs, thus penalizing some PEVs and excessively rewarding some others. Notice that the term

J^{cost} appearing in the target function is a cumulative cost, then the minimization of the target function does not guarantee that the price of the service provided to each driver remains close to the price notified at the moment of making the charging request. To take this into account, a set of constraints on the cost of charging/discharging operations is added for each PEV. Such constraints guarantee that, as required by the specifications given in Sect. 7.4.2, the cost of the charging service for each driver remains bounded iteration after iteration, without growing unpredictably. Let c_m^* denote the cost announced to the user upon arrival at the EVSE (after the first iteration of the algorithm) and $c_{m\mathrm{I}}$ the cost for the charging service provided up to time I. Then the control action can be bounded as follows

$$c_{m\mathrm{I}} + \sum_{k=\mathrm{I}}^{\mathrm{E}_m} \Delta \mathrm{P}_m \mathrm{TC}_k U_{mk} \leq (1+ \in) c_m^* \quad \forall m \in \mathrm{M} \tag{7.14}$$

where the real number $\in > 0$ is a small tolerance parameter, necessary to account for modelling inaccuracies.

A final set of control constraints is included as representative of the technical constraints that are imposed by the PEVs (constraints which are put, for example, to ensure safe charging operations and preserve the PEV energy storage and recharging systems). In particular, according to the technical specifications given in Sect. 7.4.2, a constraint is put here on the maximum allowed rate of change of the charging power from a time slot to the following one. The constraint is

$$\left| \Delta \mathrm{P}_m U_{mk} - \Delta \mathrm{P}_m U_{m,k-1} \right| \leq \delta_{\max} \quad \forall k \in [\mathrm{I}+1, \mathrm{E}_m], \quad \forall m \in \mathrm{M} \tag{7.15}$$

The maximum allowed change of charging rate $\delta_{\max}$ has to be such that $\delta_{\max} \geq \alpha_m \Delta \mathrm{P}_m$, in order to allow the termination of the charging process (when the charging power goes to zero from a previous positive value).

7.6.4 State and Termination Constraints

State constraints are related to the capacity of the batteries and the related technical limitations. In principle, the level of charge must be non negative and upper bounded by the battery capacity. In practice, for reasons related to efficiency and life cycle, as specified by the technical specifications given in Sect. 7.4.2, the battery pack is never allowed to fully charge or deplete. Then it is straightforward to establish that

$$\mathrm{X}_m^{\min} \leq x_{mk} \leq \mathrm{X}_m^{\max} \quad \forall m \in \mathrm{M}, \quad \forall k \in [\mathrm{I}, \mathrm{E}_m] \tag{7.16}$$

where x_{mk} is the SoC expressed in kWh of the mth PEV at the end of the kth time interval, $\mathrm{X}_{\mathrm{m}}^{\max}$ is the maximum allowed level of charge and $\mathrm{X}_{\mathrm{m}}^{\min}$ represents the allowed depth of discharge of the mth PEV. Interestingly, a minimum guaranteed charging profile can be established for each PEV by choosing X_m^{min} as an increasing function of time. So doing, a minimum "safety" state of charge is guaranteed at each time interval. That is relevant to remedy to real word uncertainties, among which there is the possibility that a driver terminates the charging process before the declared E_m, without giving any notification to the system.

Finally, a termination constraint must be considered in order to ensure that the final desired SoC set by the PEV user is eventually reached at time E_m (as required by the technical specifications in Sect. 7.4.2). Such set of constraints is simply given by

$$X_m^{ref} \leq x_{mE_m} \leq X_m^{max} \quad \forall m \in M \tag{7.17}$$

where the upper limit (given in the above by X_m^{max}) can be replaced by a smaller value (greater, in any case, than X_m^{ref}) in order to avoid that the PEV is recharged too much above the SoC value specified by the user.

7.6.5 Overall Problem Definition

The above detailed optimal control problem can be summarized as follows.

Problem 1 (*Optimal control of PEVs charging operations in a Load Area*)

Given a set M of PEVs plugged-in at time interval I, associated with UPs $\{X_m^{ref}, E_m\}$, technical data $\{\Delta P_m, X_m^{min}, X_m^{max}, \xi_m\}$, SoC measure X_m^0 and known market/grid data $\{C_k, P_k^{ref}, P_k^*\}$, minimize J subject to the dynamics (7.9), control constraints (7.10)–(7.15), and state and termination constraints (7.16), (7.17), where E and M_k are defined in (7.5) and (7.8) respectively.

7.6.6 Equivalent Optimization Problem

The mathematical formulation of the problem given above (presented as a classical MPC problem) is not suitable for direct implementation on a calculator (observe, for example, that variables U_{mk} are defined over the union of disjoint sets). This section is therefore dedicated to show how an equivalent mathematical formulation suitable for implementation can be derived. Eventually, the MPC problem will be written as a mixed integer quadratic programming problem, which can be interpreted and solved via well-established optimization tools.

Some additional notation is introduced. First of all, let us introduce two sets of *continuous* variables y_{mk} and z_{mk}, defined as

$$\begin{cases} y_{mk} & \text{charging rate for the } m\text{th PEV at time } k \\ z_{mk} & \text{discharging rate for the } m\text{th PEV at time } k \end{cases} \tag{7.18}$$

Recalling from the previous sections that the charging power is semi-continuous in nature (i.e. it is either zero or greater than a positive value), variables y_{mk} and z_{mk} can be further specified as follows

$$\begin{cases} y_{mk} = 0 \ \vee \ y_{mk} \in [\alpha_m, 1] \\ z_{mk} = 0 \ \vee \ z_{mk} \in [\alpha_m, 1] \end{cases} \quad \forall m \in \mathrm{M}, \quad \forall k \in [\mathrm{I}, \mathrm{E}_m] \tag{7.19}$$

A treatable definition of y_{mk} and z_{mk} can be achieved by introducing two corresponding sets of Boolean variables p_{mk} and q_{mk} defined in such a way that, when $y_{mk} = 0$ ($z_{mk} = 0$), $p_{mk} = 0$ ($q_{mk} = 0$), and when $y_{mk} \in [\alpha_m, 1]$ ($z_{mk} \in [\alpha_m, 1]$) $p_{mk} = 1$ ($q_{mk} = 1$). That can be forced by writing

$$\begin{cases} \alpha_m p_{mk} \le y_{mk} \le p_{mk} \\ p_{mk} \in \{0, 1\} \end{cases} \begin{cases} \alpha_m q_{mk} \le z_{mk} \le q_{mk} \\ q_{mk} \in \{0, 1\} \end{cases} \quad \forall m \in \mathrm{M}, \quad \forall k \in [\mathrm{I}, \mathrm{E}_m] \tag{7.20}$$

The control variable U_{mk} can be then rewritten as

$$\begin{cases} U_{mk} = y_{mk} - z_{mk} \\ |U_{mk}| = y_{mk} + z_{mk} \end{cases} \quad \forall m \in \mathrm{M}, \quad \forall k \in [\mathrm{I}, \mathrm{E}_m] \tag{7.21}$$

and the additional constraint

$$p_{mk} + q_{mk} \le 1 \quad \forall m \in \mathrm{M}, \quad \forall k \in [\mathrm{I}, \mathrm{E}_m] \tag{7.22}$$

is put in order to state that charging and discharging cannot take place simultaneously.

The aggregated controlled charging power can be rewritten in terms of y_{mk} and z_{mk} as

$$P_k = \mathrm{P}_k^s + \sum_{m \in M_k} [\Delta \mathrm{P}_m y_{mk} - \Delta \mathrm{P}_m z_{mk}] \quad \forall k \in [\mathrm{I}, \mathrm{E}] \tag{7.23}$$

Similarly, it is easy to rewrite all the constraints given in Sects. 7.6.2–7.6.4, along with the linear terms J^{cost} and J^{DER} in the objective function, in terms of the new continuous variables y_{mk} and z_{mk}. The regulation term J^{reg} can be easily rewritten as well making use of matrix notation, in which the term can be written as $J^{reg} = \left(\boldsymbol{P} - \mathbf{P}^{\mathbf{ref}} - \mathbf{P}^{\mathbf{DER}}\right)^T \boldsymbol{\Lambda} \left(\boldsymbol{P} - \mathbf{P}^{\mathbf{ref}} - \mathbf{P}^{\mathbf{DER}}\right)$, where $\boldsymbol{P}, \mathbf{P}^{\mathbf{ref}}$ and $\mathbf{P}^{\mathbf{DER}}$ are vectors of proper dimensions whose elements are, respectively, P_k, $\mathrm{P}_k^{\mathrm{ref}}$ and $\mathrm{P}_k^{\mathrm{DER}}$, while $\boldsymbol{\Lambda}$ is a diagonal matrix with diagonal entries λ_k. The quadratic term of J^{reg} is $\boldsymbol{P}^T \boldsymbol{\Lambda} \boldsymbol{P}$. It is seen from (7.23) that $\boldsymbol{P}$ is linearly dependent on the control variables

y_{mk} and z_{mk}, and therefore it can be written as $\boldsymbol{P} = [\mathbf{A}\,\mathbf{B}]\begin{bmatrix} \boldsymbol{y} \\ \boldsymbol{z} \end{bmatrix}$, where $\boldsymbol{y}$ and $\boldsymbol{z}$ are proper vectors of grouped control variables and **A** and **B** proper matrices. Finally, the quadratic term $\boldsymbol{P}^T\boldsymbol{\Lambda}\boldsymbol{P}$ is rewritten in terms of the control variables as $\boldsymbol{P}^T\boldsymbol{\Lambda}\boldsymbol{P} = [\boldsymbol{y}^T\boldsymbol{z}^T]\begin{bmatrix} \mathbf{A}^T \\ \mathbf{B}^T \end{bmatrix}\boldsymbol{\Lambda}[\mathbf{A}\,\mathbf{B}]\begin{bmatrix} y \\ z \end{bmatrix}$, being then the coefficient matrix of the quadratic term given by $\begin{bmatrix} \mathbf{A}^T \\ \mathbf{B}^T \end{bmatrix}\boldsymbol{\Lambda}[\mathbf{A}\,\mathbf{B}]$.

The mathematical problem here defined is a mixed integer quadratic programming problem (i.e. a problem with quadratic terms in the objective function and linear constraints, with both Boolean and continuous variables), for which global solution methods and related tools [32] are available (the optimization problem has been here solved via the *cplexmiqp* function of CPLEX [32], dedicated specifically to the solution of mixed integer quadratic programming problems). Further details on mixed integer quadratic programming, the related solving techniques and other applications to smart grid research field can be found in [33–36].

7.7 The State of Charge Feedback

Among the PEV users' requirements considered for the derivation of the proposed centralized charging strategy there is the fulfilment by the controller of the charging requests according to the associated UP. In particular, one of the objectives of the controller is to let the PEVs reach the final desired SoC set by the PEV users. That is the reason why SoC feedback has to be foreseen for the correct implementation of the proposed strategy (SoC feedback allows evaluating the mismatch between the SoC evolution predicted by the controller and the actual one). The SoC feedback is easily included in the problem formulation through parameter X_m^0, which can be updated according to the feedback signal at each iteration of the problem, as explained in Sects. 7.5 and 7.6.2. Currently, at the best of the authors' knowledge, there is no standardized way of automatically retrieving SoC measurements during a charging session. Such necessity is however recognized by the technical community and standards on digital communication between the EVSE and the PEV (in particular, ISO 15118) are going in this direction, making possible in the future to exchange a whole set of data crucial for enabling smart charging applications (e.g. user preferences, technical specifications of the PEV, SoC data, etc.).

For the sake of completeness, this section discusses two approaches that can be implemented in case SoC feedback is not directly available. They are based, the first, on the usage of EVSE meter readings and, the second, on the usage of detailed battery models [14].

7.7.1 Case 1: Indirect SoC Measurement Through EVSE Meter Readings

In the advanced electromobility management systems the EVSEs can be remotely controlled and monitored from the EVSE operator back–end. In particular, metering data can be retrieved from the meters inside the EVSEs. Metering data regard the instantaneous charging power supplied by the EVSE at metering time, and the charging energy supplied from the beginning of the charging session. Contrary to SoC metering, the feedback from the EVSE is already technically feasible and commonly implemented in the field. The feedback from the EVSE regarding the charging power is employed in the presented formulation to evaluate in particular the fulfilment of the tracking requirement, which is precisely related to the aggregated power supplied by the EVSEs in the LA. The feedback on the energy supplied by the EVSE up to metering time can be instead employed to monitor the correct provisioning of energy to the PEVs. In this regard, notice that the user preference related the final SoC could be replaced by a similar requirement related to the amount of energy to be provided by the EVSE. The feedback on energy would then represent a feedback precisely of the controlled variable (i.e. the energy to be provided by the EVSE). Instead, in case the user requirement is on the SoC, mathematical models of the EVSE converters and the EVSE battery pack can be employed to estimate the current SoC based on the metering data on supplied energy (the feedback would be again included in the mathematical formulation via parameter X_m^0).

In the next subsection a SoC estimation model [14] is presented as well, which is not intended however for real time usage in the present control scheme. It is a highly detailed nonlinear model used in the simulations to *emulate* the real SoC feedback from the EVSEs (feedback which is not obviously available in the simulations), thus making possible to evaluate the effectiveness of the controller.

7.7.2 Case 2: SoC Estimation Using Detailed Simulation Model

The employed simulation model [14] works according to the scheme presented in Fig. 7.4. The two converters in the PEV are assumed to be characterized by constant efficiency, being the losses modelled by parameter η. The battery is modelled as the series of an internal resistance and a controlled voltage source, whose voltage depends, according to a nonlinear relation, on the charge stored in the battery [14]. Referring to the scheme in the figure, v is the voltage of the battery pack, i the current, $\tilde{q}$ the charge (measured in Ah), e the voltage of the controlled voltage source, Q the nominal capacity of the battery (Ah), Z the polarization voltage and, finally, a and b are two model parameters. The charge $\tilde{q}$ (not to be confused with the Boolean variable q_{mk}) is computed according to the (discrete time) relation

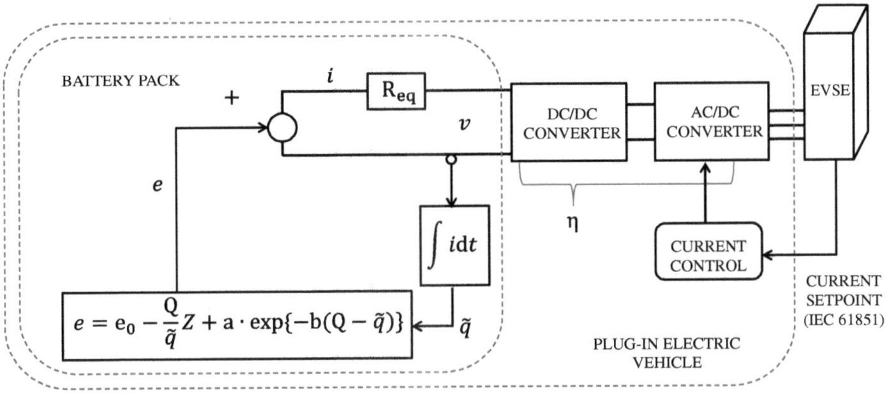

Fig. 7.4 Block scheme of the accurate simulation model

$$\tilde{q}_{m,k+1} = \tilde{q}_{mk} + \mathrm{T}\left(i_{mk}p_{mk} - i_{mk}^{\mathrm{K}}q_{mk}\right) \tag{7.24}$$

in which K is the *Peukert* coefficient amplifying the loss of charge during discharge (an hysteresis phenomenon). Also the equivalent resistance varies depending on the direction of the power flow ($\mathrm{R_{eq}} = \mathrm{R}(1.5p_{mk} + q_{mk})$).

7.8 Simulation Results

This section reports an explicative simulation study aimed at showing the effectiveness of the proposed charging controller. First of all, in the next subsection the simulation setup is detailed. Then in the subsequent section it is shown how the LAC is able to dynamically control the charging sessions in order to fulfil all the posed requirements (from the PEV users, the DSO and the DER operator), while respecting all the constraints. That is achieved by updating the computed charging load profile over the time in reaction to the asynchronous events. Among these events, a reaction to a DSM volume signal is also simulated, showing how the LAC can successfully react to the request of reducing the aggregated charging power level according to proper DSM volume signal specifications (on the amount of power reduction and the associated timelines).

7.8.1 Simulation Setup

Both single-phase and three-phase charging are considered in the following. The former refers to charging with currents limited between 6 and 16 A (resulting in a value of α_m equal to 0.375). Considering a constant grid voltage level of 230 V, the maximum charging power is therefore equal to 3.68 kW. In three-phase case instead, they are considered phase currents limited between 6 and 32 A (resulting in $\alpha_m = 0.1875$), and being thus the maximum charging power equal to 22.08 kW. It is worth recalling that it has been decided to let the LAC modulate the charging processes with higher power withdrawal, while the other charging processes are controlled by the LAC in an on-off fashion. For the sake of simplicity, all the simulated PEVs are characterized by the same following parameters: $\mathrm{X}_m^{\max} = 26$ kWh, $\mathrm{X}_{\mathrm{m}}^{\min} = 2.6$ kWh $\xi = 0.04$. Also, the same simulation model (see Sect. 7.7.2) is considered for all the simulated PEVs. As suggested in [14], the values of the PEV model's parameters can be deduced from manufacturer's data-sheet by achieving an accurate matching of the experimental charging and discharging curves (a Lithium-ion battery block [37] specifically designed for PEV applications has been considered and the following values resulted: $\eta = 1 - \xi = 0.96$, $\mathrm{Z} = 0.14$ V, $\mathrm{a} = 10$, $\mathrm{b} = 0.007$, $\mathrm{K} = 1.05$, $\mathrm{Req} = 0.01\ \Omega$, $\mathrm{Q} = 297.3$ A h and $\mathrm{e}^0 = 74$ V). The kth diagonal entry of matrix $\mathbf{\Lambda}$ is chosen as $\Lambda_{kk} = \lambda_k = 1/k^2$, the power reference $\mathbf{P}^{\mathbf{ref}}$ is taken constant for simplicity ($\mathrm{P}_k^{\mathrm{ref}} = 25$ kW). $\mathrm{P}_{\mathrm{k}}^{*}$ is taken as $\mathrm{P}_k^* = 1.2\mathrm{P}_k^{\mathrm{ref}}$. A daily profile of the Italian day-ahead tariff "prezzo unico nazionale" (PUN) has been considered for C_k. All PEVs are assumed subject to smart charging (i.e. $\mathrm{P}_k^{\mathrm{s}} = 0$). Real profiles for $\mathbf{P}^{\mathbf{DER}}$ have been taken by measurements of specific PV plant outputs [38].

Finally, simulations have been performed on an INTEL Core i5-3230 CPU, 2.40 GHz, 8 GB RAM computer, running the MS WINDOWS 8 64-bit operating system. The simulation environment has been built in MATLAB. The mixed integer quadratic programming problem defined in Sect. 7.6.6 has been solved by calling from MATLAB the *cplexmiqp* function, made available by the CPLEX for MATLAB feature of the IBM ILOG CPLEX Optimizer. The CPLEX for MATLAB module allows a user to define optimization problems and solve them within MATLAB (via the *cplexmiqp* function in this case).

7.8.2 Simulations

The simulated charging requests are reported in Table 7.1. In the following simulations V2G has been disabled for the sake of simplicity. Figure 7.5 reports the aggregated power profile resulting from uncontrolled charging, i.e., when charging starts at maximum power as soon as the charging sessions are authorized. In particular, Fig. 7.5a reports a bar chart visualization of the charging profiles (a different color is associated to the different charging profiles). Figure 7.5b reports the power reference, the threshold set by the DSO, the DER profile ($\mathbf{P}^{\mathbf{DER}}$, positive values are injections into the LA) and the net power profile of the LA (positive values mean the LA absorbs power from the main grid). It is seen that the net power profile shows large fluctuations and peaks, which is highly undesirable. As obvious, the DSO-defined reference load curve is not tracked, since no PEV load management is implemented.

Table 7.1 Simulated charging scenario

PEV ID	Arrival time (hh:mm)	Departure time (hh:mm)	Initial SoC (kWh)	Desired final SoC (kWh)	Single phase/three phases charging
1	10:00	14:00	5	20	Three phases
2	10:10	14:25	7	14	Single phase
3	10:15	14:25	9	15	Single phase
4	10:20	15:00	10	26	Three phases
5	10:40	16:00	10	20	Single phase
6	11:10	14:35	5	15	Three phases
7	11:35	17:00	7	15	Three phases
8	12:00	16:45	5	17	Three phases

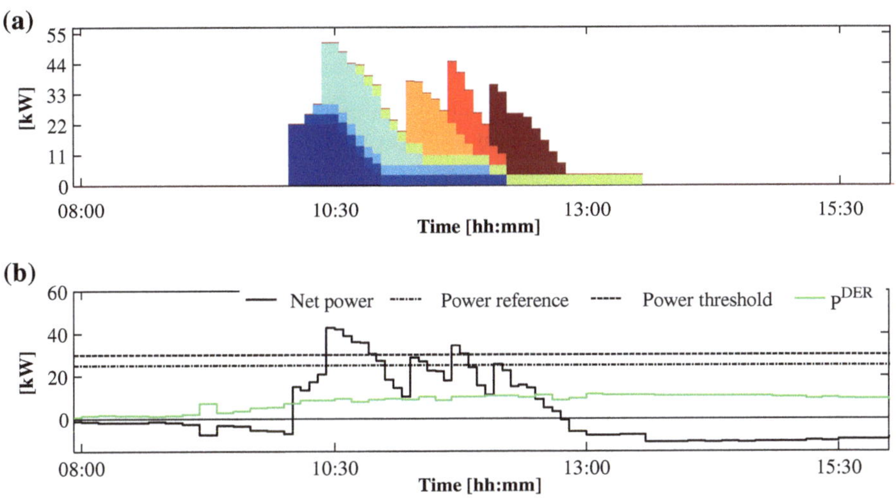

Fig. 7.5 Uncontrolled recharging: **a** load profiles of the different charging sessions and resulting aggregated demand, **b** relevant resulting power profiles, including the net power profile, the reference profile, the power threshold and the DER forecast power output

Secondly, in the same charging scenario as above, the proposed controller is tested considering a high weight μ ($\mu = 10$) of the tracking term J^{reg} of the objective function. In other words, it is simulated the case in which the remuneration offered by the DSO for the tracking of the DSO-defined reference $\mathbf{P}^{\mathbf{ref}}$ is considerably higher than the other sources of revenues (i.e. charging cost optimization and DER energy self-consumption). It is therefore expected that the net load profile accurately tracks the power reference $\mathbf{P}^{\mathbf{ref}}$. As a matter of fact, Fig. 7.6 clearly shows that accurate tracking is achieved (notice from Fig. 7.6 how the peaks present in the previous Fig. 7.5 have been shaved thanks to a proper control of the

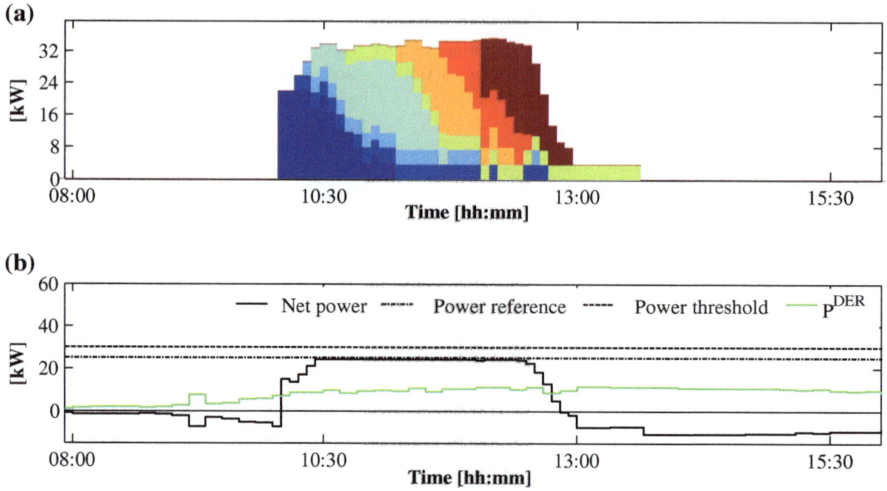

Fig. 7.6 Simulation with $\mu = 10$ (i.e. high weight given to the revenue coming from the DSO, associated to the tracking of the DSO-defined reference)

charging sessions resulting from the presented control approach). From Fig. 7.6a it is easy to distinguish single-phase charging sessions from three-phase charging sessions, due to the different nature of control (on-off control for the formers, power modulation for the latters). Also, all the posed constraints are satisfied. In particular, it can be noticed from the figure that the constraint (7.15) on the maximum rate of change of the charging power is respected. Finally, the achieved solution is such that all the charging preferences expressed by the PEV users are met (i.e. all the charging sessions end within the time specified by the users and with the desired final SoC values, as it will be explicitly shown, for the sake of brevity, in the next simulation only for the case of one of the simulated charging sessions).

In the next simulation, the weight of the tracking term is decreased ($\mu = 0.001$) in order to give more relevance to the revenue coming from the maximization of self-consumption of the energy from RES. As expected, the aggregated charging curve is increasingly flattered in order to match the PV profile on a longer horizon with respect to the two previous cases (notice from Fig. 7.7a how the PEV load is shifted ahead in time with respect to what reported in Fig. 7.6a). In this case, the tracking of the DSO-defined reference load profile is not accurate as in the previous simulation, since the controller now pursuit more the objective of maximizing self-consumption of RES power, which ensures greater revenues compared to the objective of DSO-defined load tracking. Also in this case all the constraints and, in particular, the user preferences, are met.

After having reported and discussed about load profiles at PEVs fleet level, the following discussion deals with the profiles (i.e. control signal and SoC evolution) at single PEV level, related to the last simulation presented in Fig. 7.7. Figure 7.8 reports the evolution of the charging control signal associated to PEV #1 at three

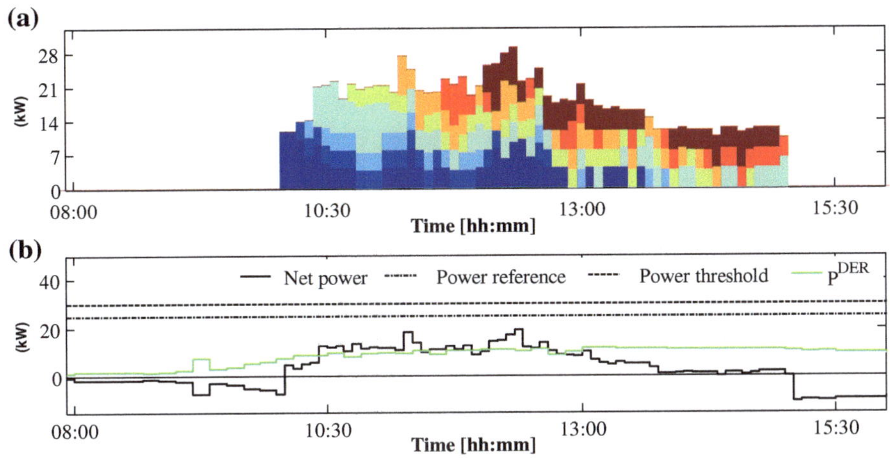

Fig. 7.7 Simulation with μ = 0.001 (i.e. high weight given to the revenue coming from DER self-consumption)

different optimization times. The control signal is either zero, or it is modulated between a minimum positive value α_m and a maximum one, as prescribed by the standard. It is evident how the control signal is updated at each optimization time in order to let the LAC accommodate the new charging requests arriving sequentially. The resulting evolution of the SoC, as computed at three different time instants via both the control model (continuous line) and the simulation model (dashed line), is reported in Fig. 7.9. Notice how the desired final SoC is reached within the time specified by the PEV user, and the bounds on SoC are respected.

Finally, a simulation of the LAC reaction to a DSM volume signal is given in the following. The DSM signal is notified to the LAC at 10:25 [hh:mm], and consists of a reprofiling of the DSO-defined reference power profile. The shape of the DSM volume signal is the typical one (see Fig. 7.10) considered for the composition of active demand services [12], and is characterized by a service time interval (i.e. the time interval during which the active demand service—i.e. the DSM volume reduction—is performed), and a "payback" zone (in the opposite direction) regulating the power profile in the immediate aftermath of the DSM service, when the impact of the payback effect is greater [12].

Making reference to Fig. 7.10, the simulated DSM signal is characterized by the following technical specifications: $V_{ser} = 8$ kW, $V_{pb} = 3$ kW, $T_{ser} + T_{pb} = 1$ h. Figure 7.11 reports the relevant LA load profiles immediately before the DSM signal is notified. From the figure it can be noticed how the LAC control is such that the aggregated load profile accurately tracks the reference only close to time I, while, ahead of time I the charging power is shifted so as to optimize costs and exploit available RES.

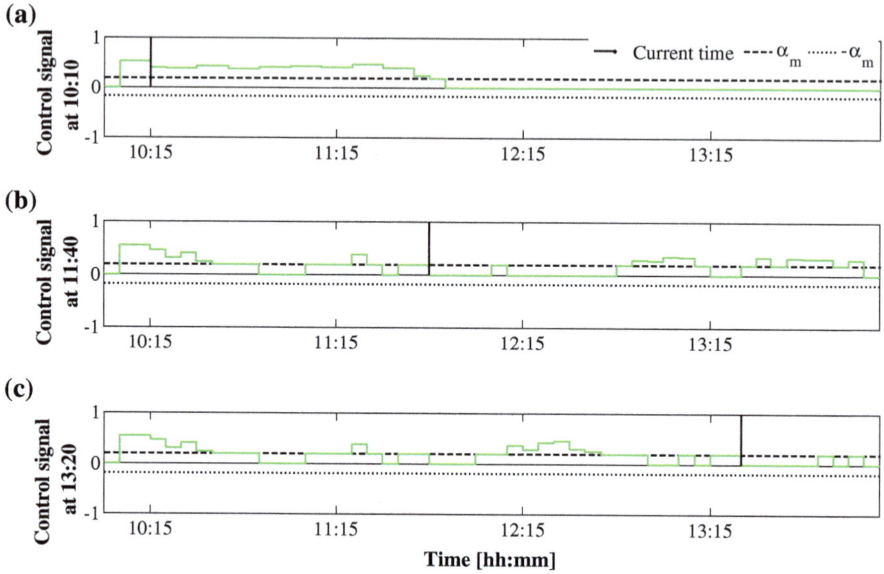

Fig. 7.8 Evolution of the control signal U_{mk} at three different LAC optimisation times

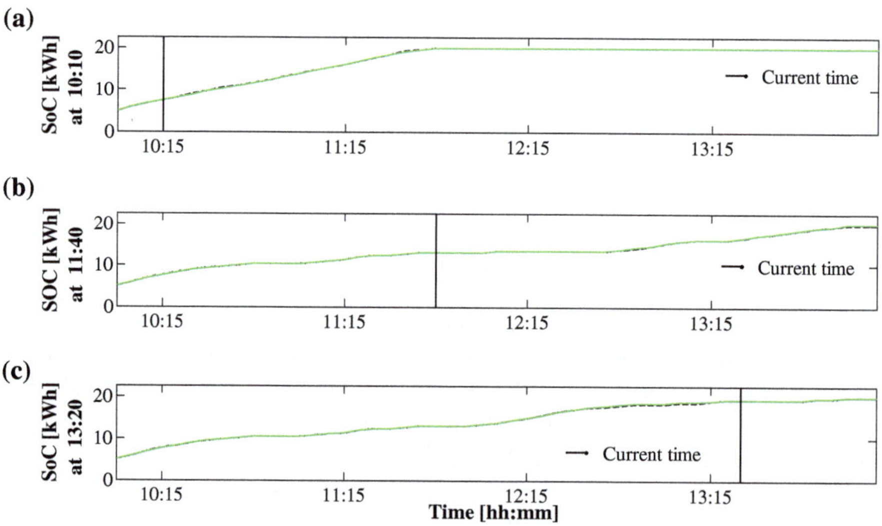

Fig. 7.9 Controlled evolution of the SoC of PEV#1 at three different time instants

Then, Fig. 7.12 reports the load profiles shortly after the notification of the DSM signal (at 10:40 [hh:mm]). It is seen how the LAC is able to properly reacting to the volume signal by dynamically rescheduling the ongoing charging sessions.

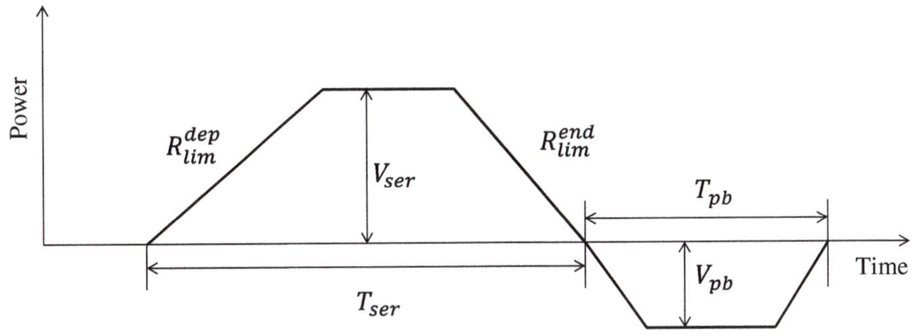

Fig. 7.10 DSM signal shaped as a typical active demand product (positive values are power reductions)

Finally, Fig. 7.13 reports the final evolution of the relevant LA load profiles. The DSM volume signal is met almost perfectly. The power threshold is respected and all the charging requests are fulfilled according to the user preferences. This simulation is relevant since it shows the potential of controlling via the LAC the aggregated load profile at LA level. As a matter of fact, control of demand will be more and more a tool by which active demand services will be provided to interested grid actors, especially for balancing purposes. As an example, the proper coordination and composition of a number of LA reprofiling actions (as the one reported in Fig. 7.13), by a higher level coordinating entity, could allow to compose and trade to interested upper level grid actors (i.e. a retailer, a DSO, a large DER operator, etc.) large volume active demand products, which could be employed by

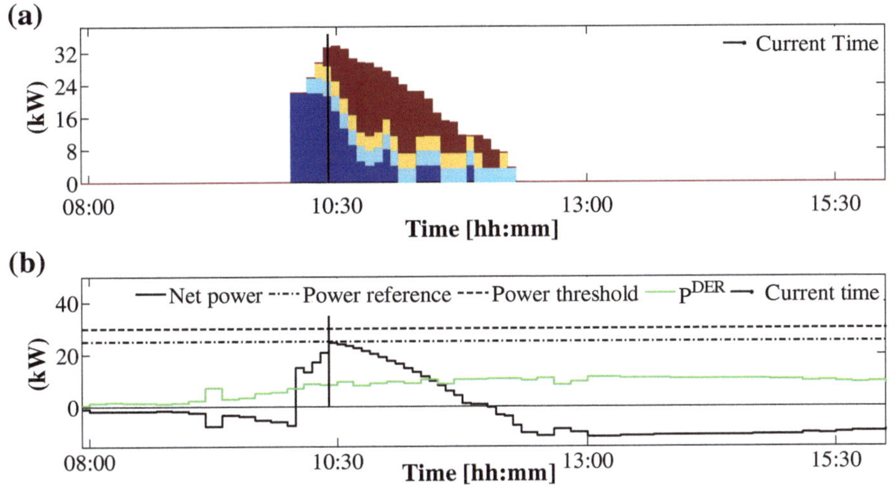

Fig. 7.11 Relevant controlled load profiles at 10:20 [hh:mm], immediately before the DSM signal notification (simulation performed with $\mu = 10$)

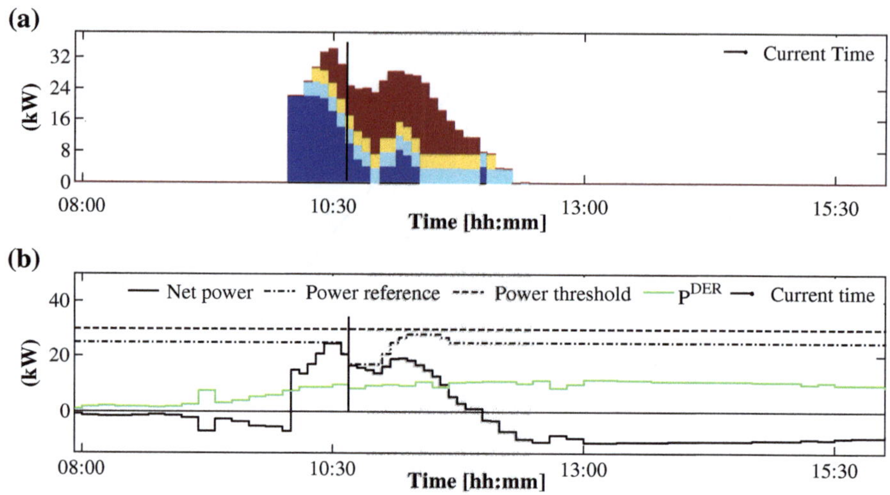

Fig. 7.12 Relevant controlled load profiles at 10:40 [hh:mm], shortly after the DSM signal notification

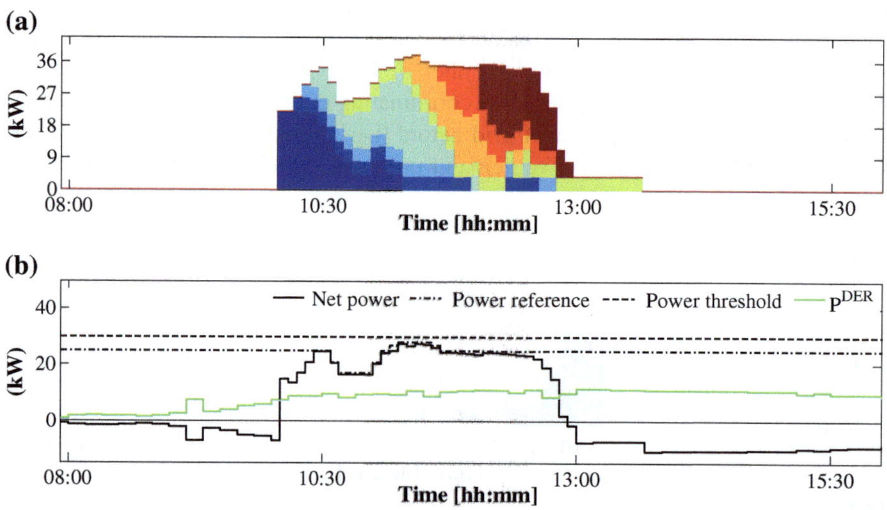

Fig. 7.13 Final relevant controlled load profiles

such actors to, e.g., remedy to short term grid imbalances (in case of the DSO) or to large variation in the RES output schedule, thus avoiding to incur in penalties.

Significantly, the capability of the LAC of dynamically rescheduling the charging sessions can be also effectively exploited to mitigate the effects of RES intermittency, guaranteeing a flattered net profile at LA level. That can be achieved by simply updating control each time a new RES forecast is available and communicated to the LAC. The maximization of RES self-consumption (for which the term J^{DER} in (7.1) is responsible), and the mitigation of the effect of RES intermittency (achieved also via the term J^{reg}) are two key factors for increasing RES integration into the distribution grid, and they are fully taken into account by the presented LAC control approach to smart charging.

7.9 Conclusions

In this chapter a MPC approach for the management of PEVs charging in distribution grids has been presented. The work has moved from an in depth analysis of the state of the art of smart charging control. Then the relevant actors and components making part of the PEV charging scenario have been identified, and their role in the provisioning of the smart charging service has been discussed. Four different use cases related to smart charging have been then introduced, being them the most relevant ones that the proposed controller (named Load Area Controller—LAC), aims to support. They are: (i) LAC reaction to a charging request, (ii) LAC reaction to an update of the user preferences, (iii) LAC reaction to a DER forecast update and, (iv) LAC reaction to a demand side management signal. All these use cases rely on the solution of the smart charging control problem addressed in this chapter (i.e. on the smart charging control functionality provided by the LAC). Based on the analysis of the charging scenario and the review of the relevant use cases, requirements and specifications for the presented LAC charging controller have been given. The LAC working logic has been then discussed, giving a motivation for the adoption of the presented MPC-based charging control strategy (basically, the need of optimizing a set of key performance indicators, the presence of constraints, the need of reacting to asynchronous events from the environment and, finally, the availability of a simple control model of the plant). The resulting MPC formulation has been then presented and discussed. In particular, the proposed control framework has been designed with the aim of: (i) optimizing charging costs, thus seeking a saving for the PEV users in a scenario characterized by time variant tariffs, (ii) seeking integration of local RES, via their balancing with the controlled charging demand and, (iii) providing a load management service to the DSO, consisting in the tracking of a DSO-defined reference power curve at load area level. Proper constraints have been introduced to make the control action compliant with the technical limitations imposed by the relevant standards (i.e. IEC 61851) and with the technical and economic requirements posed by the PEV users, the DSO and the DER operator. The natural formulation of the open-loop optimal control problem at the basis of the MPC approach has been then handled in order to achieve an equivalent mixed integer quadratic programming problem, which can be solved in near real-time to provide the EVSEs with the charging load profiles and the driver with the notification of the expected cost for charging.

Explicative simulations have been presented to show how effectively the charging requests are managed and fulfilled by the LAC, according to the three aforementioned objectives (i.e. cost minimization, RES integration and load tracking). It has been shown how the LAC can effectively update time after time the charging controls in order to react to asynchronous events coming from the environment, being them new charging requests, updates of the user preferences, notifications of DSM signals, updates of the DER forecast, etc. Such a capability, here relevant for an effective matching between PEVs charging load and RES in respect of DSO and drivers' needs, proves fundamental in all the active demand and demand side management applications so crucial in the smart grid concept.

Acknowledgments This work was supported in part by the European Union "SMARTV2G" and "MOBINCITY" projects, under grant agreements no. 284953 and no. 314328, respectively. The authors gratefully acknowledge F. Caleno, G. Coppola and T. Casacchia, from Enel Distribuzione, Rome, Italy, and F. Delli Priscoli, L. Zuccaro and A. Mercurio, from the Sapienza University of Rome, Rome, Italy, for the fruitful discussions on the reference scenario considered in the chapter.

References

1. Pieltain Fernandez L, Roman TGS, Cossent R, Domingo CM, Frias P (2011) Assessment of the impact of plug-in electric vehicles on distribution networks. IEEE Trans Power Syst 26:206–213. doi:10.1109/TPWRS.2010.2049133
2. Clement-Nyns K, Haesen E, Driesen J (2010) The impact of charging plug-in hybrid electric vehicles on a residential distribution grid. IEEE Trans Power Syst 25:371–380. doi:10.1109/TPWRS.2009.2036481
3. Putrus GA, Suwanapingkarl P, Johnston D, Bentley EC, Narayana M (2009) Impact of electric vehicles on power distribution networks. Paper presented at the 5th IEEE vehicle power and propulsion conference, 7–10 Sept 2009. doi: 10.1109/VPPC.2009.5289760
4. Richardson P, Flynn D, Keane A (2010) Impact assessment of varying penetrations of electric vehicles on low voltage distribution systems. Paper presented at the power and energy society general meeting, Detroit, 24–28 July 2011. doi: 10.1109/PES.2010.5589940
5. Kempton W, Steven L (1997) Electric vehicles as a new power source for electric utilities. Trans Res Part D: Transp Environ 2:157–175. doi:10.1016/S1361-9209(97)00001-1
6. Liu R, Dow L, Liu E (2011) A survey of PEV impacts on electric utilities. Paper presented at the 2nd IEEE PES international conference and exhibition on "innovative smart grid technologies" (ISGT Europe), Manchester, 5–7 Dec 2011
7. Falvo, MC, Foiadelli F (2010) Preliminary analysis for the design of an energy-efficient and environmental sustainable integrated mobility system. Paper presented at the power and energy society general meeting, Minneapolis, 25–29 July 2010. doi: 10.1109/PES.2010.5589545
8. EU Commission (2007) Communication of European commission to European parliament: an energy policy for Europe. Available via EUROPA. http://europa.eu/legislation_summaries/energy/european_energy_policy/l27067_en.htm. Accessed 15 May 2014
9. Deutsche E, GmbH (2006) Planning of the grid integration of wind energy in Germany onshore and offshore up to the year 2020. http://docs.wind-watch.org/dena-integratingwind2020.pdf. Accessed 15 May 2014
10. Kempton W, Tomic J (2005) Vehicle-to-grid power implementation: from stabilizing the grid to supporting large-scale renewable energy. J Power Sources 144:280–294. doi:10.1016/j.jpowsour.2004.12.022

11. Dallinger D, Wietschel M (2011) Grid integration of intermittent renewable energy sources using price-responsive plug-in electric vehicles. Renew Sustain Energy Rev 16:3370–3382. doi:10.1016/j.rser.2012.02.019
12. ADDRESS Consortium (2011) Deliverable 3.1: Prototypes and algorithms for network management, providing the signals sent by the DSO to aggregators and the markets, enabling and exploiting active demand. http://www.addressfp7.org/config/files/ADD-WP3-D3%201-ENELDISTR-ActiveDistributionGrid.pdf. Accessed 15 May 2014
13. Deilami S, Masoum A, Moses P, Masoum M (2011) Real-time coordination of plug-in electric vehicle charging in smart grids to minimize power losses and improve voltage profile. IEEE Trans Smart Grid 2:456–467. doi:10.1109/TSG.2011.2159816
14. Tremblay O, Dessaint LA, Dekkiche AI (2007) A generic battery model for the dynamic simulation of hybrid electric vehicles. Paper presented at the 3rd vehicle power and propulsion conference, 9–12 Sept 2007. doi: 10.1109/VPPC.2007.4544139
15. SmartV2G Consortium (2014) The FP7 SmartV2G project website. http://www.smartv2g.eu. Accessed 15 July 2014
16. MOBINCITY consortium (2014) The FP7 MOBINCITY project website. http://www.mobincity.eu. Accessed 15 July 2014
17. Rahimi F, Ipakchi A (2010) Demand response as a market resource under the smart grid paradigm. IEEE Trans Smart Grid 1:82–88. doi:10.1109/TSG.2010.2045906
18. Ashok S (2006) Peak-load management in steel plants. Appl Energy 83:413–424. doi:10.1016/j.apenergy.2005.05.002
19. Ashok S, Banerjee R (2000) Load-management applications for the industrial sector. Appl Energy 66:105–111. doi:10.1016/S0306-2619(99)00125-7
20. Middelberg A, Zhang J, Xia X (2009) An optimal control model for load shifting—with application in the energy management of a colliery. Appl Energy 86:1266–1273. doi:10.1016/j.apenergy.2008.09.011
21. Newsham GR, Bowker BG (2010) The effect of utility time-varying pricing and load control strategies on residential summer peak electricity use: a review. Energy Policy 38:3289–3296. doi:10.1016/j.enpol.2010.01.027
22. Di Giorgio A, Pimpinella L (2012) An event driven smart home controller enabling consumer economic saving and automated demand side management. Appl Energy 96:92–103. doi:10.1016/j.apenergy.2012.02.024
23. Matallanas E, Castillo-Cagigal M, Gutirrez A, Monasterio-Huelin F, Caamao-Martn E, Masa D, Jiménez-Leube J (2012) Neural network controller for active demand-side management with PV energy in the residential sector. Appl Energy 91:90–97. doi:10.1016/j.apenergy.2011.09.004
24. Richardson P, Flynn D, Keane A (2012) Optimal charging of electric vehicles in low-voltage distribution systems. IEEE Trans Power Syst 27:268–279. doi:10.1109/TPWRS.2011.2158247
25. Studli S, Crisostomi E, Middleton R, Shorten R (2012) A flexible distributed framework for realising electric and plug-in hybrid vehicle charging policies. Int J Control 85:1130–1145. doi:10.1080/00207179.2012.679970
26. Bashash S, Fathy H (2011) Robust demand-side plug-in electric vehicle load control for renewable energy management. Paper presented at the American control conference, San Francisco, 29 June–1 July 2011. doi: 10.1109/ACC.2011.5990856
27. Fan Z (2012) A distributed demand response algorithm and its application to PHEV charging in smart grids. IEEE Trans Smart Grid 3:1280–1290. doi:10.1109/TSG.2012.2185075
28. Green eMotion Consortium (2014) The FP7 Green eMotion project website. http://www.greenemotion-project.eu. Accessed 15 July 2014
29. Mercurio A, Di Giorgio A, Purificato F (2013) Optimal fully electric vehicle load balancing with an ADMM algorithm in Smartgrids. Paper presented at the 21st Mediterranean conference on control and automation, Chania, 25–28 June 2013. doi: 10.1109/MED.2013.6608708
30. Ehsani M, Gao Y, Emadi A (2005) Modern electric, hybrid electric, and fuel cell vehicles: Fundamentals, theory, and design. Taylor and Francis Group, New York

31. ISO/TC (2011) 23H—plugs, socket-outlets and couplers for industrial and similar applications, and for electric vehicles. IEC 62196-1 ed2.0 plugs, socket-outlets, vehicle connectors and vehicle inlets—conductive charging of electric vehicles—Part 1: general requirements
32. Bliek C, Bonami P (2013) IBM CPLEX global non-convex MIQP. http://minlp.cheme.cmu.edu/2014/papers/bliek.ppt. Accessed 15 July 2014
33. Burer S, Letchford AN (2012) Non-convex mixed-integer nonlinear programming: a survey. Surv Oper Res Manage Sci 17:97–106. doi:10.1016/j.sorms.2012.08.001
34. Misener R, Christodoulos AF (2013) GloMIQO: global mixed-integer quadratic optimizer. J Global Optim 57:3–50. doi:10.1007/s10898-012-9874-7
35. Lazimy R (1982) Mixed-integer quadratic programming. Math Program 22:332–349. doi:10.1007/BF01581047
36. Papageorgiou LG, Fraga ES (2007) A mixed integer quadratic programming formulation for the economic dispatch of generators with prohibited operating zones. Electr Power Syst Res 77:1292–1296. doi:10.1016/j.epsr.2006.09.020
37. Boston Power (2013) Boston-power data sheet: swing key 435 lithium-ion block. http://www.boston-power.com. Accessed 15 May 2014
38. PVOutput (2014) Measured PV outputs. http://www.pvoutput.org. Accessed 15 May 2014

Chapter 8
QoS Schemes for Charging Plug-in Electric Vehicles in a Smart Grid Environment

Irfan S. Al-Anbagi and Hussein T. Mouftah

Abstract Plug-in Electric Vehicles (PEVs) are expected to greatly reduce the carbon emissions from surface transport if they are widely used and efficiently charged. One of the main limitations of PEVs is their limited range and relatively long recharging times. This limitation is closely associated with the current battery technologies used in the PEVs. In order efficiently utilize the PEVs, their charging schedules and locations must be effectively integrated within the smart grid. Real-time and reliable integration of PEVs with the smart grid could solve problems related to demand response, cost and time of charging. In this chapter, we survey the state-of-the-art in wireless communication systems for PEVs integration with smart grid, different control and wireless communication strategies. We highlight the main challenges associated with the PEV-smart grid communication system. We then propose a QoS scheme for charging PEVs (QCEV) in a smart grid environment and propose a Channel Access Control (CAC) scheme that provides QoS differentiation to PEVs that are transmitting delay critical information. Unlike conventional contention based distributed QoS approaches used by the IEEE 802.11p MAC protocol, both of the QCEV and the CAC schemes provide centralized QoS differentiation in situations where immediate PEV battery charging is required. The centralization is done at the Access Point (AP) which takes an informed decision on which PEV should receive highest priority to access the channel based on the individual PEV battery levels, and also based on the availability and cost of the electricity at different locations.

I.S. Al-Anbagi · H.T. Mouftah (✉)
School of Electrical Engineering and Computer Science, University of Ottawa,
800 King Edward Ave., Ottawa, ON K1N 6N5, Canada
e-mail: mouftah@uottawa.ca

I.S. Al-Anbagi
e-mail: ialanbag@uottawa.ca

S. Rajakaruna et al. (eds.), *Plug In Electric Vehicles in Smart Grids*,
Power Systems, DOI 10.1007/978-981-287-317-0_8

8.1 Introduction

Plug-in Electric Vehicles (PEVs) and Plug-in Hybrid Electric Vehicles (PHEVs) are considered as "green" as the electricity used to charge their batteries [1]. Unlike PHEV, Pure Battery Driven PEVs (we refer to them as PEVs) depend solely on the available battery charge. Different PEV models have different features and options; they vary mostly on the range (i.e. km/charge) and the engine power in addition to other comfort features. Most compact to mid-size PEVs available in the market today can have an average of up to 160 km/charge [2]. However, luxury PEVs can provide double that range. Hence, among other factors we notice that there is a trade-off between battery cost and additional range, which explains the high cost of some PEVs that provide extended range. Based on the above, the availability of charging stations, cost and time of charging the PEVs are considered some of the main reasons why PEVs are not widespread in today's market.

Security issues, the availability of charging stations and the power required to charge multiple PEVs at the same time is a major concern for electrical utility operators and PEV owners. For example, in today's power grid, the cost of electricity is dependent on the time of the day and the instantaneous load on the grid. This is obviously done to allow the "smart grid" to manage the load on a micro-level. This is basically achieved by enabling communication between PEVs and electricity suppliers so that load and generation can be scheduled and distributed in an optimum manner. Efficient and reliable integration and communication will mitigate load issues resulting from recharging batteries, where the smart grid would match generation to electricity use and manage loading on different PEVs charging infrastructures. In doing so, the smart grid becomes an optimum solution to multiple problems which enables dynamic real-time integration of information exchange between the electricity network and the electrical transportation system [1].

Communication systems are considered as the backbone of any smart grid system. In addition to that, efficient communication and data management between the smart grid and PEVs to optimize charging cost and durations calls for an efficient, reliable and low communication latency integration between these two sub-systems. One of the main determining factors for successful integration of these systems depends on the ability of the communication system to provide Quality of Service (QoS) guarantees. QoS in a wireless communication environment can have several meanings. For example, QoS can indicate the capability to provide assurance that the service requirements of a specific application can be met. QoS can also be defined as the ability of the network to adapt to specific classes of data. QoS is a challenging issue in wireless networks in general and in mobile networks in specific due to the highly dynamic nature and dense communication environment between PEVs and Access Points (APs) (i.e. the road side units).

Wireless vehicular communication can be achieved in several different technologies such as IEEE 802.11 (WiFi), IEEE 802.16 (WiMAX), 3G cellular, and satellite technologies. PEV-smart grid communication can be supported through Vehicle-to-Infrastructure (V2I) and Vehicle-to-Vehicle (V2V) communications.

V2I communications involve vehicular nodes and APs. In these scenarios, there are several technologies that can support this communication, such as IEEE 802.11 (WiFi), IEEE 802.16 (WiMAX), and Dedicated Short Range Communications (DSRC). V2V communication scenarios include PEVs on a road which forms a Vehicular Ad hoc Network (VANET). V2V communication is basically used for transmitting safety information, warning messages and traffic information system. In addition to that, V2V communication can also be used in PEVs scenarios such as relaying battery charge information in multi hop communication network.

Among the many available wireless communication standards in vehicular environments, the IEEE 802.11p standard is the most popular standard for such environments. The IEEE 802.11p standard is an amendment to the IEEE 802.11 standard which is proposed for Wireless Access in Vehicular Environments (WAVE) [3]. The standard is basically designed to enable communication in mobile environments e.g. V2V and V2I communication. The IEEE 802.11p physical layer is similar to that of the IEEE 802.11a standard. However, to cope with the highly mobile environment, the bandwidth of the IEEE 802.11p is reduced to the half. Its Medium Access Control (MAC) protocol is inspired from the MAC of the IEEE 802.11e standard [4] with some modifications to make it more suitable for mobility (e.g. the use of 4 ACs instead of 8). Similar to the IEEE 802.11e, the IEEE 802.11p uses the Enhanced Distributed Channel Access (EDCA) mechanism. The EDCA mechanism allows the high-priority traffic to have a higher chance of acquiring the channel access and being transmitted compared to the low-priority traffic. However, the QoS differentiation used in both standards is based on the traffic type which is predefined by the application layer.

The integration of PEVs and the smart gird system in applications such as PEV charging optimization requires real-time collaboration between the AP and the PEVs to make proper QoS differentiation decisions. Conventional QoS approaches used in the IEEE 802.11p standard may not be efficient for such applications, since an uninformed QoS differentiation decision by one PEV may lead to affecting the entire energy optimization process or even affecting the overall network performance especially in dense networks.

In this chapter, we survey the state-of-the-art in wireless communication systems for PEVs integration with smart grid, different control and wireless communication strategies. We highlight the main challenges associated with the PEV-smart grid communication system. We then propose a QoS scheme for charging PEVs (QCEV) in a smart grid environment and propose a Channel Access Control (CAC) scheme that provides QoS differentiation to PEVs that are transmitting delay critical information. Unlike conventional contention based distributed QoS approaches used by the IEEE 802.11p MAC protocol, the QCEV and the CAC schemes provide centralized QoS differentiation in situations where immediate PEV battery charging is required. The centralization is done at the AP which takes an informed decision on which PEV should receive highest priority to access the channel based on the individual PEV battery levels, and also based on the availability and cost of the electricity at different locations.

8.2 Background

In the past few years, the integration of PEVs with the smart grid has been the focus of several research papers and technical reports. PEVs charging infrastructure in a smart grid environment has been extensively discussed in [5]. Furthermore, the impact of the number of the connected EVs on the stability of the electrical system has been presented in [6]. In addition to that, automated energy management between EVs and the smart grid has been studied, where multiple objective functions have been formulated to maximize customer benefits and maintain acceptable power grid operation levels [7].

In the literature, there has been many publications that review the electrification of the transportation system, vehicle area networks and the applications of the IEEE 802.11p standard.

In [8], the authors have presented an overview of the electrification of transportation system where they have considered different aspects of the integration of the EVs and PHEVs with the smart grid including some communication requirements. However, the authors have not considered the possibility of using either the IEEE 802.16 and the IEEE 802.11p protocols as possible communication protocols for this integration. The consideration of the IEEE 802.11p protocol for the integration of the ITS with the smart grid is of considerable importance since the IEEE 802.11p protocol is specifically designed to provide wireless access in vehicular environments. Furthermore, one of the main features of the protocol is that it can provide QoS which is considered as one of the main requirements identified in [9].

Faezipour et al. [10], have surveyed the main innovations in vehicle area networks featuring driver safety. They have focused on recent developments of intelligent transportation systems for intra-vehicle and inter-vehicle-area-networks to assist driver safety. Furthermore, Huang et al. [11], have discussed issues related to the challenging smart grid deployment environments and existing solutions for such environments. They have also proposed to deploy cognitive radios for smart grid communication infrastructures in challenging environments.

Msadaa et al. [12], have compare IEEE 802.16e standard and the IEEE 802.11p standard. They have investigated the potential and limitations of both technologies as a communication media for V2I communications. The authors have evaluated the performance of the two systems for different vehicle speeds, traffic data rates, and network deployments.

In this section, we discuss the state of the art in wireless communication architecture for the communication of EVs and the smart grid and the communication architecture in ITS in general. We then review the work that discuss medium access control techniques in ITS environments. We finally discuss the state of the art in QoS mechanisms in ITS scenarios and in EV-smart grid communication.

8.2.1 PEV-Smart Grid Wireless Communication Architecture

The requirements of the data communication systems used in smart grid has been widely discussed in the literature. A comprehensive overview of smart grid communication needs has been highlighted in [9], the report has discussed the challenges and opportunities in implementing EVs in the smart grid. The use of real-time wireless networks for the automation of the smart grid has been widely discussed in the literature [13, 14]. In addition to that, Zafalon et al. [15], have discussed how electric mobility influences the development of PEV in automotive industry by integrating the PEVs into the Internet of Energy (IoE) and Smart grid infrastructure by providing novel business models and requiring new semiconductor devices and modules.

A number of publications have presented wireless communication scenarios for managing the charging and discharging of EVs in smart grid scenarios. Conti et al. [16], have presented an architecture using Bluetooth for the wireless communication between driver's smart phone, vehicle and charging infrastructure. Ferreira et al. [17], have presented a mobile information system. Their application gives relevant information to Full Electric Vehicle (FEV) drivers, by supporting the integration of several sources of data in a mobile application. Their application provides recommendations to the drivers about the FEV range autonomy, location of battery charging stations, information of the electricity market, and also a route planner taking into account public transportations and car or bike sharing systems. Jansen et al. [18], have outlined an architecture of EV based vehicle-to-grid (V2G) integrating virtual power plant. They have provided the overall system architecture, a sketch of the trip-prediction algorithm, and the associated optimization problem. The authors have derived the communication requirements for their architecture and considered reliability, responsiveness, security, and application-level behaviour. Cespedes et al. [19], have specified block components, protocols and technologies required for IPv6 support in V2G communications. The authors have evaluated the performance of their proposed the framework for different V2G applications data rates and channel conditions.

The use of wireless communication to schedule the charging of EVs in smart grid architecture has been considered in [20] and [21]. Dong et al. [20], have adopted a battery replacement strategy for charging process in a charging station. The authors have developed the scheduling scheme for PHEV charging. An optimization problem is formulated and solved for an optimal charging policy that minimizes the cost due to power consumption and performance degradation. They have shown that if multiple data communication channels are available for transferring the power price information, the scheduling performance can be improved significantly. Yuan et al. [21], have proposed a coordinating control charging management system for EVs charging station consisting of single-node charging controllers and a centralized management system. Their system obtains charging data and transmits control instructions via ZigBee and GPRS wireless communication networks technology. The authors have included a model of battery

State-Of-Charge (SOC) for calculating the charging current that each EV needs and the exchange of other data through CAN-Bus.

On the other hand the issue of demand response and load management using wireless communication between EVs and the smart grid has been considered in [22, 23, 24]. Erol-Kantarci et al. [22], have proposed a communication-based PHEV load management scheme to control the load of the PHEVs. They have utilized provisioning a certain amount of energy for each distribution system based on the predicted supply level. The provisioned energy is communicated to the substation control center where each charging request is either accepted or rejected based on the utility set limits. The information is sent to the smart charging stations through a Wireless Mesh Network (WMN) that uses IEEE 802.11s. Xu et al. [23], have proposed an integrated vehicle-to-grid, grid-to-vehicle, and renewable energy sources coordination algorithm. They have implemented their algorithm over a broadband fiber-wireless communications infrastructure by co-simulating both power and communications perspectives. The authors have shown that their scheme could reduce the peak demand and enhance the performance in flattening the overall demand profile and maintaining network constraints. In addition to that, their proposed communication system shows an improvement to the throughput and the end-to-end latency. Kovacs et al. [24], have presented results from the PowerUp project, where they have specified and developed an end-to-end EV to Grid communication system, respecting the applicable communications standards. They have presented the system architecture for the end-to-end integration of the V2G communications interface and described critical Smart-Grid integration aspects for each protocol layer within the V2G communications protocol stack. The authors have also presented prototype test observations. They have shown that their ISO/IEC 15118-based solution would not interfere with the stability of the electric grid.

The use of wireless communication to control the cloud infrastructure and resources for managing the EV has been also considered in the literature. Yu et al. [25], have proposed a hierarchical cloud architecture for vehicular networks. Figure 8.1 shows their proposed cloud-based vehicular network architecture. The authors have created a pervasive cloud environment for mobile vehicles by integrating redundant physical resources in ITS infrastructures, including data centers,

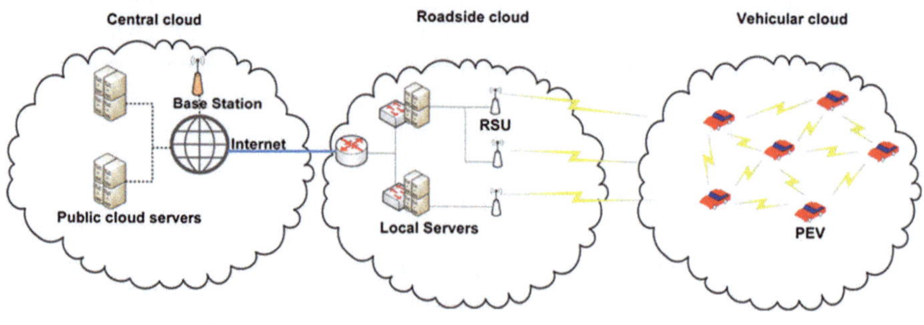

Fig. 8.1 The proposed cloud-based vehicular network architecture [25]

roadside units, and vehicles. The authors have aggregated these resources to compose cloud resources for vehicles. They have also proposed a three-layered architecture to organize the cloud resources. Their layered structure allows vehicles to select their cloud services resiliently. Their central clouds have sufficient cloud resources but large end-to-end communications delay and roadside and vehicular clouds have limited cloud resources but satisfy communications quality.

8.2.2 Medium Access Techniques

There are not many publications that discuss medium access techniques for the IEEE 802.11p standard in PEV-smart grid communication scenarios. Although, there are several studies that discuss new and improved MAC protocols for IEEE 802.11p-based communications. Cheung et al. [26], have studied the random access in a in situations where roadside APs are installed on a highway to provide temporary Internet access for vehicles. Figure 8.2 shows their drive-thru vehicle-to-roadside (V2R) communications with multiple APs. The authors have considered the problem of finding the optimal transmission policy with a single AP and random vehicular traffic arrivals. They have formulated the problem as a finite-horizon sequential decision problem, solved it using dynamic programming and designed a general dynamic optimal random access (DORA) algorithm. The authors have derived the conditions under which the optimal transmission policy has a threshold

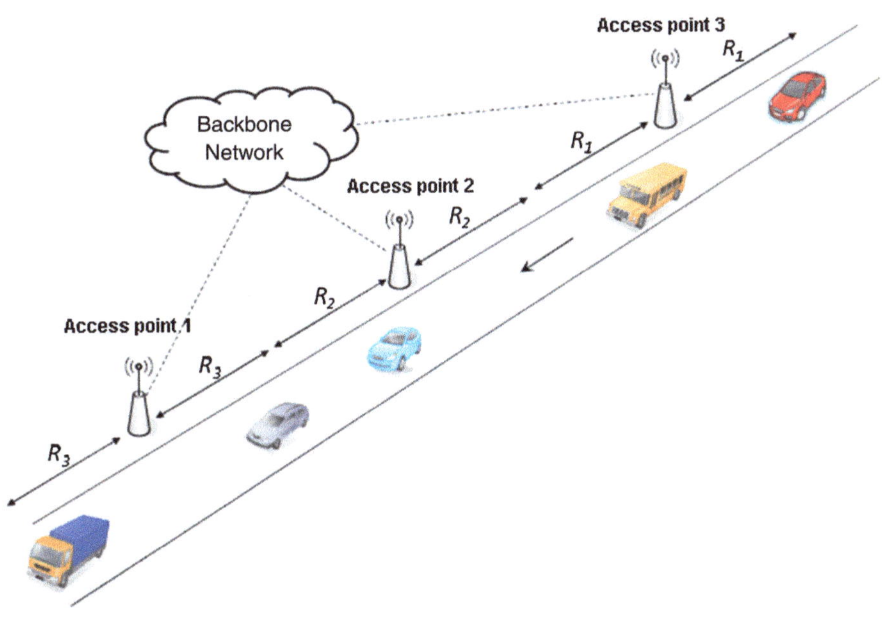

Fig. 8.2 The drive-thru vehicle-to-roadside (V2R) communications with multiple APs [26]

structure, and propose a monotone DORA algorithm with a lower computational complexity for this special case. They have also considered the problem of finding the optimal transmission policy with multiple APs and deterministic vehicular traffic arrivals.

On the other hand, Rezgui et al. [27], have proposed a Reliable Broadcast for EV Charging Assignment (REBECA) scheme. They have considered the time a transmitting vehicle on the road requests to be served, up until the time the service has been completed and the efficiency of the required service. They have shown that their scheme could determine how many EVs can be efficiently served by a number of electrical vehicle supply equipment without increasing the probability of overload on supply equipment or latency time on EVs. The authors have also we propose a random access allocation algorithm for a feasible/initial charging process solution of the proposed model. Their algorithm chooses a supply equipment location randomly among the set of available equipment. They have proposed a best access allocation algorithm, in which the EV selects the supply equipment with the smallest free slots which are able to contain the EV demand. Finally they have proposed a power balancing access allocation algorithm which takes into consideration the power balancing between supply equipment to keep a minimal variance of electricity usage between them while providing a short latency time for EVs and then guarantees service efficiency.

8.2.3 QoS in IEEE 802.11p Communication

The consideration of the IEEE 802.11p protocol in critical applications has been extensively discussed in the literature [12, 28, 29]. Furthermore, several studies have considered the use of analytical models to evaluate the performance of IEEE 802.11p-based networks [30, 31] and the use of simulation model have also been considered in [32, 33]. The use of a centralized provisioning mechanism to provide QoS differentiation to EVs communicating to an AP in a smart grid environment has not been considered in the literature. Furthermore, the concept of enforcing real-time communication between EVs and the smart grid to optimize the charging schedule and location has not addressed in such environments.

Erol-Kantarci et al. [34], have proposed a QoS-aware admission control scheme for the PHEV charging infrastructure. Their scheme operates on the energy management system of the smart grid distribution system and relies on a wireless communication network that delivers the demands of PHEVs to the energy management system and delivers the admission decisions of energy management system to PHEVs. Figure 8.3 shows there wireless packet format for both the charge request packet and the charge serve packet. They have proposed that PHEV users pay more to charge faster than the best-effort users similar to the internet traffic service differentiation mechanisms. The authors have provided mathematical analysis and simulation results for the proposed scheme.

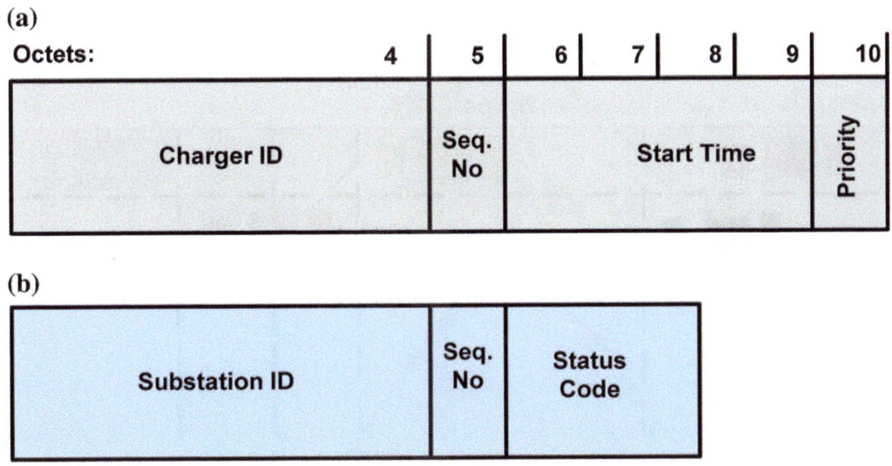

Fig. 8.3 The proposed packet formats; **a** CHARGE-REQ packet, **b** SERVICE-STAT packet [34]

Bohm et al. [28], have propose a deterministic MAC scheme for V2I communication by extending the IEEE 802.11p standard with a collision-free communication phase controlled by an AP. Their scheme schedules safety-critical, real-time data traffic in a collision-free manner by the AP. The remaining bandwidth is available to other services according to the contention-based random access scheme defined in IEEE 802.11p. The authors have used real-time scheduling analysis to adapt the bandwidth dedicated to safety critical real-time traffic to the current number of communicating vehicles and their communication needs, while maximizing the possible amount of best effort traffic in the network.

Herrera et al. [35], have introduced a platform for real-time simulation for PHEV charging stations. Their system can simulate in real-time key elements of a smart grid such as: high speed power electronics, distributed energy resources, and communication networks. The authors have presented and integrated a description of the platform for real time simulation and communication emulation. They have also presented an introduction to networked control systems and a case study of PHEV charging stations which displays the latest results accomplished with the current setup.

Analytical studies of real-time communication in WAVE scenarios has been considered in [36, 37, 38]. Li et al. [36], have presented an analytical model that evaluates the end-to-end delay against the transmission range in 802.11p-based VANETs. They have analyzed the transmission delay in saturated and non-saturated VANETs and derived the queuing delay by modeling a vehicle as a M/M/1/N model. Ran et al. [37], have studied the time requirements for last-mile communication in an low voltage smart grid by considering a futuristic neighborhood scenarios with large penetration of distributed energy sources and EVs. Luan et al. [38], have investigated the provisioning QoS ensured multimedia applications to in-motion vehicles. Figure 8.4 shows their proposed Vehicle to infrastructure

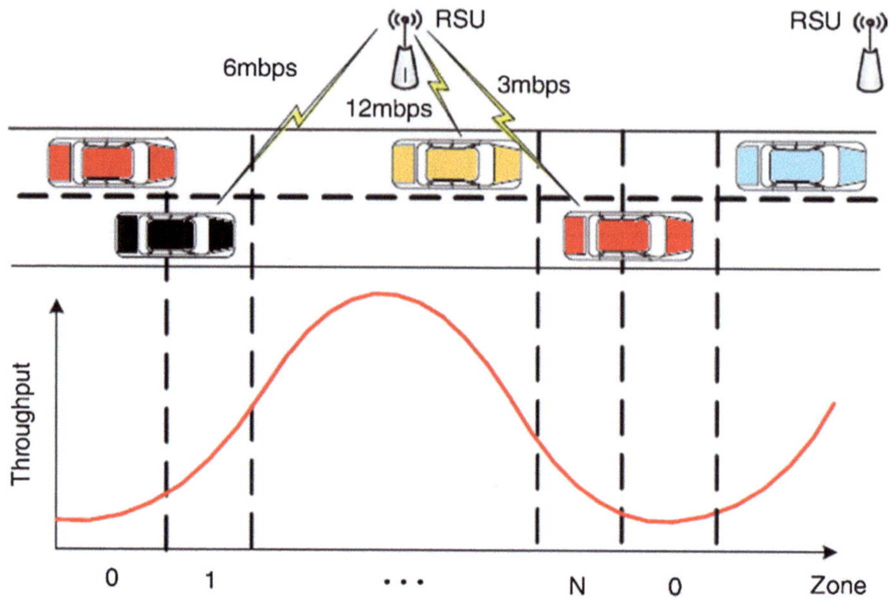

Fig. 8.4 The proposed vehicle to infrastructure road-side unit communication [38]

road-side unit communication. The authors have established a mathematical model to evaluate the performance of EDCA, the fundamental MAC scheme of 802.11p. Their model have considered the node mobility, represented by the velocity, in the modeling of MAC and unveils the impacts of mobilities on the resultant QoS performance provisioned to vehicles. The authors have reinforced the QoS provision by adjusting the QoS parameters in EDCA in tune with the mobility of vehicles.

8.3 Challenges of Implementing Wireless Networks in V2G Applications

Reliable and real-time two-way communication in V2G scenarios is one of the major issues that need be addressed to have real-world implementation of an efficient PEV-smart grid system. Due to the nature of the V2G environment and the challenges for system deployment, a communication protocol needs to consider several issues. In addition to that, in V2G communication, QoS is important to guarantee the normal operations of the smart grid. These challenges can be summarized as follows.

8.3.1 Mobile Environment

Obtaining information about real-time power price and peak hours in addition to en-route charging station reservation in highly mobile environment is one of the main challenges for these applications. Furthermore, the application of PEV-smart grid collaboration introduces a new class of applications. For example, PEVs can be used regulate the frequency and smooth the power flow of the smart grid. Therefore, in such applications a PEV needs to initiate and terminate the communication with an AP while the PEV is still within its range, especially if an PEV is traveling at an average speed of 80 km/h. Therefore, the integration of PEVs into the smart grid to satisfy such application requirement is not an easy task.

The IEEE 802.11p standard is required to handle not only communication at high mobility but also in real-time fashion. Thus, the IEEE 802.11p standard must handle the two-way communications before the communicating PEVs leave the coverage area of their AP. In addition to that, mobile PEV environment makes reliable transmission in smart grid more challenging.

The charging information of PEVs is exchanged with the AP through wireless transmission to enable real-time decision making and to improve the PEV charging efficiency. However, in a mobile environment, the charging of a PEV is becoming more demanding and challenging due to the mobility, changing power levels and limited number of APs. Furthermore, in a smart grid environment, QoS communication is essential for the entire system to adapt system behavior and provide real-time pricing. The communication delay is critical for PEV charging process decision making. Many studies consider the battery charging schemes and mechanisms of PEV in a smart grid environment, and developing the communication standard. However, the consideration of QoS and mobility in PEV-smart grid scenarios has not been considered in the literature.

8.3.2 Cooperative Communication

Cooperative (i.e. cognitive) communication among PEVs in a smart grid environment is a challenging aspect and still needs to be developed for such environments and such applications. The main aim of cognitive communication for such applications is to facilitate data exchange between PEVs and create a highly informative smart grid network. The development of a heterogeneous network architecture that enables real-time, reliable, and secure communication in V2G environment is an open research issue. The main issue in these applications is security; for instance, the information that can or cannot be broadcasted or received by PEVs and the message format associated with that. The challenge becomes even bigger when PEVs are required to provide QoS guarantees along with cognitive communication.

8.3.3 Harsh Environment

V2G communications may exist in situations where severe weather, such as snow storms, hurricanes, heavy rain and extreme heat, which may impact the QoS performance of such systems. Satisfying QoS requirements for V2G in such environments is another challenging issue that needs to be studied using field testing or using realistic models. In addition to that, during extreme weather conditions, power may be interrupted and hence the APs and the RSU need to maintain operational to support the V2G infrastructure, this issue has to be considered in such scenarios. Another challenge which is related to the nature of the implementation environment is the existence of metal structures, buildings and other natural obstacles; these in turn may cause multipath propagation, fading, path loss and other radio wave propagation effects. The effects of the propagation environment on the QoS performance of V2G systems is another open issue in this domain that needs to be studied.

8.3.4 Security Issues

In addition to the nature of bursty traffic, mobility and harsh environment, security issues also need to be considered when implementing wireless networks in V2G applications. For instance, the security and privacy of billing, payment pre-authorization and online payments from PEVs directly to the utility administrator or the charging station owner is a major concern for PEV drivers and smart grid administrators alike. In addition to that, the actual PEV location needs to be kept confidential for user privacy. Another security concern is the unauthorized transaction. Therefore, V2G security is one of the most important issues and challenges that need to be addressed. Furthermore, any security and privacy protocol should consider the QoS guarantees of such applications.

8.4 An Overview of the IEEE 802.11p MAC Protocol

The IEEE 802.11p MAC implements the EDCA mechanism adopted by IEEE 802.11e MAC. Like the IEEE 802.11e the IEEE 802.11p is designed to enable QoS support for contention-based communications. This is basically done by allowing the MAC protocol to categories each frame to a different class, which is also known as AC. The basic idea behind this categorization is that these packets are placed into different queues based on their AC number. The IEEE 802.11p defines four queues to specific ACs, each AC provides data categorization based on the characteristics and requirements of each traffic. To provide data differentiation, each AC queue is configured with different minimum Contention Window size (CW_{min}), maximum

Contention Window (CW_{max}), Arbitration Interframe Space (AIFS) and Transmission Opportunity (TXOP).

The IEEE 802.11p MAC layer uses an internal scheduler to prevent internal collisions between traffic of different ACs. This is basically done by selecting which AC has the priority to get medium access. Table 8.1 shows an example of different ACs used for different applications.

Based on the above description, each AC queue will independently contend for channel access. To physically achieve this, each (ACi) is allowed to contend for the medium access after an AIFS period. Each AC has a different AIFS duration as shown in Fig. 8.5. Therefore, each AIFS[i] can be given by the following relation:

$$AIFS[i] = T_{SIFS} + AIFSN[i] \times T_{TS} \tag{8.1}$$

where T_{TS} and T_{SIFS} are the duration of the timeslot and SIFS respectively. $AIFSN[i]$ is the AIFS slot count for priority class i for each AC_i. The IEEE 802.11p [3] defines the values of $AIFSN[i]$, CW_{max} and CW_{min} for each AC, Table 8.2 shows these defined values.

By examining the values presented in Table 8.2, we notice that low AIFSN values are assigned to high-priority ACs. Furthermore, ACs with higher priority have smaller CW. In addition to that, a larger TXOP means more packets can be exchanged between the device and its base station. Each AC_i enters into a backoff period after waiting for $AIFS[i]$, if it finds an idle channel, it initiates its packet transmission. The backoff counter is an integer which is selected randomly in the range $(0, CW_i)$. Where CW_i is defined as the maximum window size at stage i and is defined according to the following relation:

Table 8.1 Different ACs and application requirements

Access category	Application
AC[3]	Emergency information
AC[2]	Information broadcasted by vehicles
AC[1]	Inter-vehicle information exchange
AC[0]	Non-safety-related connections using SCHs

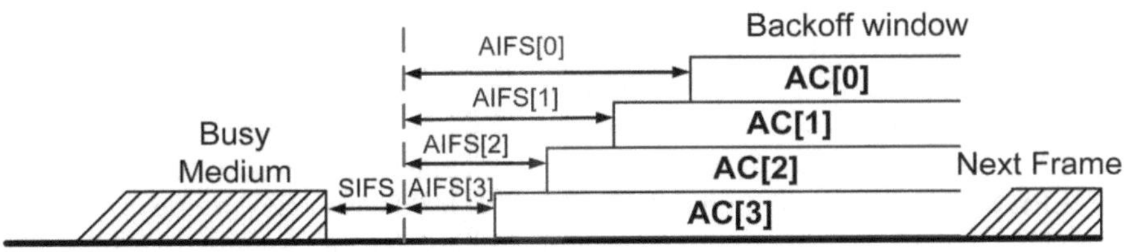

Fig. 8.5 Channel access prioritization in IEEE 802.11p

Table 8.2 IEEE 802.11p parameters

AC	$CW_{min}[i]$	$CW_{max}[i]$	$AIFSN[i]$
AC[3]	$\frac{(CW_{min}+1)}{4}-1$	$\frac{(CW_{min}+1)}{2}-1$	2
AC[2]	$\frac{(CW_{min}+1)}{4}-1$	CW_{min}	3
AC[1]	$\frac{(CW_{min}+1)}{2}-1$	CW_{max}	6
AC[0]	CW_{min}	CW_{max}	9

$$CW_i = \begin{cases} CW_{\min} + 1, & i = 0 \\ 2^i \times (CW_{i-1} + 1) - 1, & 1 \le i \le M - 1 \\ CW_{\max}, & M \le i \le M - f \end{cases} \tag{8.2}$$

Therefore, the window size W_i

$$W_i = \begin{cases} CW_{\min} + 1, & i = 0 \\ 2^i \times W_0, & 1 \le i \le M - 1 \\ CW_{\max} + 1, & M \le i \le M + f \end{cases} \tag{8.3}$$

The value of CW_{min} and CW_{max} are defined as 15 and 1,023 as reported by the draft standard of IEEE 802.11p [3]. For every time slot sensed idle, the backoff counter is decremented by one unit, otherwise it keeps the same value until the channel becomes idle again for a period of $AIFS[i]$. When the counter reaches zero, the packet in the AC_i queue is transmitted. In case of a collision, the backoff window size doubles, and the backoff procedure is re-initiated. After reaching the maximum number of retries (R_{max}) which is equal to $M + f$, the packet is dropped and the system returns back to the idle state and waits for a new packet.

8.4.1 EDCA for 802.11p

In addition to the adoption of the IEEE 802.11e EDCA in the IEEE 802.11p protocol, the protocol also uses additional enhancement due to the nature of IEEE 802.11p applications. One of the major enhancements at the MAC layer is the use of multi-channel operation. These different channels (i.e. Control Channel (CCH) and Service Channel (SCH)) provide different traffic categories to achieve additional QoS guarantees to critical traffic.

Both the CCH and SCH provide four ACs labelled from AC[0] to AC[3]. Whereas AC [3] has the highest priority and AC[0] has the lowest priority. Table 8.1 shows examples of different ACs with typical application requirements in the IEEE 802.11p protocol. In the IEEE 802.11p, for delay critical traffic, more time is assigned to CCH, and for regular traffic more time is assigned to SCH [3]. Therefore, a balance between the time allocated to CCH and SCH channels is required depending on the traffic type and the environment [3]. In addition to different ACs and channel type, the size of the Contention Window (CW) (where CW_{min} is 15 and CW_{max} is 1,023) along with the combination of the backoff window are used to enforce different QoS guarantees [3].

8.4.2 Multi-channel Operation

The IEEE 802.11p protocol divides the MAC layer of the Dedicated Short Range Communication (DSRC) into two sub-layers. The upper MAC layer (defined by IEEE 1609.4 standard), which uses channel coordination, channel routing and UP. The lower sub-layer is responsible for wireless medium access. The standard provides guidelines for a single station (PEV) under one CCH and SCH by providing alternating channel access mechanism. The combination of these alternating CCH and SCH intervals is referred to as Sync interval [3]. The CCH and SCH intervals are 50 ms long each [3], with a guard interval of 2 ms to separate them.

Network management messages such as WAVE Short Message (WSM) and Service Advertisements (SAs) are exchanged over CCH. Therefore, initially all PEVs are required to instantaneously listen to the CCH to achieve synchronisation and network management. This listen period has to be considered when distributing the channel allocation between the CCH and the SCH channels especially in heavy data transmission [3]. The IEEE 1609.4 standard provides four possible access choices, these namely are; continuous, alternating, immediate and extended channel access [39]. These different channel access choices depend very much on the application traffic and the QoS requirements. Figure 8.6 shows the four channel access mechanisms defined in [3].

From Fig. 8.6, we notice that during CCH time channel activities on SCH are terminated and during SCH activities CCH activities are terminated [3].

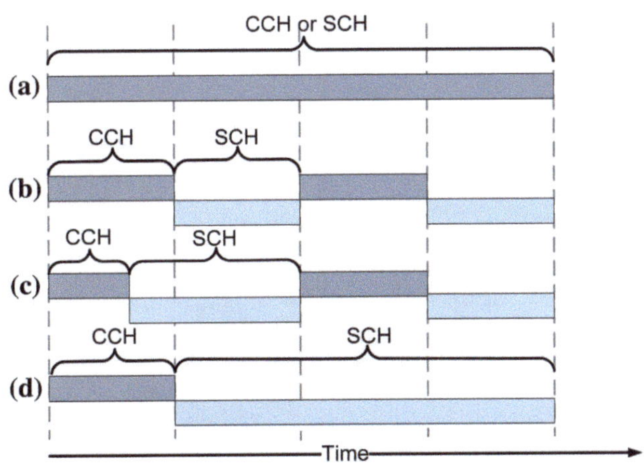

Fig. 8.6 Channel access mechanisms; **a** continuous, **b** alternating, **c** immediate, **d** extended

8.5 Scenario Description

There are multiple options to charge PEVs [1], the choice of one charging option over the other depends very much on the cost of charging, the availability of the charging, time required to charge and the distance from the PEVs to the charging stations. In this scenario, we assume that there are three options to charge the PEV. In our scenario we categorise these options based on cost, availability and time. Therefore, we assume that the Change Station (CS) is the best option followed by the fast charging station then finally charging at home as shown in Fig. 8.7.

In our scenario, we assume that there are (N) PEVs on the road at any moment of time and that these PEVs travel at an average speed of 60 km/h. Furthermore, all of these PEVs have established a link with one or more APs (within their transmission range) using the IEEE 802.11p protocol as shown in Fig. 8.7. These PEVs communicate with the AP to exchange various safety information and also to exchange available charge information. Each AP is connected to a Smart Grid Server (SGS) to acquire information for optimizing the cost of charge (CC), distance to charging locations (DC) and time to charge (TC). During the initial network set-up phase, all PEV send their State-of-Charge (SoC), distance to arrival (d) and their destination (D) to the AP. The AP responds by providing optimum charging options to these PEVs

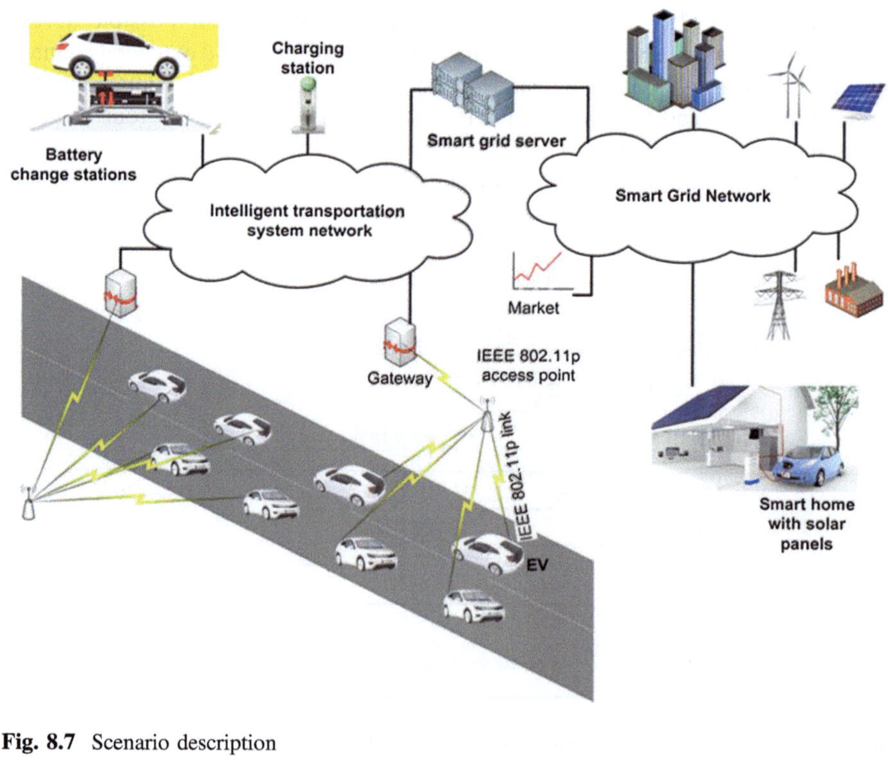

Fig. 8.7 Scenario description

and also provides QoS differentiation to an PEV that requires immediate charging. The optimization of charging options is done in a near real-time manner at the AP by communicating with the SGS. The PEV driver can decides which option to take and then makes the reservation to charge at the selected option through the AP.

8.6 The QCEV Scheme

The QCEV scheme provides QoS differentiation to PEVs that require immediate charging due to the low battery level available at the time of link establishment with the AP. The aim of QCEV scheme is to give the PEV with critical battery charge higher probability in accessing the channel and hence reducing the end-to-end delay and increasing the throughput.

The QCEV scheme works as follows; a group of (N) PEVs within the WAVE Basic Service Set (WBSS) broadcast their charge status to the AP during the network association phase using the normal DCF procedure described in [3]. At the same time, the AP has access to available charging infrastructure and current charging costs through communicating with smart grid servers. After the association phase is complete, the AP evaluates the available charge in all PEVs in its WBSS and then decides which PEV will have higher priority to have channel access.

The QCEV scheme modifies the default prioritization of transmission in EDCA which is implemented by an Arbitrary Inter-Frame Space (AIFS). AIFS is an extension of the backoff procedure in DCF by assigning new AIFS values for different ACs. The duration AIFS [AC] is a duration derived from the AIFS Number (AIFSN) and is given in the Eq. 8.1.

Duration (BD) is given by:

$$BD = CW_{\min} \times Slot_{Duration} \tag{8.4}$$

where, $CW_{\min}$ is the minimum CW described in [st].

According to [3], different ACs are associated with different AIFSNs. Therefore, by using Eqs. 8.1 and 8.2, we see that the AC with a smaller AIFS has higher priority to access the channel. Furthermore, different CW sizes are assigned to different ACs which is also used to enforce propriety. This procedure is used by the default IEEE 802.11p MAC protocol which uses EDCA. Based on Fig. 8.6 and Table 8.2, we see that both the CCH and SCH channels support four traffic classes each having different priorities. Therefore, allocating more time to CCH will considerably reduce the delay of critical message since medium access is faster.

In the application proposed in this chapter, a more centralized QoS differentiation is required. The centralization is enforced by the AP based on evaluation of the remaining charges in all of the PEVs and based on the information fed back form the smart grid at that moment. It is essential to highlight that a distributed QoS differentiation described in the IEEE 802.11p protocol is not an optimum solution for this scenario due to the need for learned decision based on available battery in all PEVs in a BSS. In the QCEV scheme, when the AP determines that a specific PEV (we refer to it as xPEV) needs to have higher priority compared to other PEVs, it broadcast an "alarm signal" to all PEVs in its WBSS indicating the address of the xPEV. Upon the reception of this broadcast, all of the PEV except xPEV will use AC[0] and the alternating channel access mechanism. xPEV on the other hand, will be the only station allowed to use AC[3] and use the CCH extended channel access mechanism. The CCH extended channel access mechanism is more efficient in terms of delay (since it is used for emergency short messages) than the alternating channel access mechanism because half the length of channel is allocated for short message exchange. Algorithm 1 shows the QCEV algorithm run at the AP.

Algorithm 1 QCEV Algorithm at the AP.

//Network establishment phase
//For all STAs
$AP \leftarrow (SoC), (d) \; \& \; (D)$
//For N STAs find Min (SoC) & Max (d)
$STA_{Min_{(SoC)}, Max_{(d)}} \leftarrow xSTA$
$AP \leftarrow (CC), (DC) \; \& \; (TC)$ from SGS
//Find Min (CC) & Min (TC) for $xSTA$
AP broadcast $xSTA_{IP_{add}}$, $xSTA_{CC_{Min}}$ & $xSTA_{TC_{Min}}$
//Find Min (CC) & Min (TC) for $\lceil (N\text{-}1)\; STA \rceil$
Transmit Min $[CC_{(N-1)}]$ & $[TC_{(N-1)}]$
(EDCA Algorithm)

8.7 The CAC Scheme

In the second proposed scheme, we assume that there are two traffic classes being transmitted across the network. The first is the high priority traffic for delay critical data such as transmission by electrical emergency vehicles or transmission of critical incidences on the road. The second class is the low priority traffic which is transmission of routine traffic such as exchange of road traffic information, weather, SoC, location of BCS and etc. We assume that all of the vehicles (PEVs) connect to an AP using a star topology and that the communication between PEVs is done through the AP. Therefore, a message transmitted from one PEV to the rest of the PEVs connected to the same AP should be done before the PEVs leave the coverage area of the AP. This assumption is made to ensure real-time delivery of delay critical messages. We assume that APs are interconnected through an intelligent transportation system network and that the propagation delay from one AP to another is negligible.

The proposed centralized Channel Access Control (CAC) scheme provides service differentiation to vehicles that have high priority and need immediate channel access. Therefore, we expect a reduction in the delay when the CAC scheme is used. The CAC uses the same technique used by the QCEV, but it uses further QoS differentiation.

The CAC scheme works as follows; during the network establishment phase, a group of (N) vehicles within the WBSS exchange their network ID (i.e. MAC address) and the priority level with the AP using the normal DCF procedure described in [3]. Based on the network ID and the priority level of the received handshake, the AP classifies the traffic into two categories, namely, low priority and high priority traffic. High priority traffic is allocated to vehicles that are transmitting delay critical data such as emergency vehicles (i.e. data with an alarm flag). Whereas, low priority traffic is assigned to all other vehicles (i.e. no alarm flag). Based on that, if within a group of N vehicles there is one or more high priority vehicles, the AP will broadcast a message to all vehicles in its WBSS indicating a high priority transmission and the source of this transmission. In the CAC scheme, in addition to the differentiation technique used by QCEV, low priority vehicle use $[\frac{R_{max}}{2}]$ and high priority vehicles use $[R_{max} \times 2]$. In doing so, low priority vehicles have less number of transmission retries and hence give up channel access and give up transmission sooner than high priority vehicles. In addition to that, different CW sizes are assigned to different ACs which is also used to enforce propriety. This procedure is used by the default IEEE 802.11p MAC protocol which uses EDCA. Based on Fig. 8.5 and Table 8.2, we see that both the CCH and SCH channels support four traffic classes each having different priorities. Therefore, allocating more time to CCH will considerably reduce the delay of critical message since medium access is faster. Therefore, when the broadcast is received from the AP, the high priority vehicle will be the only vehicle allowed to use AC0 and use the CCH extended channel access mechanism. Similar to QCEV, the CCH extended channel access mechanism is more efficient in terms of delay than the other channel access mechanisms. It is important to mention here that if we permanently allow high priority vehicles to use $[\frac{R_{max}}{2}]$ and low priority vehicles to use $[R_{max} \times 2]$, this will lead to deterioration in network performance. Because, when high priority vehicles do not exist or do not communicate in the WBSS, low priority vehicles will suffer from unnecessary transmission delay and drop in throughput. Therefore, the CAC scheme allows the AP to adaptively control when this data differentiation should take place.

8.8 Simulation Results and Analysis

To evaluate the performance of the QCEV and the CAC schemes, we use QualNet [40] network simulator to simulate the scenario described in Fig. 8.7. We assume that there are (N) PEVs on the road at a given time and that these PEVs travel at an average speed of 60 km/h. We run each simulation for 1 h and average 10 simulation runs to obtain the results.

Figure 8.8 shows the end-to-end delay from the xPEV to the AP. We show that when the default QoS defined in the IEEE 802.11p MAC protocol is used, the end-to-end delay is always higher compared to the situation where the QCEV is used. This reduction in delay takes place since the distributed QoS mechanisms used by

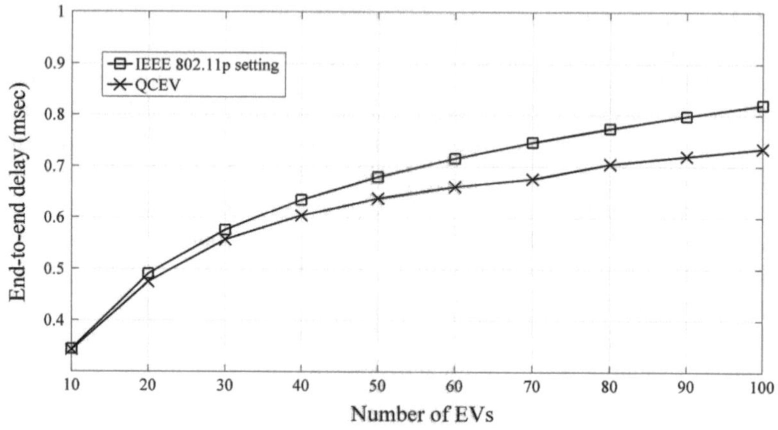

Fig. 8.8 QCEV end-to-end delay

the IEEE 802.11p MAC allow different PEVs to use contention based on internal decisions of remaining battery charge levels and hence making independent priority levels. Therefore, if multiple PEVs decide that their battery levels are low, they all consider this situation as a high priority, and use the same AC leading to higher contention. In the QCEV scheme on the other hand, since PEVs send their battery levels, the AP decides which PEV deserves the highest channel access priority. We show that the reduction in delay becomes more obvious when the number of PEVs increase due the reduction in the contention level.

Figure 8.9 shows the throughput when the IEEE 802.11p protocol and the QCEV scheme are used. We show that the throughput is always higher when the QCEV mechanism is used; this is due to the reduction in the contention when centralized QoS mechanism is used.

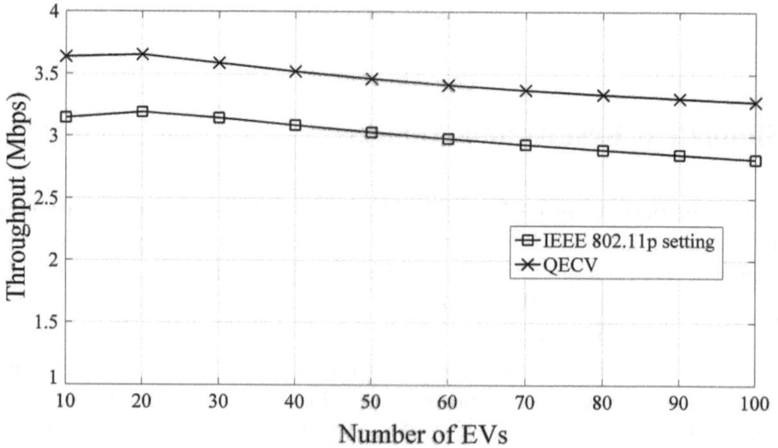

Fig. 8.9 QCEV throughput

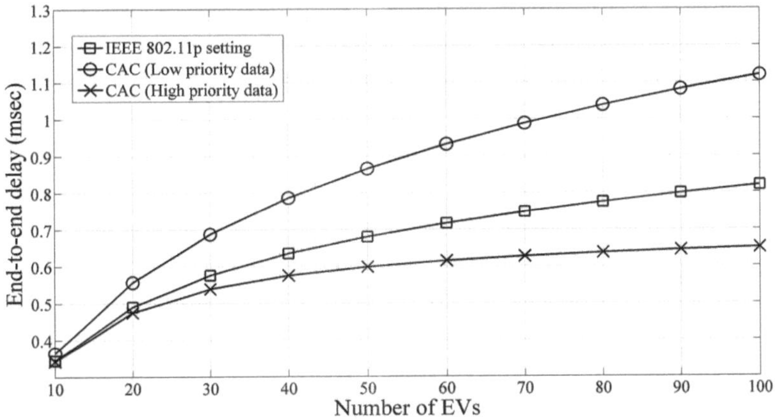

Fig. 8.10 CAC throughput

Figure 8.10 shows the end-to-end delay from the two different classes of vehicles to the AP. We show that when the default QoS defined in the IEEE 802.11p MAC protocol is used, the end-to-end delay is always higher compared to the situation where high priority data is transmitted. Furthermore, we show that when low priority data is transmitted the delay is higher than the default situation. If multiple PEVs decide that their traffic is critical, they all consider this situation as a high priority, and use the same AC leading to higher contention. In the CAC mechanism on the other hand, since PEVs send their network ID and alarm level to the AP, the AP decides which PEV gets the highest priority. We show that the reduction in delay becomes more significant as the number of PEVs increase.

Figure 8.11 shows the throughput when the IEEE 802.11p, low priority and the high priority data are transmitted. We show that the throughput is not affected when

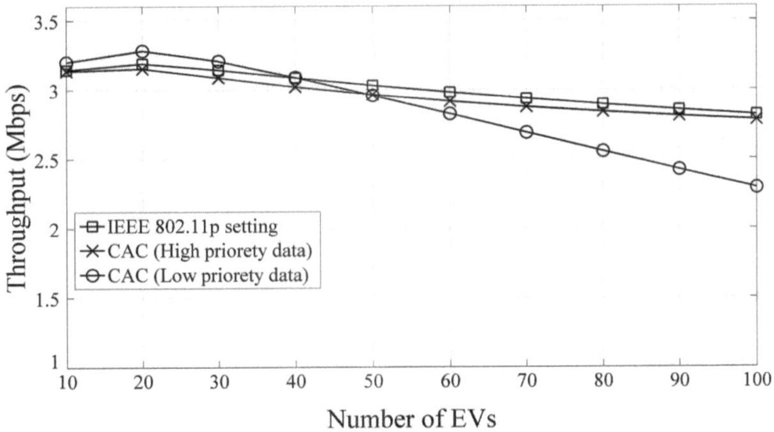

Fig. 8.11 CAC throughput

the high priority is transmitted and that it remains very close to the IEEE 802.11p throughput values even with very high number of PEVs. On the other hand, when PEVs are transmitting low priority data, the throughput remains comparable to the other two situations. However, when the number of PEVs increase beyond 60, the throughput begins to slightly decrease due to the reduced channel access rates and the limited number of retries.

8.9 Conclusions

Quality of Service (QoS), Reliable and real-time communication between PEVs/ PHEVs on one side and the smart grid on the other side has been identified as one of the main challenging issues to successfully integrate these two technologies. In fact electrical utility operators have set up specific delay and reliability requirements for certain V2G applications.

In this chapter, we presented a survey of the state-of-the-art in wireless communication systems for PEVs integration with smart grid. We highlighted the main challenges associated with the PEV-smart grid communication system. We also presented two QoS schemes for charging PEVs (QCEV) in a smart grid environment and a centralized Channel Access Control (CAC) scheme.

Our schemes considers the charging requirements of PEVs communicating with the Access Point (AP) using the IEEE 802.11p protocol and centrally provides Quality of Service (QoS) differentiation to PEVs that require immediate access to a charging infrastructure. The QCEV scheme controls the channel access of contending PEVs by forcing PEVs with less priority to use lower Access Categories (ACs) with longer channel access mechanism. The CAC scheme builds on the QCEV scheme, but it provides further enhancements by forcing PEVs with less priority to spend more time in backoff. Our simulation results show that the QCEV scheme and the enhanced CAC scheme could effectively reduce the end-to-end delay and increase the throughput for PEVs with high priority. Furthermore, simulation results show that as the number of PEVs communicating to an AP increase, the effectiveness of the QCEV and CAC schemes becomes more pronounced.

As a future work we intend to investigate the performance of the two schemes in multihop environment where the data packets are routed through multiple PEVs before reaching the AP.

References

1. Kemp R, Blythe P, Brace C, James P, Parry-Jones R, Thielens D, Thomas M, Wenham R, Urry J (2010) Electric vehicles: charged with potential. In: Royal academy of engineering report. http://www.raeng.org.uk/publications/reports/electric-vehicles. Accessed Nov 2014
2. Owning an electric car/electric car range (2014) http://www.owningelectriccar.com/electric-car-range.html. Accessed Feb 2014

3. IEEE standard for information technology, local and metropolitan area networks, specific requirements part 11: wireless LAN medium access control (MAC) and physical layer (PHY) specifications amendment 6 (2010) Wireless access in vehicular environments. IEEE STD 80211p-2010 (amendment to IEEE STD 80211-2007 as amended by IEEE STD 80211k-2008, IEEE STD 80211r-2008, IEEE STD 80211y-2008, IEEE Std 80211n-2009, and IEEE STD 80211w-2009), pp 1–51. doi:10.1109/IEEESTD.2010.5514475
4. IEEE standard for information technology local and metropolitan area networks specific requirements part 11 (2005) Wireless LAN medium access control (MAC) and physical layer (PHY) specifications—amendment 8: medium access control (MAC) quality of service enhancements. IEEE STD 80211e-2005 (amendment to IEEE STD 80211, 1999 edition (Reaff 2003), pp 1–212. doi:10.1109/IEEESTD.2005.97890
5. Salem, Corvallis, Eugene (2010) Electric vehicle charging infrastructure deployment guidelines for the oregon i-5 metro areas of Portland. Technical report, Electric Transportation Engineering Corporation
6. Pieltain Fernandez L, Roman T, Cossent R, Domingo C, Frias P (2011) Assessment of the impact of plug-in electric vehicles on distribution networks. IEEE Trans Power Syst 26 (1):206–213. doi:10.1109/TPWRS.2010.2049133
7. Palensky P, Dietrich D (2011) Demand side management: demand response, intelligent energy systems, and smart loads. IEEE Trans Ind Inf 7(3):381–388. doi:10.1109/TII.2011.2158841
8. Su W, Eichi H, Zeng W, Chow MY (2012) A survey on the electrification of transportation in a smart grid environment. IEEE Trans Ind Inf 8(1):1–10. doi:10.1109/TII.2011.2172454
9. Oldak M, Kilbourne B (2014) Communications requirements comments of utilities telecom council. http://www.energy.gov. Accessed Feb 2014
10. Faezipour M, Nourani M, Saeed A, Addepalli S (2012) Progress and challenges in intelligent vehicle area networks. Commun ACM 55(2):90–100
11. Huang J, Wang H, Qian Y (2012) Smart grid communications in challenging environments. In: Proceedings of the IEEE third international conference on smart grid communications (SmartGridComm), pp 552–557. doi:10.1109/SmartGridComm.2012.6486043
12. Msadaa I, Cataldi P, Filali F (2010) A comparative study between 802.11p and mobile WiMAX-based V2I communication networks. In: Proceedings of the 4th international conference on next generation mobile applications, services and technologies (NGMAST), pp 186–191. doi:10.1109/NGMAST.2010.45
13. Al-Anbagi I, Erol-Kantarci M, Mouftah H (2013) Priority and delay-aware medium access for wireless sensor networks in the smart grid. IEEE Syst J PP(99):1–11. doi:10.1109/JSYST.2013.2260939
14. Al-Anbagi I, Erol-Kantarci M, Mouftah H (2013) A reliable IEEE 802.15.4 model for cyber physical power grid monitoring systems. IEEE Trans Emerg Topics Comput 1(2):258–272. doi:10.1109/TETC.2013.2281192
15. Zafalon R, Vermesan O, Coppola G (2013) e-mobility the next frontier for automotive industry. In: Proceedings of the design, automation test in europe conference exhibition (DATE), pp 1745–1748. doi:10.7873/DATE.2013.351
16. Conti M, Fedeli D, VirgultiM (2011) B4V2G: bluetooth for electric vehicle to smart grid connection. In: Proceedings of the of the ninth workshop on intelligent solutions in embedded systems (WISES), pp 13–18
17. Ferreira J, Monteiro V, Afonso J (2014) Vehicle-to-anything application (v2anything app) for electric vehicles. IEEE Trans Ind Inf pp 10(3):1927–1937. doi:10.1109/TII.2013.2291321
18. Jansen B, Binding C, Sundstrom O, Gantenbein D (2010) Architecture and communication of an electric vehicle virtual power plant. In: Proceedings of the first IEEE international conference on smart grid communications (Smart-GridComm), pp 149–154. doi:10.1109/SMARTGRID.2010.5622033
19. Cespedes S, Shen X (2011) A framework for ubiquitous IP communications in vehicle to grid networks. In: Proceedings of the IEEE GLOBECOM workshops (GC Wkshps), pp 1231–1235. doi:10.1109/GLOCOMW.2011.6162378

20. Dong Q, Niyato D, Wang P, Han Z (2013) An adaptive scheduling of PHEV charging: analysis under imperfect data communication. In: Proceedings of the IEEE international conference on smart grid communications (SmartGrid-Comm), pp 205–210. doi:10.1109/SmartGridComm.2013.6687958
21. Yuan Z, Xu H, Han H, Zhao Y (2012) Research of smart charging management system for electric vehicles based on wireless communication networks. In: Proceedings of the IEEE 6th international conference on information and automation for sustainability (ICIAfS), pp 242–247. doi:10.1109/ICIAFS.2012.6419910
22. Erol-Kantarci M, Sarker J, Mouftah H (2011) Communication-based plug-in hybrid electrical vehicle load management in the smart grid. In: Proceedings of the IEEE symposium on computers and communications (ISCC), pp 404–409. doi:10.1109/ISCC.2011.5983871
23. Xu DQ, Joos G, Levesque M, Maier M (2013) Integrated V2G, G2V, and renewable energy sources coordination over a converged fiber-wireless broadband access network. IEEE Trans Smart Grid 4(3):1381–1390. doi:10.1109/TSG.2013.2253337
24. Kovacs A, Marples D, Schmidt R, Morsztyn R (2013) Integrating evs into the smart-grid. In: Proceedings of the IEEE 13th international conference on ITS telecommunications (ITST), pp 413–418
25. Yu R, Zhang Y, Gjessing S, XiaW, Yang K (2013) Toward cloud-based vehicular networks with efficient resource management. IEEE Netw 27(5):48–55. doi:10.1109/MNET.2013.6616115
26. Cheung MH, Hou F, Wong V, Huang J (2012) DORA: dynamic optimal random access for vehicle-to-roadside communications. IEEE J Sel Areas Commun 30(4):792–803. doi:10.1109/JSAC.2012.120513
27. Rezgui J, Cherkaoui S, Said D (2012) A two-way communication scheme for vehicles charging control in the smart grid. In: Proceedings of the 8th international wireless communications and mobile computing conference (IWCMC), pp 883–888. doi:10.1109/IWCMC.2012.6314321
28. Bohm A, Jonsson M (2008) Supporting real-time data traffic in safety-critical vehicle-to-infrastructure communication. In: Proceedings of the 33rd IEEE conference on local computer networks, LCN 2008, pp 614–621. doi:10.1109/LCN.2008.4664253
29. Luan TH, Ling X, Shen XS (2012) Provisioning QoS controlled media access in vehicular to infrastructure communications. Elsevier Ad Hoc Netw 10(2):231–242
30. Han C, Dianati M, Tafazolli R, Kernchen R, Shen X (2012) Analytical study of the ieee 802.11p mac sublayer in vehicular networks. IEEE Trans Intell Transp Syst 13(2):873–886. doi:10.1109/TITS.2012.2183366
31. Misic J, Badawy G, Rashwand S, Misic V (2010) Tradeoff issues for CCH/SCH duty cycle for IEEE 802.11p single channel devices. In: Proceedings of the IEEE global telecommunications conference (GLOBECOM 2010), pp 1–6. doi:10.1109/GLOCOM.2010.5684264
32. Grafling S, Mahonen P, Riihijarvi J (2010) Performance evaluation of IEEE 1609 WAVE and IEEE 802.11p for vehicular communications. In: Proceedings of the 2nd international conference on ubiquitous and future networks (ICUFN), pp 344–348. doi:10.1109/ICUFN.2010.5547184
33. Islam T, Hu Y, Onur E, Boltjes B, de Jongh J (2013) Realistic simulation of IEEE 802.11p channel in mobile vehicle to vehicle communication. In: Proceedings of tje conference on microwave techniques (COMITE), pp 156–161. doi:10.1109/COMITE.2013.6545061
34. Erol-Kantarci M, Sarker J, Mouftah H (2012) Quality of service in plug-in electric vehicle charging infrastructure. In: Proceedings of the IEEE international electric vehicle conference (IEVC), pp 1–5. doi:10.1109/IEVC.2012.6183227
35. Herrera L, Murawski R, Guo F, Inoa E, Ekici E, Wang J (2011) PHEVs charging stations, communications, and control simulation in real time. In: Proceedings of the IEEE vehicle power and propulsion conference (VPPC), pp 1–5. doi:10.1109/VPPC.2011.6043167
36. Li J, Chigan C (2010) Delay-aware transmission range control for VANETs. In: Proceedings of the IEEE global telecommunications conference (GLOBECOM 2010), pp 1–6. doi:10.1109/GLOCOM.2010.5684168

37. Ran B, Negeri E, Baken N, Campfens F (2013) Last-mile communication time requirements of the smart grid. In: Proceedings of the sustainable internet and ICT for sustainability (SustainIT), pp 1–6. doi:10.1109/SustainIT.2013.6685207
38. Luan TH, Ling X, Shen XS (2012) Provisioning QoS controlled media access in vehicular to infrastructure communications. Elsevier Ad Hoc Netw 10(2):231–242
39. Sun N (2011) Performance study of ieee 802.11 p for vehicle to vehicle communications using opnet: a thesis presented in partial fulfilment of the requirements for the degree of master of engineering in telecommunications and network. PhD thesis, Massey University, Auckland, New Zealand
40. Qualnet network simulator (2014) http://web.scalablenetworks.com/content/qualnet. Accessed Feb 2014

Chapter 9
Centralized Charging Control of Plug-in Electric Vehicles and Effects on Day-Ahead Electricity Market Price

Pavan Balram, Le Anh Tuan and Lina Bertling Tjernberg

Abstract Global policy targets to reduce greenhouse gas emissions have led to increased interest in plug-in electric vehicles (PEV) and their integration into the electricity network. Existing electricity markets, however, are not well suited to encourage direct participation of flexible demand from small consumers such as PEV owners. The introduction of an aggregator agent with the functions of gathering, aggregating, controlling and representing the energy needs of PEV owners in the electricity market could prove useful in this regard. In this chapter, a mathematical model of PEV aggregator for participation in the day-ahead electricity market is described. The modeling is done by treating each of the individual vehicle batteries as a single large battery. The centralized charging and discharging of this battery is then scheduled based on the traveling needs of the PEV owners determined by an aggregated driving profile and the cumulative electrical energy needs of vehicles over the optimization horizon. Two methods for scheduling PEV demand named as joint scheduling method (JSM) and aggregator scheduling method (ASM) are presented. The two methods are subsequently used to observe the effects of introducing flexible scheduling of PEVs on the day-ahead market price in an IEEE test system and a Nordic test system. Results from the IEEE test system case studies will indicate that the scheduling of PEV energy through direct centralized control at high PEV penetration levels of 50 % or greater could lead to potential lowering of day-ahead market prices as compared to an indirect control method such as the use of fixed period charging. Results from the Nordic test

---

P. Balram (✉) · L.A. Tuan
Division of Electric Power Engineering, Department of Energy and Environment,
Chalmers University of Technology, Gothenburg, Sweden
e-mail: pavan.balram@chalmers.se

L.A. Tuan
e-mail: tuan.le@chalmers.se

L.B. Tjernberg
Department of Electromagnetic Engineering, School of Electrical Engineering,
KTH Royal Institute of Technology, Stockholm, Sweden
e-mail: linab@kth.se

S. Rajakaruna et al. (eds.), *Plug In Electric Vehicles in Smart Grids*,
Power Systems, DOI 10.1007/978-981-287-317-0_9

system case study shows that controlled scheduling of PEV demand could lead to only a small increase in day-ahead market price of electricity.

Keywords Aggregator · Demand scheduling · Electricity markets · Day-ahead market · Plug-in electrical vehicles

9.1 Introduction

Today's electricity markets are typically designed based on the operational characteristics of large, conventional production units. The structural and operational rules of electricity markets are continuously being adapted to changes that are occurring due to the pro-active environmental policies in the energy sector [1]. In line with these policies, electricity generation from renewable energy sources has received much focus in the recent past and is only expected to increase in the coming decades as reported by the International Energy Agency [2]. Most of the newly installed generating capacity from wind and solar power sources are intermittent in nature, thereby giving rise to limited control over power generation. To maintain energy balance within the power system at all instances, it could become imperative to have controllability from other resources, which could be either non-renewable energy sources from the generation side, or control of resources on the demand side.

The introduction of smart metering systems at households, and integration of information and communication technologies (ICT) with power system components, gives rise to a potential for control of virtually any type of demand. This means that even the lowest power consuming devices within households could be collectively controlled to provide services during power system operation. Currently, many demand response programs delving into the control of household appliances have been launched and are being researched upon. It is imperative to understand the implications of investing in ICT and smart meters for end consumer demand control. The main question of whether demand control on small consumer level would benefit in any way would need to be answered. A starting point would be to categorically observe the effects of demand control on consumers. Two categories that could have a direct effect are:

i. *Social effects*—including the behavioral changes needed to be adopted by consumers to perform demand control, and consumer feeling of performing a common good by promoting and supporting environmentally friendly resources and participating in programs that could prove to be good for the society in general.
ii. *Economic effects*—the customers would need to know how demand response programs would affect their electricity bills. Investment in smart meters could result in better awareness for the consumers on their consumption patterns that could, in the long term, result in overall energy demand reduction. To observe the effects of demand response on the electricity market price would require further research to study the system level impacts.

Electrification within the transportation sector is considered to provide good opportunities for demand control in future power systems [3]. With battery energy storage systems, PEVs provide flexibility regarding the sources of electrical energy for charging. Hence, if global policies are driven towards tapping renewable resources such as wind, solar, biomass, biogas, wave, tidal for power production, then power sources with lower carbon footprint could be used to charge the vehicles. With battery systems in PEVs, greater flexibility could be achieved by storing renewable energy when it is available, and then re-using this energy during times of higher power imbalance. Hence, PEVs could also be utilized to offset some of the intermittency in power production from renewable sources.

9.2 The Nordic Electricity Market

The electricity market is an arrangement for sale of electrical energy as a commodity between various free players—producers, consumers, retailers and traders. Additional players such as transmission system operators (TSO) and DSO facilitate the functioning of electricity markets and the subsequent delivery of electrical energy to end consumers. The power generated by the producers is delivered to consumers through transmission and distribution networks. As the electricity network acts as a backbone for the delivery of energy, the network owners are generally established monopolies that are independent and neutral. The producers and consumers pay a fee known as the *point tariff* to the network owners for every kWh of electric energy produced into or consumed from the network [4]. This ensures that the market mechanism is facilitated, while assuring financial compensation to the TSO/DSO for managing network related operations.

The electricity markets within the European Union (EU) as well as and other parts of the world are constantly evolving. The basic structure is however similar to that of the Nord Pool, which was the first international electricity market [5]. Considering the EU level plan of a harmonized electricity market to facilitate cross-border trading [1], it could be reasonably assumed that future developments would not drastically change the framework of electricity markets from the present. Currently, market players can enter into various power contracts that are further described below in the context of the Nordic electricity market. Figure 9.1 shows an overview of market participants along with the various types of contracts they could enter into. A description of some important contracts that the players could enter into is described in the following subsections.

9.2.1 Bilateral Contracts

The market participants can enter into conventional bilateral contracts that involve a direct trade between a buyer and seller of electrical energy. Considering around

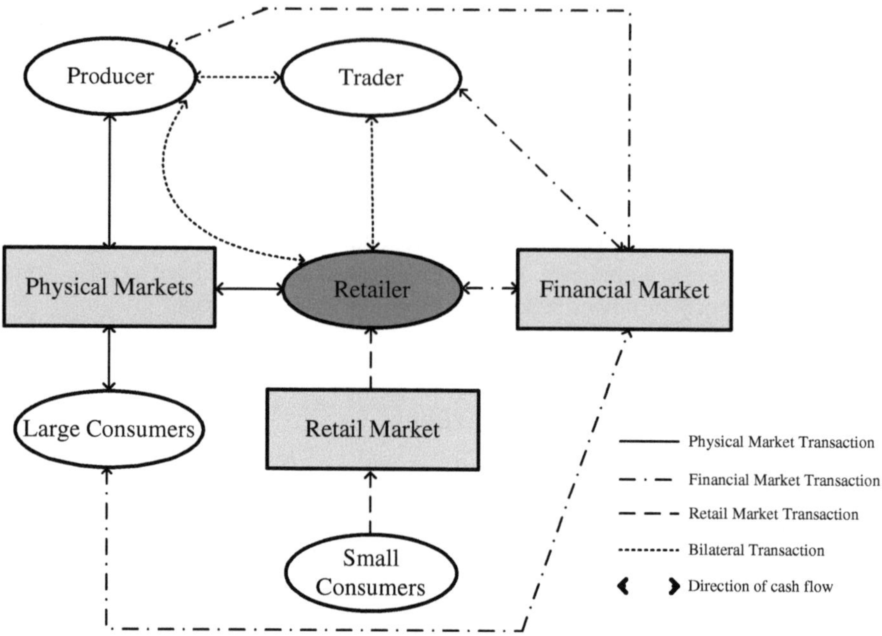

Fig. 9.1 Market players and their interactions within an electricity market framework

84 % of power consumption in the Nordic and Baltic countries are bought at the day-ahead market (DAM) [5], it could be observed that electricity market trading is becoming more appealing to the players. The physical markets including the DAM are described in the following subsection.

9.2.2 Physical Electricity Markets

Like many other commodities, electricity could also be traded within a wholesale market framework. A common DAM called *Elspot* exists for the Nordic and Baltic countries where the market players trade bulk of the electricity production and consumption. The clearing of Elspot results in a production and/or consumption plan for each market player with a delivery obligation, which requires the players to abide by their individual plans.

Electrical energy is however, dynamic, in the sense that electricity has to be instantaneously available when there is demand with few economically viable storage options. This singular characteristic along with the fact that Elspot is cleared ahead of the delivery time of electricity necessitates the use of forecasting methods by the players to estimate their production and consumption. The resulting power deviations that could occur due to forecasting errors, component failures etc., need to be rectified. Players are provided an opportunity to do this through the continuously traded *Elbas* market that is available to balance out the players' individual deviations from their Elspot plans.

It is still possible that last minute imbalances could occur due to failure of components or various faults within the power system. The responsibility of maintaining power balance within the power system during delivery period rests on the TSO who jointly operate the *regulating power market* (RPM) to provide a mechanism for correcting the resulting imbalance during delivery period and ensure the desired level of security of supply within the power system. This market is cleared retroactively as opposed to the Elbas market.

There is a physical obligation associated with the electricity markets, i.e., it has to be ensured that the energy traded in the market is delivered to the end consumers during the specified delivery period. Hence, the Elspot, Elbas and regulating power markets are collectively addressed as physical markets.

9.2.3 Financial Electricity Markets

It is imperative that the market players are able to quantify and hedge the financial risks associated with their participation in the physical markets. Financial markets provide a platform to manage risks by hedging against price fluctuations in the wholesale markets. Common contracts made available in financial markets are [6]: Futures, Forwards, Options and Contracts-for-difference.

It is also important to mention that the financial and physical markets have a specific time-line over which they are operated. Contracts in financial markets are cleared days, weeks, months or years ahead of delivery as opposed to physical markets that are generally cleared 45 min to 1 day-ahead. E.g., an overview of the time-line for Nordic electricity markets operation is shown in Fig. 9.2.

9.3 Demand Response in Electricity Markets

Demand response could be defined as the independent variation in consumption made by consumers as a reaction to different possible incentives. An incentive could, e.g., be price signals from a market for electricity or a network signal

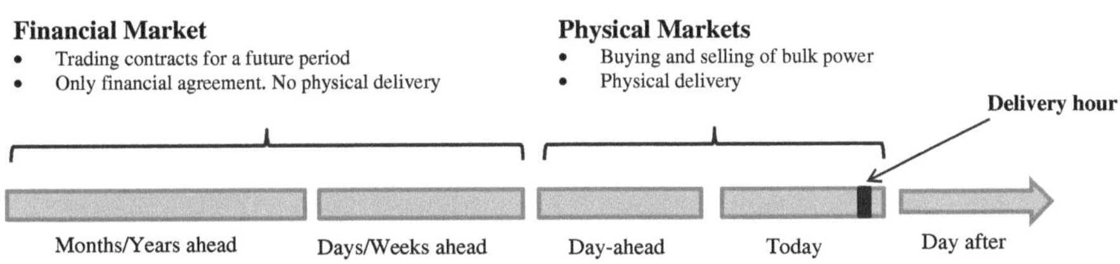

Fig. 9.2 Overview of timeline of Nordic physical and financial markets

provided by the distribution system operator (DSO) in order to maintain the security and reliability of the power system during emergency conditions. Regardless of the type of incentive used, the end result of demand response could be represented as one the load shaping objectives described in [7, 8]:

a. *Peak clipping*—implies the reduction of peak load by using direct load control over customers' appliances. This form of control could be used to reduce the overall cost and dependence on peaking generating units. A good example of peak clipping is the use of interruptible or curtailable tariffs for industrial customers in many vertically integrated power system architectures.
b. *Valley Filling*—implies the process where incentives are provided to increase new demand during off-peak hours. This could be accomplished, e.g., by providing price incentives to new space heating or electric vehicle demand to consume during off-peak hours.
c. *Load Shifting*—implies shifting part of the load from on-peak to off-peak periods. This could involve displacing loads during a particular hour that would otherwise normally be served by electricity.
d. *Flexible Load Shaping*—implies a combination of various incentives. These incentives could include interruptible load, integrated energy management systems or individual customers' load control, etc.

The consumers who are willing to respond to incentive-based signals could be referred to as active consumers. An active consumer could be either a large industrial consumer or a small domestic consumer. Since, the consumption levels of domestic consumers are small compared to the volumes traded in electricity markets, an agent similar to a retailer could become essential to represent the needs of domestic consumers in electricity markets. With controllable resources, however, this agent generally referred to as an 'aggregator' could possibly assume new functions that might require it to control consumer appliances in real-time.

Demands could respond to the price of electricity in the market to consume during low price hours as opposed to high price hours. Alternatively, response from demand could also be used to provide ancillary services to network operation such as frequency response, voltage and reactive power control, black start capability, voluntary load shedding etc. Such programs involving control of demand-side have historically been utilized but limited to large industrial consumers. With the roll out of smart meters, real-time control of domestic consumers' consumption could also be achieved. This could result in greater demand side participation in electricity markets possibly leading to more efficient use of generation resources while also reducing the stress on transmission network during peak consumption periods. Realizing its importance, many electricity markets have begun to open up for greater demand side participation—notable are the area-price based Nordic electricity market and the locational marginal pricing (LMP) based PJM market [9, 10].

9.3.1 Nordic Electricity Market

Features: The Nordic physical markets provide opportunities for price dependent demand to compete directly with price dependent generation. This is especially the case with large scale industrial consumers who have the flexibility to bid for energy directly on the market on an hourly basis and to adjust their consumption in order to prevent being exposed to very high prices. When it comes to small and medium sized consumers, there is a plan to move towards a common retail electricity market with the Nordic region that offers the option of variable retail pricing for consumers directly based on the wholesale price of electricity. In this regard, the installation of smart metering systems has been adapted to measure real-time consumption pattern of domestic and commercial consumers of electricity.

Challenges: Notable challenges exist for domestic consumer participation in the Nordic electricity markets. Though an aggregator agent could be a legal entity in current Nordic day-ahead, intra-day and financial electricity markets, barriers arise when the aggregator would want to participate in the regulating power market. This is due to the fact that aggregator would need to assume the role of a balance responsible party (BRP) in order to participate in RPM, or contract with another BRP. There could be further limitations due to the rules and regulation regarding aggregation of demand in general and also, regarding a new market player assuming the role of a BRP. Another barrier that could hinder the participation of an aggregator is the minimum bid volume requirement by the TSO in RPM, which is 5 MW. This could prove to be a large volume for aggregators, especially in bidding areas with surplus production resources.

9.3.2 Pennsylvania-Jersey-Maryland Market

Features: In the Pennsylvania-Jersey-Maryland (PJM) electricity market, the end-users can participate in PJM's day-ahead energy, capacity, reserves and regulation markets by reducing their demand for electricity. Currently, the mechanism provided through demand response programs only attempts to replicate electricity market price signals instead of exposing them directly to end-users. This is done through curtailment service provider (CSP). Specifically, the role of CSP is defined by PJM as [10], "the entity responsible for demand response activity for electricity consumers in the PJM wholesale markets." Some notable features about a CSP are:

- It may be a company that solely focuses on a customer's demand response capabilities, a local electricity utility, an energy service company or other type of company that offers these services.
- It identifies demand response opportunities for customers and implements the necessary equipment, operational processes and/or systems to enable demand response both at the customer's facility and directly into the appropriate wholesale market.

- It should have appropriate operational infrastructure and a full understanding of all the wholesale market rules and operational procedures.

Challenges: Some barriers [11] that limit customer exposure to wholesale electricity prices could be due to the inadequate metering infrastructure, lack of jurisdictional clarity among regulatory authorities, lack of clear business rules etc. Furthermore, retail prices are set by regional authorities whose operations are decoupled from federal agencies.

9.4 PEV Aggregator

Small and medium level consumers do not have means to directly trade in electricity markets due to the multi-MW size of minimum bid requirement imposed. In order to trade their flexibility, they would require the services of an aggregator agent that gather the flexibility offered by many consumers and builds or pools in active demand capacity to be traded as a single resource. Example of loads that could be aggregated include: fans, electric cooling and heating, electric boilers, refrigerators etc. The aggregators could also generate agreements with consumers to adjust their electricity consumption at moment's notice. A dedicated aggregator for trading flexibilities offered by electric vehicles is the 'PEV aggregator'. Within the context of electricity markets, the functions of a PEV aggregator are similar to that of an electricity retailer. Figure 9.3 describes the interaction of an aggregator with other market participants. The aggregator could purchase electricity from wholesale markets and subsequently sell them to PEV owners through retail contracts. However, it is imperative for the DSO to ensure that the electricity traded by the aggregator is within the capacity of the distribution network. Hence, there could be an additional need for the DSO to communicate and subsequently validate the energy trading performed by an aggregator.

There could be additional functions that need to be accommodated in order to include the concept of aggregator more efficiently within the electricity market [12]. Some of these have been listed as follows:

Fig. 9.3 Overview of the aggregator and its interaction in the physical markets

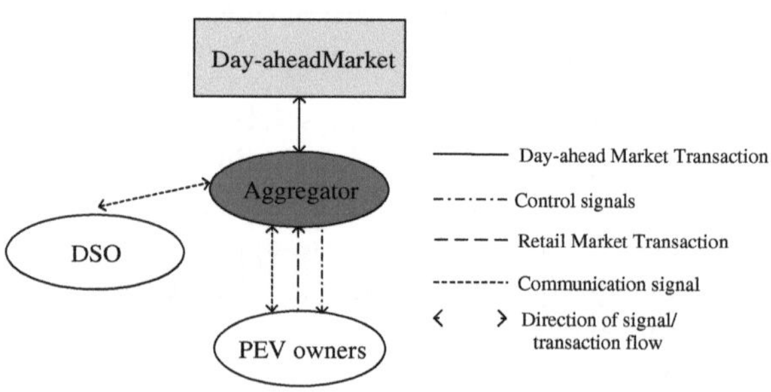

i. There should be necessary communication infrastructure in place for the aggregator to obtain near real-time electricity consumption measurement, vehicle battery state and consumption needs of PEV owners [13].
ii. There should be a mechanism in place for the control of PEV owner batteries. The batteries could be controlled directly by the aggregator with energy schedule validation by the DSO if the necessary automatic control infrastructure is established and market and power system operational rules permit the same. If the rules impose separation of the operational aspect and business aspect of the aggregator, then it could be possible for the DSO to take over the PEV battery control function based on the energy scheduling plan communicated to the DSO by the aggregator [12].
iii. For higher participation from small consumers, it could become essential to reduce minimum bid size in the market to values lower than 1 MW [14].
iv. It might become necessary to introduce shorter time periods of around 30 min or less between market closure and operating hour in order to reduce forecast errors by the aggregator [14].

In this chapter, (i) and (ii) described above are assumed to be available to the aggregator. It is also possible to incorporate (iii) could be incorporated within a RPM model by reducing the minimum bid size to be submitted by the aggregator to the market and (iv) within the DAM by modifying the time resolution for scheduling by the aggregator.

The following section presents two methods for incorporating PEV aggregator and their charge scheduling in DAM [15]. Firstly, the *Joint Scheduling Method* (JSM), where PEV energy is scheduled simultaneously with the generation units—the objective function being minimization of total generation cost. Secondly, the *Aggregator Scheduling Method* (ASM), where the PEV battery energy is first scheduled independently by an aggregator agent based on the estimated electricity market price. The charging schedule, which represents the PEV energy demand, is then submitted to the market in the same way as other conventional loads. Consequently, it is possible to assess the effects of PEV energy demand on electricity market price and compare the impacts among cases with and without PEVs, as well as among cases with the two different scheduling approaches.

9.5 Methods for PEV Energy Scheduling

9.5.1 Joint Scheduling Method

In JSM, the PEV energy scheduling is considered to be performed by a central entity like a system operator that also plans for the dispatch of the generators. The central operator is assumed to receive data related to the generators and PEV batteries. The operator could then schedule both the generators and the PEV charging energy demand by minimizing the total cost of generation. In a scenario where advanced methods of communication and control are feasible, individual PEV owners could directly interact with the market by submitting the necessary PEV data. In this scheduling method, the central operator is assumed to receive the following three sets of information:

(1) Generator costs along with its technical constraints.
(2) Daily PEV driving energy requirements, driving pattern data and aggregated PEV battery energy limits.
(3) Hourly conventional load data, which represents the inflexible demand from all other loads other than PEV demand.

Using these three sets of information, the market model jointly schedules the generators and PEV load to minimize the total generation cost within a unit commitment framework [16]. This is shown in Fig. 9.4.

The generators are assumed to bid their true marginal cost of generating electricity and the market is settled with the minimum generation cost objective [17]. Loads except PEVs are considered to be perfectly forecasted a priori, and are fixed for each hour.

The objective function of the market model is to minimize the total cost of generation to supply the load over the time horizon T. This cost also includes the start-up cost of generating units. This is formulated as shown in (9.1).

$$Minimize \quad DAMC = \left(\sum_{t \in T} \sum_{i \in N} VC_i p_{i,t} + y_{i,t} SC_i \right) \tag{9.1}$$

where, $DAMC$ is the total cost of scheduling the generators in the day-ahead market, VC_i is the variable cost of power production of generating unit i and $p_{i,t}$ is the power produced by unit i at time t, $y_{i,t}$ is a binary variable indicating the starting up of unit i at time t and SC_i is the start-up cost of unit i.

The objective function $DAMC$ in JSM as shown in (9.1) is subject to constraints (9.2)–(9.15) imposed by the generating units, (9.16)–(9.19) imposed by the aggregated PEV batteries and power balance constraint (9.20).

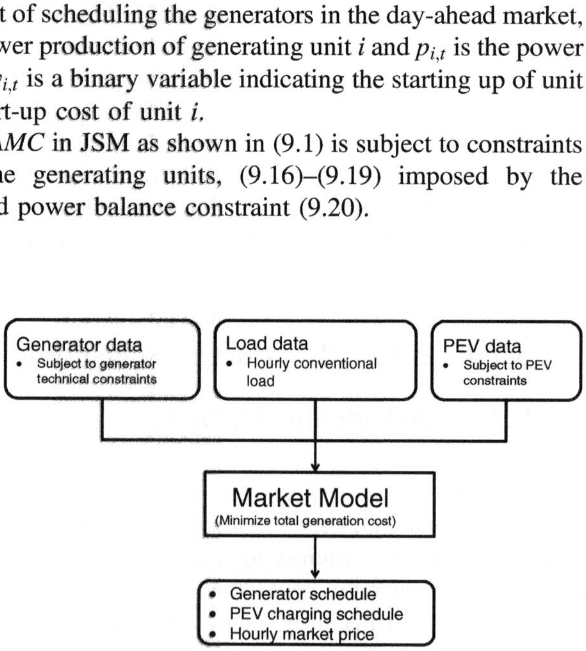

Fig. 9.4 Overview of the JSM

9.5.1.1 Generating Unit Constraints

The generating units should generate power greater than their minimum limits at all times t as shown in (9.2)

$$p_{i,t} \geq v_{i,t} P_i^{\min}; \quad \forall i \in N,\ t \in T \tag{9.2}$$

where, $p_{i,t}$ is the active power output of generator unit i at time t, $v_{i,t}$ is a binary variable indicating the online status of generator unit i at time t and $P_i^{\min}$ is the minimum active power output of generator unit i. The decision of whether the generating unit generates power at time t is taken using a binary variable $v_{i,t}$. The value of $v_{i,t} = 1$ indicates that the unit i is committed to generate power at time t whereas a value of $v_{i,t} = 0$ indicates that the unit i is de-committed from generating power at time t.

The constraints for maximum available power from the generating unit and its ramp rate limit are formulated as shown in (9.3) and (9.4). These constraints account for the generating unit capacity, start-up ramp rate limit, shut-down ramp rate limit and the ramp-up limit of the unit. The maximum available output from the generator becomes zero when $v_{i,t} = 0$, i.e., the unit is offline.

$$p_{i,t} \leq P_i^{\max} \left(v_{i,t} - z_{i,t+1} \right) + z_{i,t+1} SD_i; \quad \forall i \in N,\ t \in T \tag{9.3}$$

$$p_{i,t} - p_{i,t-1} \leq RU_i v_{i,t-1} + SU_i y_{i,t}; \quad \forall i \in N,\ t \in T \tag{9.4}$$

where, $P_i^{\max}$ is the maximum active power output of generator unit i, SD_i is the shut-down ramp limit of generating unit i, RU_i is the ramp-up limit of generating unit i, SU_i is the start-up ramp limit of generating unit i and $z_{i,t}$ is a binary variable indicating shut down status of generator unit i at time t (the unit is shut down if value of $z_{i,t}$ is 1 and online if its value is 0).

The constraint in (9.5) enforces the ramp-down rate limit and the shut-down ramp rate limit for the unit.

$$p_{i,t-1} - p_{i,t} \leq RD_i v_{i,t} + SD_i z_{i,t}; \quad \forall i \in N,\ t \in T \tag{9.5}$$

where, RD_i is the ramp-down ramp limit of generating unit i.

Expressions (9.6)–(9.9) impose minimum up time constraints on the generating units.

$$\sum_{t=1}^{GU_i} \left[1 - v_{i,t} \right] = 0; \quad \forall i \in N \tag{9.6}$$

$$GU_i = Min\left[T, \left(UT_i - U_i^0 \right) V_i^0 \right]; \quad \forall i \in N \tag{9.7}$$

$$\sum_{k=t}^{t+UT_i-1} v_{i,t} \geq UT_i y_{i,t}; \quad \forall i \in N,\ t \in \{GU_i + 1, \ldots, T - UT_i + 1\} \tag{9.8}$$

$$\sum_{k=t}^{T} \left[v_{i,t} - y_{i,t}\right] \geq 0; \quad \forall i \in N,\ t \in \{T - UT_i + 2, \ldots, T\} \tag{9.9}$$

where, GU_i is the number of period that generating unit i must be online at the beginning of the day-ahead market optimization horizon due to its minimum up time constraint, UT_i is the minimum up time of generating unit i, U_i^0 is the time periods that generating unit i has been online at the beginning of the day-ahead market optimization horizon, V_i^0 is the initial commitment status of generating unit i (1 if it is online and 0 if it is offline), $y_{i,t}$ is a binary variable indicating start-up status of generator unit i at time t (the unit is started up if its value is 1 and offline if it is 0).

Constraint (9.6) accounts for the initial status of the units. GU_i is the total number of initial periods during which the unit i must be online and is calculated as shown in (9.7). The constraint in (9.8) ensures that the minimum up time constraint during all the possible sets of UT_i consecutive periods is satisfied for each period following GU_i. If a generating unit is started up in one of the last $(UT_i - 1)$ periods, (9.9) ensures that it remains online during the rest of the periods until $t \in \{T\}$.

The set of expressions in (9.10)–(9.13) impose the minimum down time constraints on the generating units. These are similar to the minimum up time constraints with the difference that $(1 - v_{i,t})$, $y_{i,t}$, UT_i, U_i^0 in (9.6)–(9.9) are replace by $v_{i,t}$, $z_{i,t}$, DT_i, S_i^0 in (9.10)–(9.13), respectively.

$$\sum_{t=1}^{GD_i} v_{i,t} = 0; \quad \forall i \in N \tag{9.10}$$

$$GD_i = Min\left[T, \left(DT_i - S_i^0\right)\left(1 - V_i^0\right)\right]; \quad \forall i \in N \tag{9.11}$$

$$\sum_{k=t}^{t+DT_i-1} \left[1 - v_{i,t}\right] \geq DT_i z_{i,t}; \quad \forall i \in N,\ t \in \{GD_i + 1, \ldots, T - DT_i + 1\} \tag{9.12}$$

$$\sum_{k=t}^{T} \left[1 - v_{i,t} - z_{i,t}\right] \geq 0; \quad \forall i \in N,\ t \in \{T - DT_i + 2, \ldots, T\} \tag{9.13}$$

where, GD_i is the number of period that generating unit i must be offline at the beginning of the day-ahead market optimization horizon due to its minimum down time constraint, DT_i is the minimum down time of generating unit i and S_i^0 is the time periods that generating unit i has been online at the beginning of the day-ahead market optimization horizon.

The constraints in (9.14) and (9.15) are necessary to model the start-up and shut-down status of the units and avoid simultaneous commitment and de-commitment of a unit.

$$y_{i,t} - z_{i,t} = v_{i,t} - v_{i,t-1}; \quad \forall i \in N,\ t \in T \tag{9.14}$$

$$y_{i,t} + z_{i,t} \leq 1; \quad \forall i \in N,\ t \in T \tag{9.15}$$

9.5.1.2 PEV Battery Constraints

In the developed mathematical model, the individual batteries are assumed to be aggregated and treated as a single battery. The constraints essentially reflect the charging and discharging operation of the aggregated vehicle battery while accounting for the traveling energy needs of PEV owners based on their aggregated driving pattern. It is further assumed that the vehicles are available to the grid for charging at all times when they are not traveling.

Minimum Energy Requirement

It is assumed that the PEV owner gives the aggregator information about when and how much of traveling is intended for the next day. Based on this information, the battery is charged only that amount of energy necessary over its initial state.

$$SOC^{ini} + \sum_{t=1}^{T} E_t \geq SOC^{\min} + \sum_{t=1}^{T} E_t^{next} \tag{9.16}$$

where E_t^{next} is the MWh energy required by the PEV for next day travel during hour $t \in T$, SOC^{ini} is the initial state of energy in the battery in MWh and $SOC^{\min}$ is the minimum energy requirement imposed by the PEV owner on the battery in MWh. An example of minimum energy requirement input data of a single PEV provided to the aggregator is shown in Table 9.1.

Charging Period Limit

The PEV aggregator should ensure that the charging of the PEV occurs in such a way that the battery is charged during hours $tf = 1, 2, \ldots, (t-1)$ before it travels

Table 9.1 PEV related data [15, 19]

Battery capacity	24 kWh
Energy consumption	0.192 kWh/km
Distance travelled	40 km/day
Energy consumption per day	7.68 kWh/day
Charging power	3.7 kW

during hour t for all values of $t \in T$. This is formulated mathematically as shown in (9.17).

$$\sum_{tf=1}^{t-1} \left(E_{tf} - E_{tf}^{next} \right) \geq E_{t}^{next} \tag{9.17}$$

Battery State
Charging and discharging of the battery during consecutive hours results in a change in its energy level for all $t \in T$. This is formulated as shown in (9.18).

$$SOC_t = SOC_{t-1} + E_t - E_t^{next} \tag{9.18}$$

Battery Energy Limits
The energy state in the battery should not deviate from its minimum and maximum limits, $SOC^{\min}$ and $SOC^{\max}$, respectively for all $t \in T$ as shown in (9.19).

$$SOC^{\min} \leq SOC_t \leq SOC^{\max} \tag{9.19}$$

It could be noted that the aggregator could face uncertainty by obtaining input data from PEV owners. One method to plan for PEV charging scheduling while considering uncertainty in PEV demand as well as electricity price in the day-ahead market is proposed in [18] using a stochastic programming framework.

9.5.1.3 Power Balance Constraints

The power balance between generation and supply must be maintained. This is mathematically formulated as shown in (9.20).

$$\sum_{i} p_{i,t} = CL_t + E_t \tag{9.20}$$

The total demand consists of the conventional demand CL_t and the demand from the PEV charging energy E_t. The PEV charging energy is an endogenous variable when PEV scheduling is performed using the JSM. However, it is provided as an input parameter to the DAM model when the ASM is utilized.

9.5.2 Aggregator Scheduling Method

In ASM, the PEV aggregator is assumed to function similar to an electricity retailer in the market. The aggregator plans for DAM participation by independently scheduling PEV energy based on its objective of minimizing the total cost of

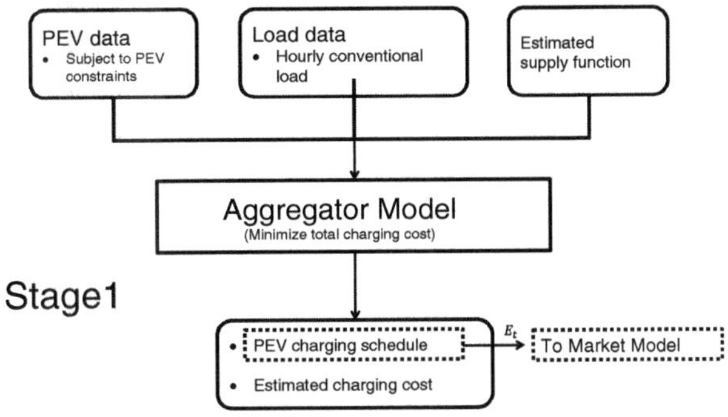

Fig. 9.5 Overview of the ASM: Stage-1

charging. For the scheduling, the PEV aggregator is assumed to have the following three sets of information:

(1) Daily PEV driving energy requirements, driving pattern data and aggregated PEV battery energy limits.
(2) Hourly conventional load data, which represents the inflexible demand from all other loads other than PEV demand.
(3) Estimated supply function to model the PEV aggregator as a price maker. If the aggregator needs to be modeled as a price taker, then the estimate supply function input data could be replaced directly with the forecasted electricity price profile.

Using the above sets of data, the aggregator schedules the charging energy of PEVs such that the total cost of charging is minimized as shown in Fig. 9.5.

9.5.2.1 The PEV Aggregator Model

The PEV aggregator ensures that the charging and discharging events of the vehicle's aggregated battery is scheduled considering the unavailability of PEVs due to driving needs. Batteries within electric vehicles are essentially loads that are required to be charged with sufficient energy to ensure smooth operation of the vehicle according to the driver's needs. Hence, it could be reasonable to assume that the main position held by the PEV aggregator is as a consumption entity within the electricity market. Considering such a stance, the objective function of the aggregator would then be to make sure that the cost from energy purchased for charging of all the PEVs is minimized while accounting for the driving needs of the PEVs. Due to its participation in the day-ahead market, the charging energy price would depend on the market price of electricity. If hourly charging costs are directly imposed on the PEV owners, the objective function could then be represented using (9.21).

$$Minimize \quad ACC = \left(\sum_{\forall t \in T} \pi_t^m * E_t \right) \tag{9.21}$$

where, ACC is the total cost of charging estimated by the PEV aggregator, π_t^m is the estimated market price from the estimated supply function and E_t is the charging energy to be scheduled at time t over the optimization horizon T. It is possible that the charging price used by the PEV aggregator is either an endogenous variable or an exogenous parameter. If the market is such that it requires the aggregator to plan the hourly charging needs before submitting its energy requirements to the market, then the electricity price would need to be estimated and it would identify itself as an exogenous parameter within the aggregator model.

The objective function of the PEV aggregator shown in (9.21) is subject to constraints imposed by the needs of vehicle owners along with the technical limitations of the battery as described in (9.16)–(9.19).

The estimated supply function gives an approximation of how the market price varies with changes in total demand. This function is important to identify the effect of total PEV demand on the market price when it is no longer a price taker. The estimated charging price could be modeled as a function of the total demand within the system as is shown in (9.22).

$$\pi_t^m = f(CL_t, E_t) \tag{9.22}$$

where, CL_t is the aggregator forecasted conventional load. The supply function can also be estimated from historical data on price and demand level cleared in the market.

9.5.2.2 The Market Model

The PEV charging schedule E_t from the ASM: Stage-1 is then input into the market model in the ASM: Stage-2 where the generators are scheduled to meet the total demand from the conventional load and the scheduled PEV energy in a way so as to minimize the total generation cost. This is shown in Fig. 9.6.

The objective function of the market model in ASM: Stage-2 is described by *DAMC* in (9.1) and is subject to constraints (9.2)–(9.15) imposed by the generating units and the power balance constraint (9.20).

9.6 Case Studies

In this section, the presented joint scheduling and aggregator scheduling methods are applied on a modified IEEE 30-bus test system and a Nordic test system to observe the effect of PEV demand scheduling on changes in price of electricity.

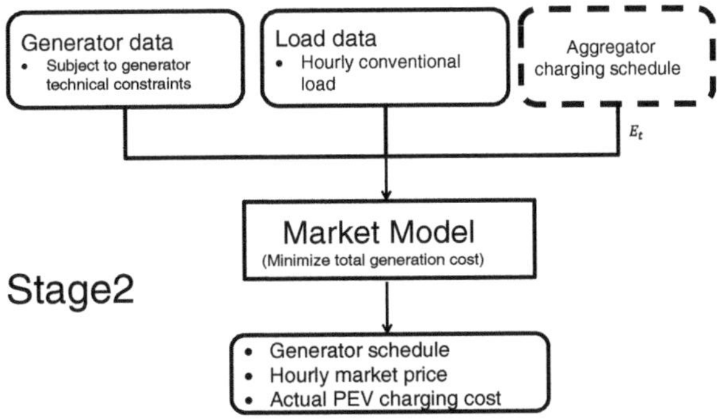

Fig. 9.6 Overview of the ASM: Stage-2

PEV related data shown in the following paragraphs is used that is common to all the case study simulations performed. The description of the IEEE 30-bus test systems and the representative Nordic system along with the results from the case studies are further described in Sects. 9.6.1 and 9.6.2, respectively.

The input data related to PEVs used for both the JSM and ASM case studies were obtained from a report published by the Grid for Vehicles (G4 V) project under the European Commission's 7th framework program [19], and are shown in Fig. 9.7 and Table 9.1.

The driving pattern in Fig. 9.7 is dependent on vehicle users and it is reasonable to assume that the driving behavior would not change drastically with the introduction of PEVs. Hence, the conventional vehicle user behavior is considered to be representative of the expected PEV user behavior.

The battery capacity and energy consumption in Table 9.1 are calculated based on the expected composition, at high penetration levels, of battery electric vehicles and plug-in hybrid electric vehicles, and represent a weighted average value [15, 19].

The battery charging and discharging characteristics are highly non-linear and depend on the type of battery. Li-ion batteries are considered as they appear to be the most promising type for PEV application [20]. Their charging curve indicates

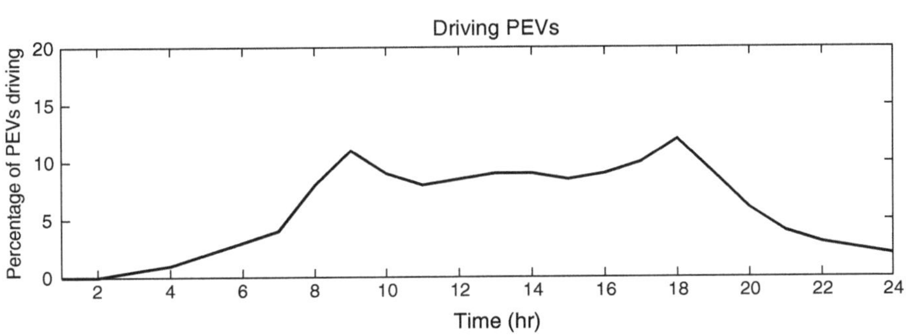

Fig. 9.7 Driving pattern of PEVs based on conventional vehicle data [15, 19]

that the charging power is nearly constant within a certain range of their state-of-charge [21]. Hence, the values of SOC^{min} and SOC^{max} are fixed at 20 and 85 % of the battery capacity for all simulations.

9.6.1 The IEEE 30-Bus Test System

The presented JSM and ASM have been applied to a modified IEEE 30-bus test system [22] to observe the effects of PEV aggregator demand scheduling on the price of electricity. The test system consists of nine generating units that are subjected to the following general technical constraints [16]:

- Min/Max generation limit
- Min up/down time
- Ramp up/down rate limits
- Start-up/Shut-down ramp rate limits

The penetration level of PEVs is defined as the ratio of total number of PEVs to the total number of vehicles in the system. It is estimated that a total of 170,000 PEVs would, in addition to the conventional load, result in energy requirements that would lead to the flattening of the daily load curve at a level corresponding to the peak demand. Since, information about vehicles in this test system is not readily available; it is assumed that there are a total of 170,000 vehicles in the system.

In ASM, the aggregator is considered to make use of the estimated supply function described in (9.23) to estimate the effect of PEV load on the market price and schedule the charging accordingly.

$$\pi_t^m = a_1 (CL_t + E_t) + a_0 \tag{9.23}$$

where, a_1 and a_0 are the constant coefficients. The estimated supply function for this system is shown in Fig. 9.8.

9.6.1.1 Fixed Period Charging

To obtain an idea of how the total load and market price will vary with the introduction of PEVs while the market has limited control over the charging, a fixed period charging method is described. A simple controlled charging mechanism of PEVs can be implemented by fixing their charging schedule during certain hours of the day when the conventional load is low. Figure 9.9 shows the variation of total hourly load at different levels of PEV penetration within the system when PEV charging is limited to hours 1–6.

Figure 9.10 shows the variation of hourly market price with different levels of PEV penetration within the system. It can be observed that at penetration levels of 20 and 50 %, increase in market price is not significant indicating that even a simple

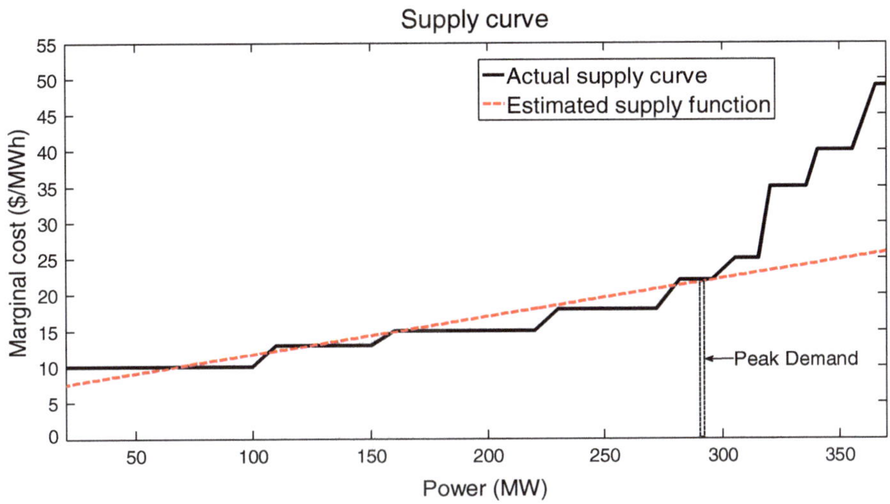

Fig. 9.8 Actual supply curve and estimated supply function used for modified IEEE 30-bus system

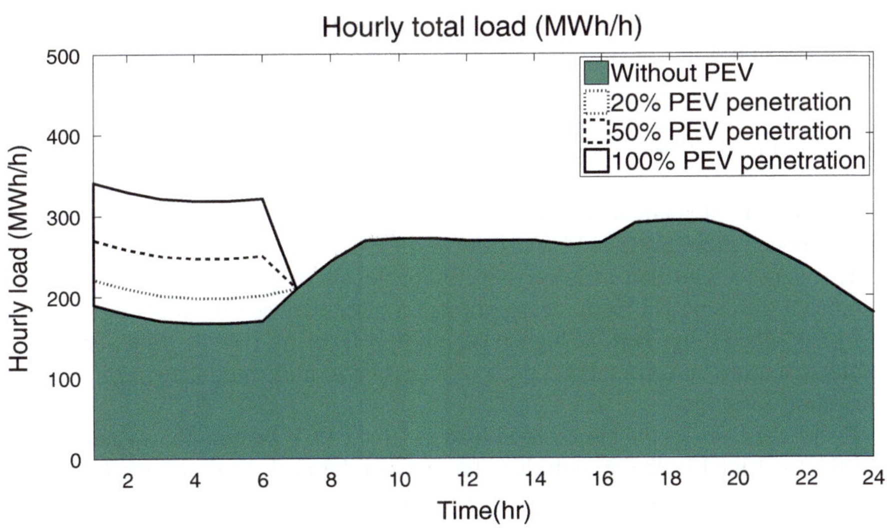

Fig. 9.9 Fixed period charging result—hourly total load

charging mechanism could be effective in maintaining an acceptable increase in market price by the introduction of PEVs. But at higher penetration levels, i.e., 100 %, Fig. 9.9 indicates that the total load during early hours exceeds the peak demand due to conventional load alone (hour-18). The increase in market price can

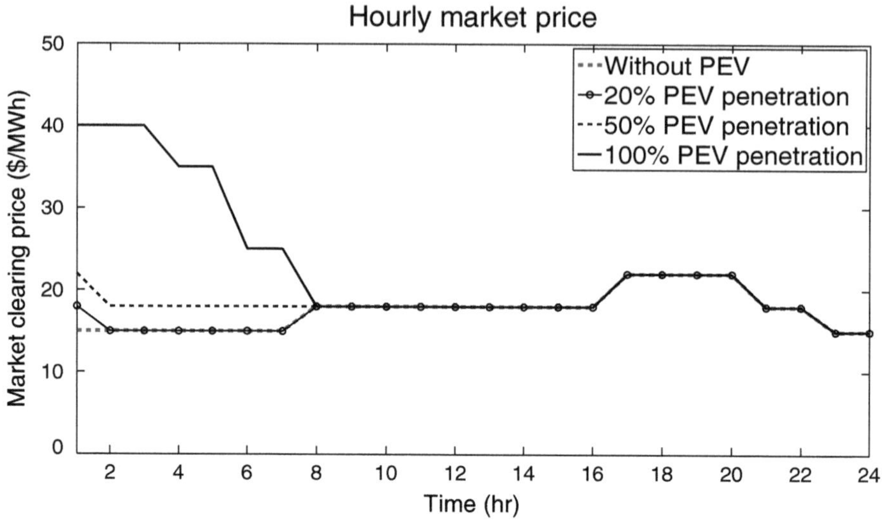

Fig. 9.10 Fixed period charging result—hourly market price

also be seen in Fig. 9.10 as more expensive generators need to be scheduled to supply the additional PEV load resulting in a market price as high as 40 $/MWh during the first 3 h.

9.6.1.2 Joint Scheduling Method

The JSM is implemented for this test system and the resulting market price for various penetration levels of PEVs is shown in Fig. 9.11. Comparing Figs. 9.10 and 9.11, it can be seen that at lower penetration levels of 20 and 50 %, there is no significant difference in the increase of market price between fixed period charging and joint scheduling. But, at higher penetration level of 100 %, joint scheduling results in a more uniform market price of 22 $/MWh, indicating better utilization of generating resources.

Figure 9.12 shows the hourly total load at 100 % PEV penetration. It can be seen that the total load in the system does not exceed the peak load at hour 18 even at 100 % PEV penetration. This can be significant in systems that are stressed and might need network reinforcement in the case of fixed period charging, but the same can be avoided using joint scheduling. It is interesting to note that little or no charging takes place during the hours 23 and 24. This could possibly be due to two reasons—one, the optimization horizon in the model is limited to 24 h and two, the PEV energy requirements need to be respected before their hour of travel.

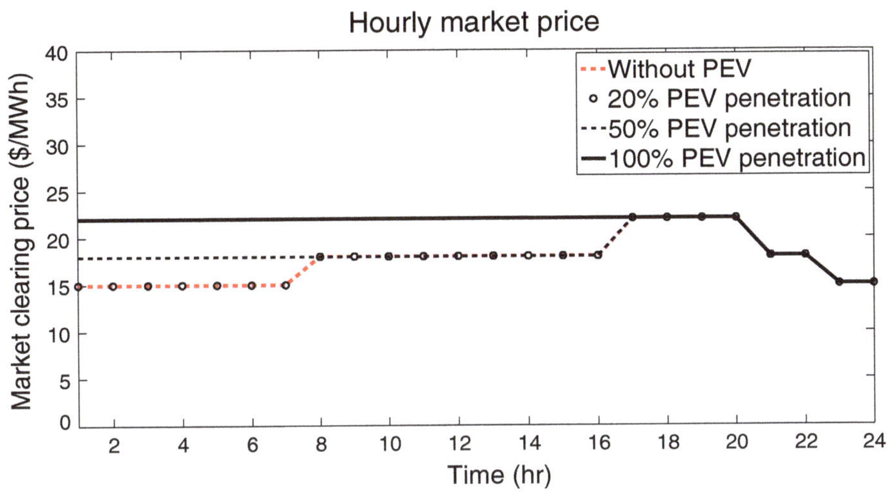

Fig. 9.11 JSM result: market price at zero and 100 % PEV penetration

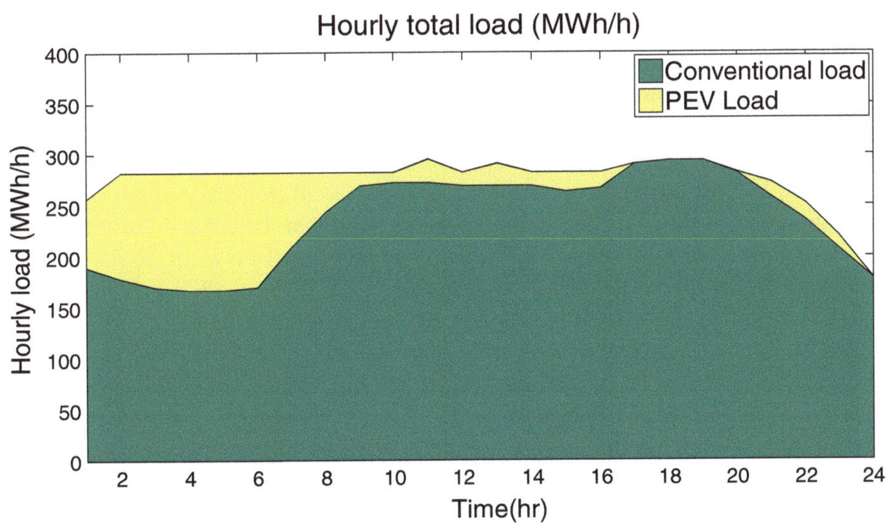

Fig. 9.12 JSM result: system demand at zero and 100 % PEV penetration

9.6.1.3 Aggregator Scheduling Method

The results obtained from ASM are shown in Fig. 9.13 for various PEV penetration levels. Comparing Figs. 9.10 and 9.13, it can be seen that at 20 % penetration, the market price during the day increases similarly in both models. However, at 50 % penetration level, the aggregator model results in an increase of market price by

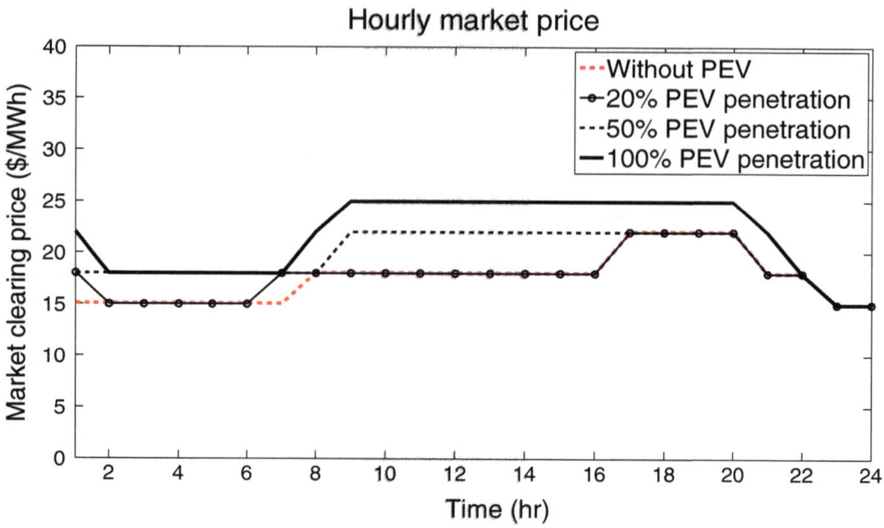

Fig. 9.13 ASM result: market price at various PEV penetration levels

4 $/MWh during hours 9–20. This could be attributed to the aggregator not being able to perfectly forecast the dependency of price on changes in demand. Since, forecasting brings about an error in the estimated price, the aggregator schedules higher or lower charging energy during an hour, depending on whether the demand dependency was underestimated or overestimated, respectively.

The hourly load for ASM is compared with the result from JSM for 100 % PEV penetration and is shown in Fig. 9.14. It can be seen that the error in estimation by the aggregator results in lower PEV load to be scheduled between hours 2 and 7 when compared with the JSM. Due to this under-scheduling of PEV load during the early hours, greater PEV load is scheduled between hours 9 and 21.

The corresponding changes in market price can be seen in Fig. 9.15. This price directly reflects the errors in forecasting by the aggregator on market price. It is lower by about 4 $/MWh during hours 2–7 but, consequently, increases by 4 $/MWh during the later hours 9–21 when compared to JSM results.

9.6.2 The Nordic Test System

The joint scheduling approach is used to simulate the participation of PEV aggregator and its charging energy scheduling in the Nordic day-ahead market, which consists of four participating countries from the Nordic region, namely—Norway, Sweden, Finland and Denmark. The market players who want to trade electricity on the Elspot market must submit their sell offers and/or buy bids for every hour of trading to the market, no later than 12:00 h, on the day before the power delivery.

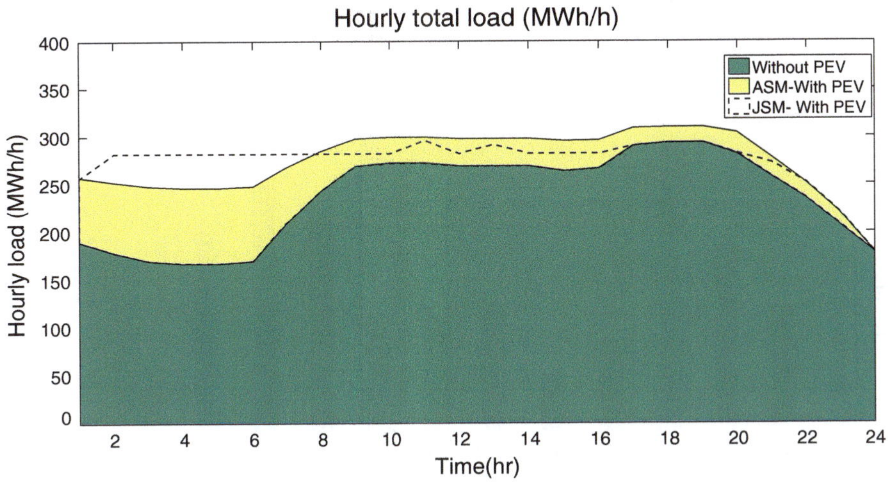

Fig. 9.14 ASM result: system demand at various PEV penetration levels

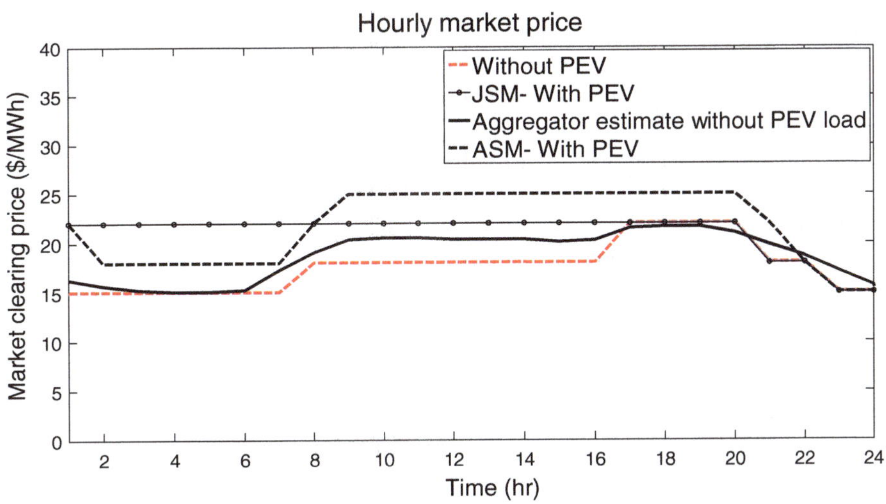

Fig. 9.15 Comparison of resulting market price with aggregator and joint scheduling methods

These bids are submitted via the internet to the website of Nord Pool Spot. The collected sell bids are cumulated in increasing order of price to form a supply curve and the buy bids are cumulated in decreasing order of price to form a demand curve —for every hour. The intersection of the two curves gives the market price of electricity for that hour. More information on the operation of Elspot can be found in [4].

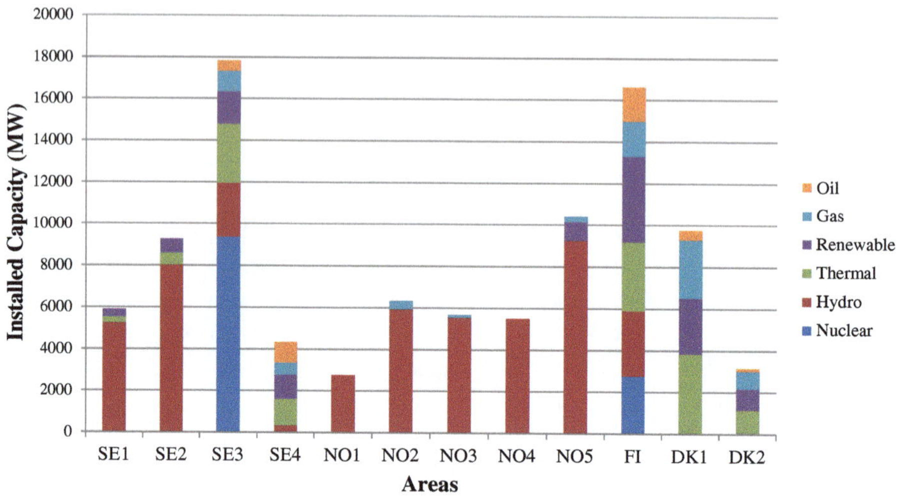

Fig. 9.16 Generating capacity distribution by bidding area in the Nordic region [23]

Due to the physical restrictions imposed on energy trading by transmission lines, the Nordic electricity market area is divided into a number of bidding areas. The TSO decides on the criteria and number of bidding areas. Since, the operation of a TSO is generally limited to one country; a bidding area does not traverse political boundaries between countries. Currently, Norway is split into 5 bidding areas—NO1 to NO5; Sweden into four—SE1 to SE4; Denmark into two—DK1 and DK2; and Finland into one—FI.

The total installed generating capacity in the Nordic region is around 96 GW [23]. Figure 9.16 shows the share of total installed generation capacity based on the bidding areas in the Nordic region (excluding Estonia, Latvia and Lithuania).

Installed generation capacity data for units greater than 100 MW including the type of generating technology for all four countries is obtained from [24]. The variable cost of power generation based on different technologies in [25] is used and scaled to reflect the average system prices in Nord Pool Spot for the year 2012 [26], after which the aggregated supply curve in the Nordic market can be obtained as shown in Fig. 9.17.

The aggregated supply curve is based on installed generation capacity in four countries—Norway, Sweden, Finland and Denmark. A normal market situation is considered, where, all the installed generation capacity is available. Two generation technologies that influence this assumption critically in the Nordic market are—hydro and nuclear power. With respect to hydro power, it reflects a situation when there is sufficient inflow to the hydro power station reservoirs in Norway and Sweden. This can further be classified as a normal winter that occurs every other year. This is in line with a study on vulnerabilities of the Nordic power system where, 90 % hydro availability is assumed in Norway and Sweden during normal

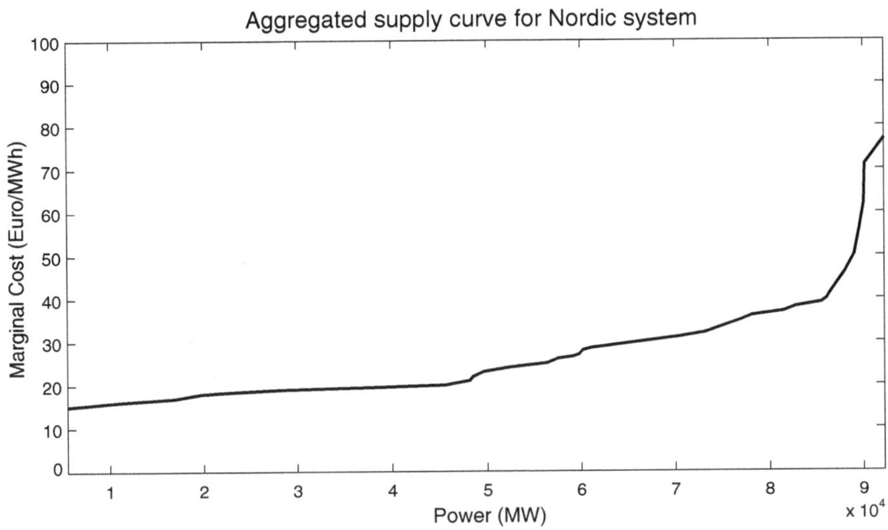

Fig. 9.17 Supply curve in the representative Nordic market [25]

hydro conditions [27]. Similarly, due to the low probability of forced outages of nuclear power generation in Sweden and Finland, 100 % availability is assumed.

The vehicle data required to simulate the participation of PEVs in the Nordic market is based on statistics available for conventional fuel driven vehicles and is obtained from [28–31]. The resolution of this data currently available is for each county present in each of these countries. The total number of conventional vehicles in the Nordic area is found to be around 12.7 million. These are approximately segmented according to bidding areas and the resulting distribution is shown in Fig. 9.18.

It is difficult to estimate the elasticity of conventional demand in the short-term since this elasticity would occur in special circumstances, where, the price of electricity is very high over a sustained period of time (days or weeks). Hence, the conventional demand is assumed to be inelastic. The impact of the assumptions made in this study on the final results and its analysis is optimistic, while at the same time, reflecting a highly expected market situation. It is imperative to mention that the simulation models are designed for a single auction market while the Nordic market is, in fact, a double auction market where a number of market players determine the outcome. A direct consequence of this may be a lower resulting market price due to better utilization of generating resources.

The external interconnection capacities between countries within and outside of the Nordic area are included as inelastic demand thereby representing an export scenario from the Nordic countries. This is indicative of an anticipated market situation, though, in reality, the complete transmission capacity may not be utilized.

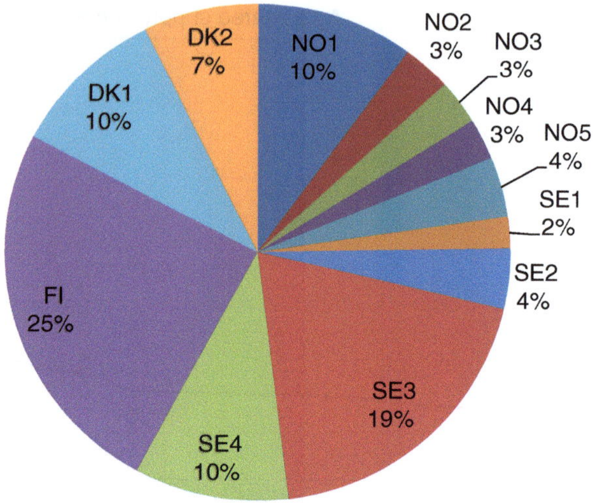

Fig. 9.18 Conventional vehicles' distribution in Nordic region based on bidding areas [28–31]

The JSM will be applied to the following two cases of the Nordic market:

(1) *Unconstrained case*: When the trading of electricity is not limited by the interconnection capacities between different bidding areas in the Nordic region.
(2) *Constrained case*: Trading of electricity is limited by the interconnection capacities between different bidding areas in the Nordic region, which are modeled based on the net transfer capacity (NTC) values [32].

Only joint scheduling is used for the case study of the Nordic market, because the aggregator scheduling is heavily dependent on the accuracy of the estimated supply function given by (9.23). The accuracy of this function could be improved by modeling the price as a polynomial function of demand, although, by doing so, the complexity of the optimization function increases and the resulting model might not necessarily provide a solution. The consequence of such an assumption is the results being more optimistic, where the available generation and flexible demand are utilized more resourcefully.

9.6.2.1 Unconstrained Case

If there were no upper limits on interconnection capacities, one supply and one demand curve could in theory be used for the clearing of the whole Nordic DAM. In a single auction market, it would translate into a single supply curve for the entire Nordic market. This would then be matched with the demand curve during that particular hour to obtain the market price for electricity.

The demand profile for this system was obtained using the data in [33] for a Tuesday during week 51 with an aggregated peak demand of 69 GW [23]. In such a context, if EVs are introduced into the system and their charging energy traded in

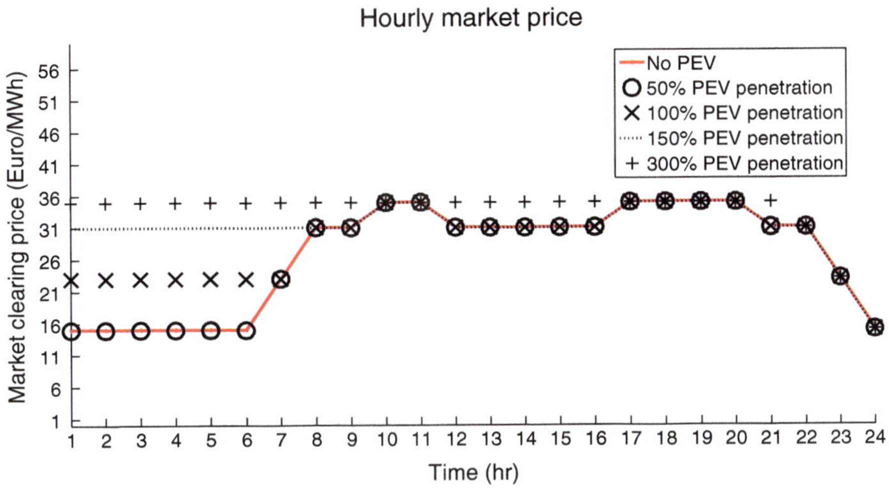

Fig. 9.19 Unconstrained case result—changes in market price by the introduction of PEVs

the Nordic market as flexible demands, the corresponding changes to the electricity price in the day-ahead market, for various penetration levels of PEVs, can be obtained as shown in Fig. 9.19.

It could be observed that even if all the 12.7 million conventional vehicles were replaced by PEVs, the market price would increase by 8 €/MWh during low demand periods. It would require an introduction of at least 37 million PEVs before the system price during most hours corresponds to the peak load price of 35 €/MWh during hour 18. Hence, the Nordic market could be considered to be highly resilient towards the introduction of PEVs. The changes in hourly total load and market price, with the introduction of 12.7 million PEVs in the Nordic region, are shown in Figs. 9.20 and 9.21, respectively. Since, only a single auction market is considered, it could be seen that JSM schedules the charging energy of PEVs during low demand periods as shown in Fig. 9.20, when the price of electricity is low as shown in Fig. 9.21. The impact of this is a minimal increase in total demand and corresponding electricity price.

9.6.2.2 Constrained Case

With interconnection capacities in place, area prices apply when power traded between at least two areas in the market exceeds the total available transmission capacity between those areas. The area market prices in the Nordic market for the constrained case are shown in Fig. 9.22. Y-axis denotes the area prices; x-axis denotes the 12 bidding areas and the colored bars from blue to red denote the 24 h under consideration for each area.

It can be seen in Fig. 9.22 that areas FI, SE4 and DK2 already suffer from high prices compared to other areas primarily due to the dominant fossil fuel-based local

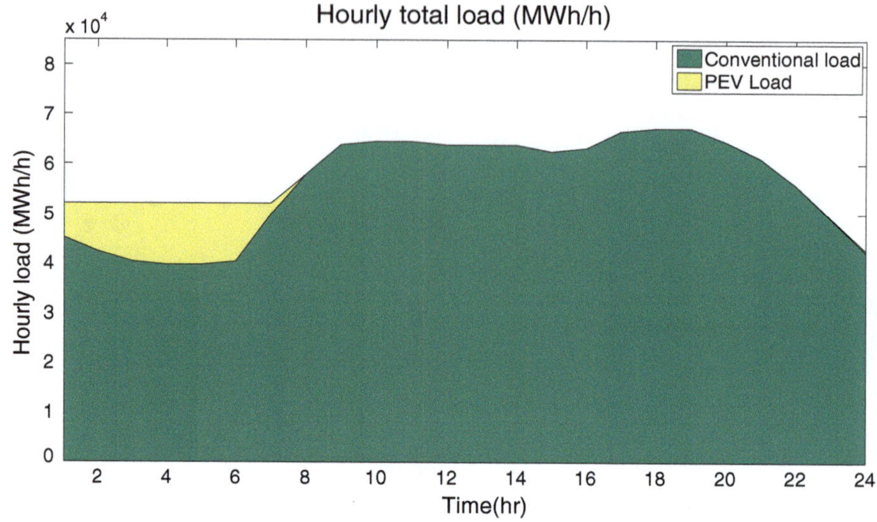

Fig. 9.20 Unconstrained case result—total load at 100 % PEV penetration

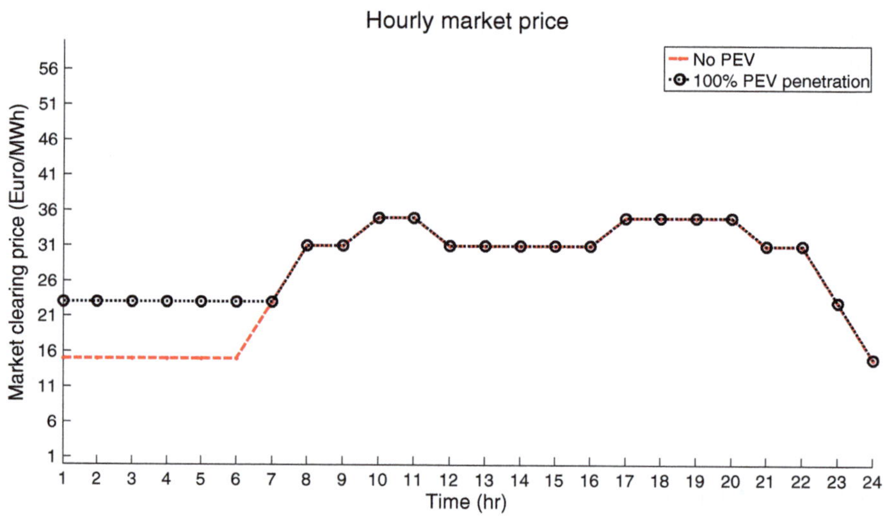

Fig. 9.21 Unconstrained case result—market price at 100 % PEV penetration

generation. Prices in most of the areas are different indicating that the interconnections between these areas have been fully utilized. In areas NO1 and NO2, it can be observed that the prices during all of the 24 h are the same indicating that the available transmission capacity is not completely utilized.

Extension of the model to include scheduling of PEV charging results in area prices as shown in Fig. 9.23, for 100 % penetration of PEVs in the market. In hydro power

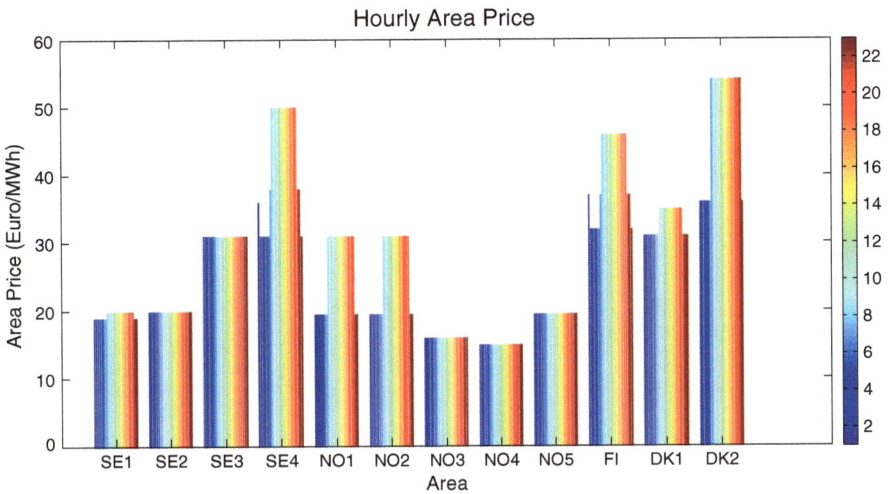

Fig. 9.22 Constrained case result—area prices with only conventional load

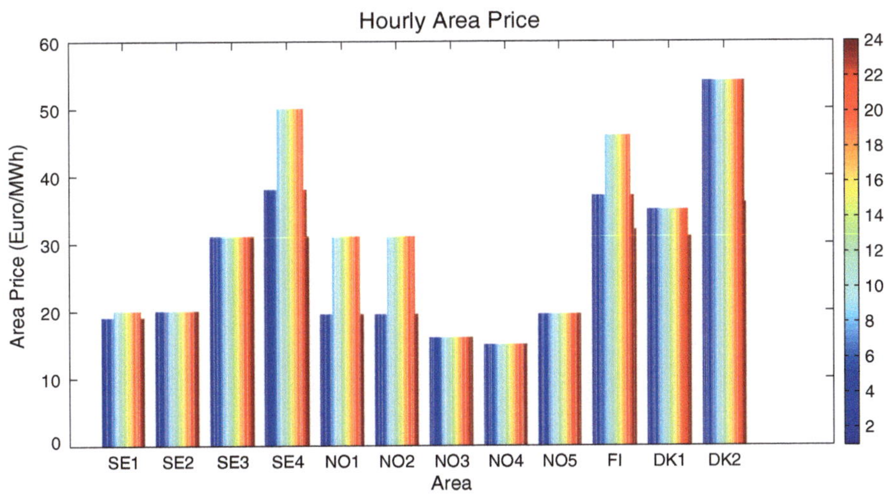

Fig. 9.23 Constrained case result—area prices with 100 % PEV penetration

dominated areas—SE1, SE2, NO1-NO5, it can be seen that the market price remains relatively the same during all hours even with a high penetration of PEVs. It is also found that mainly two areas, namely—SE4 and DK2 are affected by the high levels of PEV penetration. At 100 % PEV penetration level, the electricity price in DK2 increases to 54 €/MWh even during the low demand hours 1–7, whereas it increases to 38 €/MWh during the same hours. Further introduction of PEVs would result in a market price higher than 54 €/MWh in DK2 that corresponds to the price at peak demand with only conventional demand.

Area price for SE4 at different penetration levels is shown in Fig. 9.24a. Similarly, for the bidding area DK2, the area price at different penetration levels is shown in Fig. 9.24b. It can be seen that the area prices in SE4 and DK2 increase with an increased penetration of PEVs in the Nordic system. This may be attributed

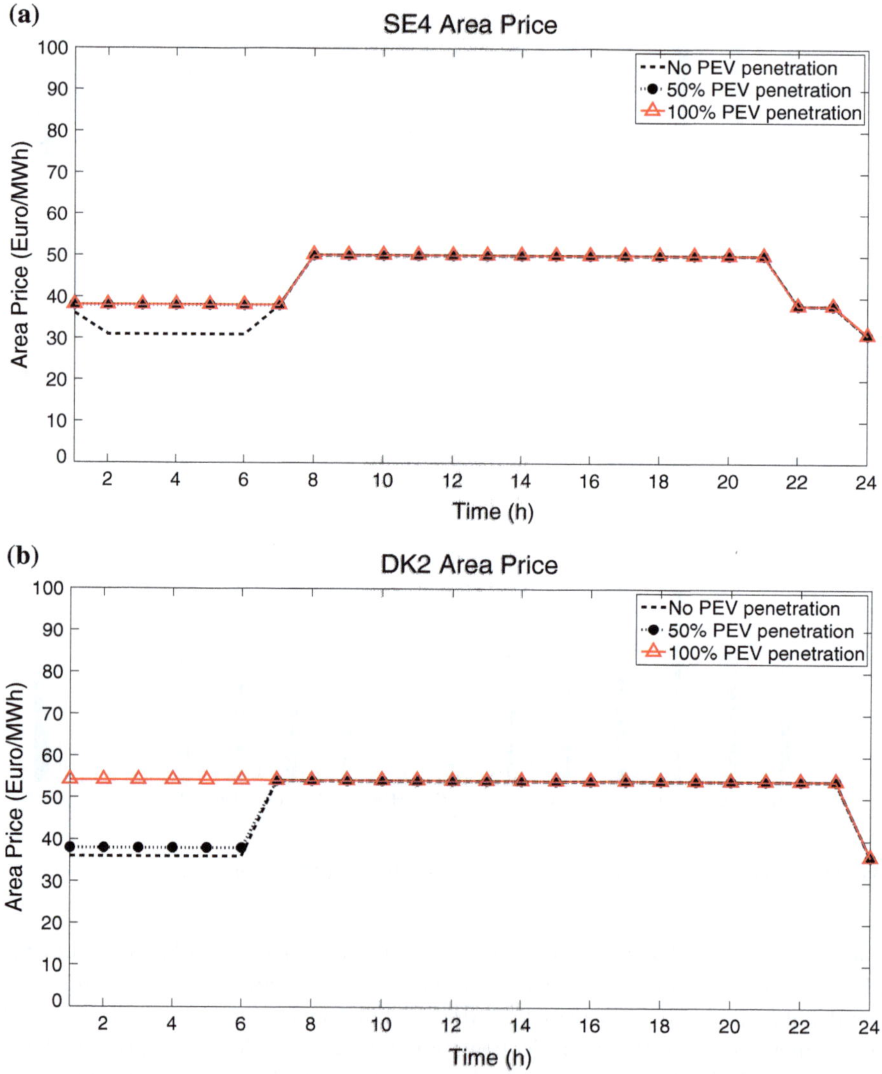

Fig. 9.24 Constrained case result: **a** SE4 area price at different PEV penetration levels, **b** DK2 area price at different PEV penetration levels

to a number of factors, e.g., these two areas are dominated by thermal generators which are generally more expensive, capacity of transmission lines connecting them to generator surplus areas are insufficient and greater population in these areas account for relatively higher number of PEVs being integrated at higher penetration levels.

9.7 Conclusions

In this chapter, two methods for scheduling PEV demand named as JSM and ASM have been presented. These methods could be used to evaluate the effects of PEVs scheduling on the overall system load shape and the effects on electricity market price. The JSM could prove useful in a market setup where there is a possibility to schedule both the generation and demand side resources; whereas the ASM could be useful where individual market players would require performing their individual energy scheduling. The two methods were applied to an IEEE 30-bus test system and the Nordic test system to find the effects of PEV energy scheduling on market price of electricity. From the case study on the IEEE 30-bus test system, it was found that market integration of PEVs might lead to an increase in market price at higher penetration levels using fixed period charging, at which point, advanced methods of scheduling of PEV fleet charging could become necessary. The JSM may require changes in the operational structure of electricity markets, but the model could result in better utilization of resources as it simultaneously schedules both the generation and demand resources. In the unconstrained case, the Nordic market was found to be highly resilient toward integration of PEVs. Transmission network constraints form an important factor on the system level which could influence the actual flow of power, and hence, directly influence the penetration level of the PEVs that could be accommodated in the system before a significant increase in market price.

Acknowledgments The authors wish to acknowledge the financial support provided by E.ON Sverige AB for the project '*E.ON A2—Future Electric Power Systems*' within the *E.ON-Chalmers* collaboration framework.

References

1. European Commission (2009) Directive 2009/72/EC of the European Parliament and of the Council of 13 July 2009 concerning common rules for the internal market in electricity and repealing Directive 2003/54/EC. Official J Eur Union
2. OECD Publishing (2010) World energy outlook 2010. Organization for Economic Cooperation and Development Ebrary, Inc. (distributor), Washington, Palo Alto
3. Boulanger AG, Chu AC, Maxx S, Waltz DL (2011) Vehicle electrification: status and issues. Proc IEEE 99:1116–1138

4. Nord Pool Spot (2013) The nordic electricity exchange and the nordic model for a liberalized electricity market. http://nordpoolspot.com/globalassets/download-center/rules-and-regulations/the-nordic-electricity-exchange-and-the-nordic-model-for-a-liberalized-electricity-market.pdf . Accessed 10 November 2014
5. Nord Pool Spot (2013) Nord Pool Spot—Europe's leading power markets. http://nordpoolspot.com/globalassets/download-center/annual-report/nord-pool-spot_europes-leading-power-markets.pdf. Accessed 10 November 2014
6. Kirschen D, Strbac G (2005) Basic concepts from economics. Fundamentals of power system economics. Wiley, New York, pp 11–47
7. Gellings CW, Barron W, Betley FM, England WA, Preiss LL, Jones DE (1986) Integrating demand-side management into utility planning. IEEE Trans Power Syst. doi:10.1109/TPWRS.1986.4334958
8. Lampropoulos I, Kling WL, Ribeiro PF, van den Berg J (2013) History of demand side management and classification of demand response control schemes. In: 2013 IEEE power and energy society general meeting (PES), pp 1–5
9. Smart Energy Demand Coalition (2014) Mapping demand response in Europe today. http://sedc-coalition.eu/wp-content/uploads/2014/04/SEDC-Mapping_DR_In_Europe-2014-04111.pdf. Accessed 10 November 2014
10. PJM (2014) Demand response. http://www.pjm.com/markets-and-operations/demand-response.aspx. Accessed 5 May 2014
11. Monitoring Analytics (2009) Barriers to demand side response in PJM. The independent market monitor for PJM. http://www.monitoringanalytics.com/reports/Reports/2009/Barriers_to_Demand_Side_Response_in_PJM_20090701.pdf. Accessed 10 November 2014
12. Lopes JAP, Soares FJ, Almeida PMR (2011) Integration of electric vehicles in the electric power system. Proc IEEE 99:168–183
13. Hay C et al (2010) Introducing electric vehicles into the current electricity market. EDISON Project. http://www.edison-net.dk/Dissemination/Reports/Report_004.aspx. Accessed 10 November 2014
14. Andersson SL, Elofsson AK, Galus MD, Göransson L, Karlsson S, Johnsson F, Andersson G (2010) Plug-in hybrid electric vehicles as regulating power providers: case studies of Sweden and Germany. Energy Policy 38:2751–2762
15. Balram P, Tuan LA, Bertling Tjernberg L (2012) Effects of plug-in electric vehicle charge scheduling on the day-ahead electricity market price. In: 2012 3rd IEEE PES international conference and exhibition on innovative smart grid technologies (ISGT Europe), pp 1–8
16. Arroyo JM, Conejo AJ (2002) Multiperiod auction for a pool-based electricity market. IEEE Trans Power Syst 17:1225–1231
17. Bhattacharya K, Bollen MH, Daalder JE (2001) Operation of restructured power systems. Kluwer, Boston
18. Balram P, Tuan LA, Bertling Tjernberg L (2013) Stochastic programming based model of an electricity retailer considering uncertainty associated with electric vehicle charging. In: 2013 10th international conference on the European energy market (EEM), pp 1–8
19. Silva V (2011) Estimation of innovative operational processes and grid management for the integration of EV. Grid for Vehicles. http://www.g4v.eu/datas/reports/G4V_WP6_D6_2_grid_management.pdf. Accessed 10 November 2014
20. Work Package 1.3: Parameter manual. Grid for vehicles (G4 V) project. http://g4v.eu/datas/Parameter_Manual_WP1_3_RWTH_101216.pdf. Accessed 3 April 2014
21. Battery university (2014) Charging lithium-ion batteries. http://batteryuniversity.com/learn/article/charging_lithium_ion_batteries
22. Shahidehpour M, Yamin H, Li Z (2002) Market operations in electric power systems: forecasting, scheduling, and risk management. Wiley Online Library, New York
23. EURELECTRIC (2009) Statistics and prospects for the European electricity sector, 37th edn, EURPROG 2009
24. Nord Pool Spot (2012) Power plant units (installed generation capacity greater than 100 MW)

25. Swedish Energy Agency (2004) The energy market 2004. https://energimyndigheten.a-w2m.se/FolderContents.mvc/Download?ResourceId=2163. Accessed 10 November 2014
26. Energy in Sweden (2014) Facts and figures (excel) 2014
27. Doorman G, Kjølle GH, Uhlen K, Huse ES, Flatabø N (2004) Vulnerability of the Nordic power system. Report to the Nordic Council of Ministers, SINTEF Energy Research
28. Statistics Norway (2012) Stock of vehicles and population, by County
29. Statistics Sweden (2013) Vehicles in counties and municipalities at the turn of year 2013/2014
30. Statistics Finland (2013) Motor vehicle stock
31. Statistics Denmark (2012) Motor vehicles registered 1 January by reporting country, vehicle and time
32. ENTSO-E (2014) Maximum NTC. European network of transmission system operators for electricity (ENTSO-E)
33. Grigg C, Wong P, Albrecht P et al (1999) The IEEE reliability test system-1996. A report prepared by the reliability test system task force of the application of probability methods subcommittee. IEEE Trans Power Syst 14:1010–1020

Chapter 10
Optimal In-Home Charge Scheduling of Plug-in Electric Vehicles Incorporating Customer's Payment and Inconvenience Costs

Mahmud Fotuhi-Firuzabad, Soroush Shafiee and Mohammad Rastegar

Abstract Plug-in electric vehicles (PEVs) are identified as one of the motivating technologies in smart grid era. However, if their highly disruptive impacts on the distribution system are left unaddressed, it may obstruct both smart grid development and PEV adoption. This chapter develops a novel in-home PEV charging control (PCC) algorithm that schedules both the time and level of charging PEVs incorporating customer's desired comfort level. This optimization-based problem attempts to achieve a trade-off between *minimizing the electricity payment* and *minimizing the waiting time to fully charge the PEVs* in presence of a time of use (TOU) pricing tariff combined with inclining block rates (IBRs). The projected algorithm is *online* in which each PEV is scheduled at its plug-in time, and the charge scheduling of plugged-in PEVs are updated when the next PEV is plugged-into the home outlet. The proposed method is applied to a smart home with different number of PEVs and various levels of customer's comfort. In addition, the impacts of solving PCC problems on the specifications of the IEEE 34-bus residential test feeder with different PEV penetration levels are investigated. The simulation results are presented to demonstrate the effectiveness and applicability of the proposed PEV charge scheduling scheme.

M. Fotuhi-Firuzabad (✉) · S. Shafiee · M. Rastegar
Department of Electrical Engineering, Center of Excellence in Power System Management and Control, Sharif University of Technology, Tehran, Iran
e-mail: fotuhi@sharif.edu

S. Shafiee
e-mail: soroush.shafiee@gmail.com

M. Rastegar
e-mail: rastegar_m@ee.sharif.edu

S. Rajakaruna et al. (eds.), *Plug In Electric Vehicles in Smart Grids*,
Power Systems, DOI 10.1007/978-981-287-317-0_10

10.1 Introduction

Power system reliability, environmental concerns, and energy efficiency are the three major challenges of smart grid (SG) implementation [1]. In the SG era, plug-in electric vehicle (PEV) technology is an emerging paradigm and a promising solution to some environmental and economical problems. However, utilities are becoming concerned about the performance degradations and overloads that may take place in distribution systems due to multiple domestic PEV charging activities. So, as PEVs become more popular, they are more threatening the reliability and security of the grids.

In recent years, several studies demonstrated that high penetration levels of PEVs have negative impacts on the distribution network [2–5]. Due to proven adverse impacts of PEV charging, it seems necessary to monitor and control these inappropriate consequences. SG provides an outstanding opportunity to intelligently manage PEV charging in the distribution network.

Several literatures concentrated on the PEV charging control methods in a smart environment. PEVs can be charged either in public places or at homes. An aggregator of retail customers or distribution company (DISCO) is responsible for public charging of PEVs at charging stations or parking lots. Various objective functions such as maximizing customer benefit [6, 7] and maximizing average state of charge (SOC) for all vehicles [8, 9] are formulated to control PEVs charging at public charging places. A number of researches also focus on the in-home charging control. To the best of authors' knowledge, in almost all the literatures associated with the in-home charging, an aggregator or distribution system operator (DSO) is responsible for charging a group of PEVs. The responsible organization designs different strategies or optimization-based programs to manage PEVs charging. Sortomme et al. [10] propose three objective functions, i.e. minimizing losses, maximizing load factor, and minimizing load variance, to achieve an optimum scheduling of charging PEVs. References [3, 11] coordinate PEVs charging in a residential distribution network to minimize power losses and voltage deviation. Also, a smart load management is proposed in [12–14] to control in-home PEVs charging in a residential network to minimize the cost of charging. In these studies [3, 10–14], the DSO or aggregator manages PEVs charging.

It can be deduced from studying these literatures that the main purpose of proposed PEVs charging algorithm is to maximize social welfare, i.e. minimize peak load, system losses, or the total cost. As mentioned before, references [3, 10, 11] minimize system losses while it might be unappealing for the customers themselves. For example, presented results in [3] show that by applying the proposed optimization method, a PEV in one load point is charged faster than a PEV in another point. This might lead to irrational different payment costs for PEV owners in a time-varying pricing environment. Payment cost is the most important concern and appealing motive for customers to control their PEVs charging. Although references [12–14] minimize charging cost of a group of PEVs, this may also cause a discrimination against customers; because in these optimization procedures, customer preferences are not considered properly. Moreover, recently, some reports and surveys have been published about demand response programs, which prove that customers do not properly respond to the price signal. These kinds of customers do not sacrifice their comfort instead of the less payment cost. If such a program, for which the customer participation is required, does not appropriately provide the customers' comfort, customers will not participate in that program. Similarly, an applicable PEV charge scheduling program should cover all aspects of customer's preferences to satisfy the customer to participate in the program. The customers' interests and comfort can be incorporated in the optimization process as constraints [15, 16] and objective function [16–18].

The most prominent factor in customer's decisions is payment cost. Therefore, similar to many recent studies, the customer payment cost should be employed in the optimization problem as an objective function. A difference between the results of controlled and uncontrolled charging, only based on the payment cost, is that in the case of controlled charging the charging time usually shifts from afternoon hours to the later ones at midnight. This may not be desirable for all customers. We should provide a situation for the customer in which the PEV can be charged as soon as the customer wants. This can be presented as the waiting cost function in the objective of optimization problem. The concept of waiting time was previously used in [18] to schedule the household appliances. The presented method in [18] is a day ahead scheduling which confronts uncertainties. The challenge of real time pricing uncertainties is probed in the paper [18]. However, in the present study, the uncertainty challenges raised by the day-ahead scheduling of PEV are mitigated by proposing an online charge scheduling algorithm.

This chapter addresses the in-home PEV charging control (PCC) problem as an optimization effort incorporating the customer's comfort. The optimization problem is solved from the customer point of view by a PEV charging scheduler (PCS) improvised in each home. The output of the optimization problem would be the charging time and the level of connected PEVs at home. The objective is to achieve a trade-off between minimizing the customer inconvenience cost and payment cost of PEV charging. The payment cost is a function of PEVs charging demand and declared time-varying tariffs. In order to minimize the payment cost, the PCS shifts the charging of PEVs to inexpensive periods which rationally coincide with the valley of consumption profile. As a matter of fact, the peak to average ratio (PAR) of the load profile would likely decrease which is desirable from the utility viewpoint. The proposed inconvenience cost function encompasses the waiting time of the customer to have a fully charged PEV battery. It is clear that the desired level of convenience is different for various customers and should be set by the PEV owners in each house. This is considered in the problem formulation and guarantees more customer satisfaction. Also, in most of previous studies, the charge scheduling problem is solved for a day-ahead period in which the customers' behavior would likely face to some uncertainties. For example, in [16], departure time, travelling time, and out-of-home energy consumption of PEVs are modeled by a normal probability distribution function due to lack of data about customer's behavior. We tackle the problem by proposing an online charge scheduling algorithm in which the

PCS gets required data from the PEV once it is plugged in, and schedules the charging time and level of all the plugged in PEVs in the horizon time of scheduling. The proposed problem is materialized in a simple linear programming (LP) fashion that can be easily integrated in the energy management system of a household.

Verification of the proposed method effectiveness is probed and presented in different cases. PCC problem is solved in a smart home with different number of PEVs and different levels of comfort. In order to investigate the proposed PCC problem impacts on the distribution networks specifications, the optimization results with random customers' comfort levels are applied to the IEEE 34-bus residential test feeder with different PEV penetration levels.

The rest of the chapter is outlined as follows. The system data structure as well as dynamic pricing model used in this chapter is treated in Sect. 10.2. The basic PCC formulation, which just minimizes charging payment cost, is presented in Sect. 10.3. In Sect. 10.4, the PCC formulation is extended in order to take into account the inconvenience cost of waiting time for PEV charging. Simulation results are presented and discussed in Sect. 10.5. The chapter is concluded in Sect. 10.6.

10.2 System Data Structure and Dynamic Pricing

In this section, the data structure and requirements utilized in smart grids to manage the PEV charging are initially reviewed and then, dynamic pricing methods, specifically time of use (TOU), inclining block rate (IBR) pricing and the combination of these two pricing techniques are briefly discussed.

10.2.1 Data Structure and Requirements

A feasible data structure for applying an optimization program to schedule PHEV charge at a home is presented here. The general wholesale electricity market scenario is shown in Fig. 10.1, in which the retailer/utility serves a number of end

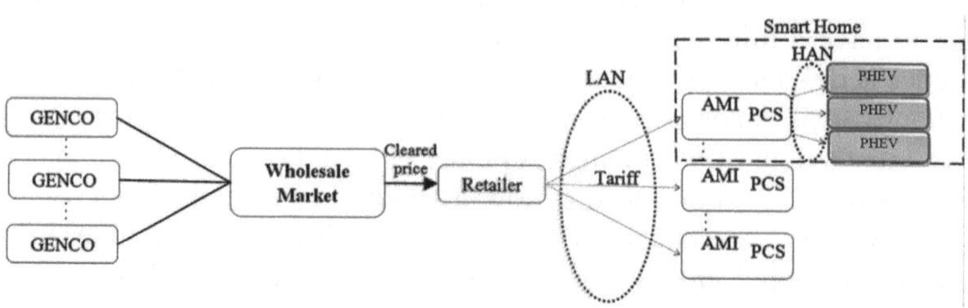

Fig. 10.1 Feasible data transfer structure

users. The market is cleared by the ISO and prices are announced to the market participants. Next, retailer declares retail price for end-users mainly based on the wholesale market prices over a digital communication infrastructure, e.g., a local area network (LAN). PCS dedicated to each advanced metering infrastructure (AMI) is responsible for controlling the charging of PEVs by setting up charging times of batteries. PCS solves the PCC problem considering declared time-varying tariffs. As shown in Fig. 10.1, control signals are transmitted from PCS to the responsive appliances over a home area network (HAN) [19].

10.2.2 Dynamic Pricing

One of the effective inputs of PCC problem is time-varying tariffs. Time-differentiated pricing models can potentially lead to economic and environmental advantages compared to the flat rates [19]. Changing the pattern of customer electricity consumption is the basic idea behind the time-varying pricing. The tariffs considered in this chapter are TOU, and a combination of TOU tariff and IBR.

In the TOU tariff, electricity price changes in definite levels during hours of the day. TOU can be described in different levels. Three-level TOU pricing can be formulated as below:

$$\gamma_t = \begin{cases} \gamma_1 & \text{if } t \in T_1 \\ \gamma_2 & \text{if } t \in T_2 \\ \gamma_3 & \text{if } t \in T_3 \end{cases} \tag{10.1}$$

where, t is the hourly time index, γ_t is the TOU electricity tariff at hour t, γ_1, γ_2 and γ_3 are respectively tariffs at off-peak periods (T_1), mid-peak periods (T_2), and on-peak periods (T_3) during a day. Obviously $T_1 \cup T_2 \cup T_3 = 24$ h and $\gamma_1 \leq \gamma_2 \leq \gamma_3$. Transferring PEV charging from on-peak tariff periods to lower tariff ones leads to peak load shaving and valley filling which are desirable from DSO point of view.

IBR pricing can lead to load balancing and reducing PAR [20]. In the IBR pricing, energy consumption more than a predetermined threshold would impose a penalty cost to the customer [21]. This penalty is such that the amount of consumed energy more than the threshold should be paid by a higher tariff than that of below the threshold. An hourly IBR tariff, $r(L_t)$, is mathematically presented as:

$$r(L_t) = \begin{cases} \alpha & 0 \leq L_t \leq \partial \\ \beta & L_t > \partial \end{cases} \tag{10.2}$$

where, L_t is the total amount of consumed energy at hour t which here is the total energy consumption of connected PEVs at hour t, and ∂ is the predetermined threshold of IBR pricing. The cost of energy consumption lower than and beyond threshold ∂ is respectively calculated by tariffs α and β where β is greater than α.

It is expected that by considering IBR pricing in PCC problem, charging level of batteries be capped to the threshold, as much as possible. Thus, the periods of PEVs charging at a home will be distributed during the allowable time interval of charging PEVs. This prevents occurrence of high peak load demand, which could take place due to overlapping the charging of PEVs at a home.

Combining these two non-flat rate pricings, TOU and IBR, can accumulate their advantages together. Combination of TOU and IBR tariff, $r_t(L_t)$, can be formulated as:

$$r_t(L_t) = \begin{cases} \alpha_t & 0 \le L_t \le \partial \\ \beta_t & L_t > \partial \end{cases} \tag{10.3}$$

where, α_t and β_t are, respectively, the tariff of consumed energy at hour t less and higher than threshold ∂. Note that, in (10.2), α and β are constant and thus, the tariff is independent of the hours of day and only depends on the level of hourly-consumed energy. Nevertheless, in (10.3), α_t and β_t depend on not only the total hourly energy consumption at each hour, but also on the time of day. Off-, mid-, and on-peak periods are determined in advance, and α_t and β_t are predefined in three levels for these periods. So, we have $\{\alpha_1, \beta_1\}$, $\{\alpha_2, \beta_2\}$, and $\{\alpha_3, \beta_3\}$ for $t \in T_1$, $t \in T_2$, and $t \in T_3$, respectively. It would be rational to have α_t less than γ_t and β_t higher than γ_t [22].

10.3 Basic PCC Formulation

In this section, the objective function for basic in-home PCC, the online algorithm of scheduling, and the associated constraints are described to control the charging periods and charging level of PEVs.

Consider that K is the set of PEVs connected to the outlet at home. t_p^k and t_d^k $(\forall k \in K)$ are respectively plug-in time and departure time of the kth PEV. The charge scheduling vector associated with the kth PEV at home (P^k) can be defined as:

$$\mathrm{P}^k = \left[p_{t_p^k}^k, p_{t_p^k+1}^k, \ldots, p_{t_d^k-1}^k\right], \quad \forall k \in K \tag{10.4}$$

The PCS specifies the optimal choice for P^k, where p_t^k, $\forall t \in [t_p^k, t_d^k)$, is charging level of the kth PEV at hour t. Note that

$$0 \le p_t^k \le p^{k,\max}, \quad \forall t \in \left[t_p^k, t_d^k\right). \tag{10.5}$$

where, $p^{k,\max}$ [kW] is maximum charging level of the kth PEV.

The PEV battery should be fully charged before departure time. This is certified by

$$\sum_{t=t_p}^{t_d-1} p_t^k = E_c^k, \quad \forall k \in K \tag{10.6}$$

where, E_c^k [kWh] is the energy required to fully charge battery of the kth PEV after arriving home. It is obvious that $t_d^k - t_p^k$ should be equal to or greater than the period needed to fully charge the kth PEV with maximum charging level.

The basic objective function is to minimize household PEVs charge payment cost, represented by:

$$\min CCF = \sum_{t=t_s}^{t_e-1} C_t(L_t) \tag{10.7}$$

where, CCF is charge cost function, t_s is the start time of scheduling, t_e is the end time of scheduling, and $C_t(L_t)$ is the payment cost at hour t based on the defined three-level TOU tariff combined with IBR. The scheduling horizon $([t_s, t_e])$ is determined based on the plug-in time and departure time of all connected PEVs, which is clearly described later in this section. The tariff is mathematically presented in (10.8).

$$C_t(L_t) = \begin{cases} \alpha_t L_t & L_t < \partial \\ \alpha_t \partial + (L_t - \partial)\beta_t & L_t \geq \partial \end{cases} \tag{10.8}$$

where,

$$L_t = \sum_{k \in K} p_t^k \tag{10.9}$$

PCS determines the charge scheduling vector of connected PEVs based on the declared pricings and PEV charging characteristics, i.e. the required energy for charging the PEVs, their maximum charging level, plug-in time, and departure time. When a new PEV arrives home and is plugged into the outlet, PCS adds this PEV to the set K and schedules the charging of this new PEV. Since the scheduling is done in an online way and the defined pricing in (10.3) depends on the level of energy consumption at each time, charge scheduling vector of connected PEVs should be updated when a new PEV is connected to the outlet at home. Thus, to have an online scheduling, an algorithm is required to update the data and re-schedule the new set of PEVs based on the objective functions. In the following, the algorithm of charge scheduling is profoundly described by an example.

Consider a smart home with two PEVs; the first PEV arrives home at 5 p.m. with 30 kWh battery capacity, $E_c^k = 20$ kWh, and maximum charging level of 4 kW. The driver plugs it to the outlet at 5 p.m. and sets 7 a.m. of the day after as the departure time. Thus, the set K has one PEV with $t_p^1 = 5$ p.m., $t_d^1 = 7$ a.m., $E_c^1 = 20$ kWh and $p^{1,\max} = 4$ kW. Also, $L_t = p_t^1$. PCS should fully charge this PEV between 5 p.m. and 7 a.m. according to (10.6). The objective function is to minimize the charging cost in the scheduling horizon. Since there is only one connected PEV, the start time of scheduling is the plug-in time of this PEV ($t_s = t_p^1 = 5$ p.m.) and the end time of scheduling is the departure time of this PEV ($t_e = t_d^1 = 7$ a.m.). Suppose that the threshold for hourly IBR pricing is 3 kWh. The objective function is as below:

$$\min CCF = \sum_{t=5\text{ p.m.}}^{6\text{ a.m.}} C_t(p_t^1) \tag{10.10}$$

Assume that solving PCC results in the PEV charge scheduling vector of:

$$\mathrm{P}^1 = [p^1_{5\text{ p.m.}}, p^1_{6\text{ p.m.}}, \ldots, p^1_{6\text{ a.m.}}] = [0, 0, 0, 0, 3, 3, 0, 0, 3, 3, 3, 3, 2, 0].$$

At 11 p.m., the second PEV comes back home with the battery capacity of 20 kWh, the remained energy of 10 kWh, and the maximum charging level of 2 kW. It is plugged into the outlet at this time. The vehicle owner also sets 9 a.m. as its departure time. As the first PEV has been charged up to 11 p.m., the charge cost up to the plug-in time of the second PEV should be considered in the cost function which is here referred to as CC^0. In this case, $CC^0 = \sum_{5\text{ p.m.}}^{10\text{ p.m.}} C_t(p_t^1)$. At 11 p.m., the new connected PEV is added to the set K. So, $t_p^2 = 11$ p.m., $t_d^2 = 9$ a.m., $E_c^2 = 10$ kWh, and $p^{2,\max} = 2$ kW. Once the second PEV is plugged in, the scheduling horizon for PCS should be updated. Also, a new charge scheduling should be effectuated for the set K in the updated forward horizon time. The start time of scheduling (t_s) is set to the plug-in time of the last connected PEV (11 p.m.). Note that, the first PEV has been charged up to this time and it is necessary for PCS to reschedule both the connected PEVs from this time. The end time of scheduling is obviously the maximum departure time of both PEVs ($t_e = 9$ a.m.). Also, $L_t = p_t^1 + p_t^2$. Therefore, the objective function for charging these two PEVs is reformulated as:

$$\min CCF = CC^0 + \sum_{t=t_s}^{t_e-1} C_t(L_t) = CC^0 + \sum_{t=11\text{ p.m.}}^{8\text{ a.m.}} C_t(p_t^1 + p_t^2) \tag{10.11}$$

subjected to (10.5), (10.6), (10.8) and (10.9). The charging vector of the first PEV from 11 p.m. to 7 a.m. is rescheduled after the plug-in time of the second PEV. For instance, solving PCC for the determined scheduling horizon may result in:

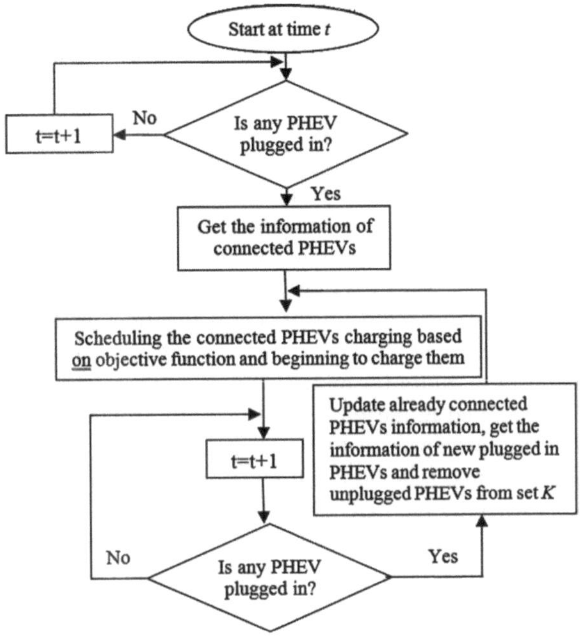

Fig. 10.2 Online algorithm of PEV charge scheduling

$$\mathrm{P}^1 = [\underbrace{p^1_{5\,\mathrm{p.m.}}, p^1_{6\,\mathrm{p.m.}}, \ldots,}_{\substack{before\,plug-\mathrm{in}\,time\\ of\,\mathrm{2nd}\,PHEV}} \underbrace{p^1_{11\,\mathrm{p.m.}}, p^1_{12\,\mathrm{p.m.}}, \ldots, p^1_{6\,\mathrm{a.m.}}}_{\substack{after\,plug-\mathrm{in}\,time\\ of\,\mathrm{2nd}\,PHEV}}]$$

$$= [\underbrace{0, 0, 0, 0, 3, 3}_{\substack{before\,plug-\mathrm{in}\,time\\ of\,\mathrm{2nd}\,PHEV}}, \underbrace{3, 3, 3, 1, 1, 1, 2, 0}_{\substack{after\,plug-\mathrm{in}\,time\\ of\,\mathrm{2nd}\,PHEV}}] \quad \text{and}$$

$$\mathrm{P}^2 = [p^2_{11\,\mathrm{p.m.}}, p^2_{12\,\mathrm{p.m.}}, \ldots, p^2_{8\,\mathrm{a.m.}}] = [0, 0, 0, 2, 2, 2, 0, 2, 2, 0].$$

The elements of P^1, before plug-in time of 2nd PEV, is the output of solving (10.10) and after that is the output of solving (10.11). Moreover, when a PEV is fully charged or unplugged, the PCS removes this PEV from the set *K*.

The above procedure for charge scheduling of the connected PEVs can be generalized with more number of PEVs at a home. Figure 10.2 concludes the online algorithm for PEV charge scheduling.

10.4 Inclusion of Customer Inconvenience Cost in PCC Formulation

PEV characteristics, depended on vehicle owners' behavior, are extracted from 2009 NHTS by analyzing four created files. These data include daily miles driven, starting time of charging, number of vehicles per house, and vehicle type of houses. References [10, 23, 24] assumed that the PEV owners plug in their vehicles when they arrive home after their last trip in a day. Therefore, the last trip arrival time of the vehicles is assumed as the starting time of charging.

Minimizing the payment cost is the most appealing requirement of customers in consequence of PEV charging control. However, customers do not scarify their comfort in return for minimizing cost. Therefore, the customer's comfort should be incorporated in the process of PCC to provide customer satisfaction and convince him to participate in PEV charge scheduling programs. As mentioned before, the waiting time to have a full charged battery is projected here as the inconvenience cost whereby the PCC objective function is extended. The concept of waiting time is proposed in [18], in which an optimization-based load control method has been proposed that both the total payment associated to all appliances as well as cost of waiting time are incorporated. Although the proposed algorithm in this chapter has a few common concepts with [18], this chapter proposes an online charge scheduling algorithm that mitigates the uncertainty challenges raised by the day-ahead scheduling. The inconvenience cost function and the extended PCC procedures and formulations are deeply described in this section.

When the kth PEV is plugged into the outlet at t_p^k, its battery has an initial state of charge equal to $SOC^k(t_p^k)$. As the PCS charges the battery, the SOC of the battery increases until it becomes 100 %. The SOC of battery of the kth PEV at hour t is:

$$SOC^k(t) = SOC^k(t_p^k) + \sum_{s=t_p^k}^{t-1} \frac{p_s^k \eta^k}{C^k} \tag{10.12}$$

where, C^k is the battery capacity of the kth PEV [kWh], η^k is its charger efficiency, and $SOC^k(t)$ is the percentage of battery capacity remained in the kth PEV battery at hour t. $(1 - SOC^k(t))$ shows the percentage of the kth PEV battery capacity which has not been charged until hour t. The faster the PCS charges the battery, the sooner $(1 - SOC^k(t))$ approaches zero. So, inconvenience function for the kth PEV (IF^k) can be reasonably the summation of $(1 - SOC^k(t))$ from the start time of scheduling (t_s) to its departure time when its owner wishes the battery to be fully charged. This reflects the waiting time of the kth PEV owner to have a fully charged battery and is formulated as below:

$$IF^k = C^k \sum_{t=t_s}^{t_d^k - 1} (1 - SOC^k(t)) \tag{10.13}$$

The total inconvenience cost function (*ICF*) shows the inconvenience cost of all connected PEVs of the house and is presented in (10.14).

$$ICF = \sum_{k \in K} \xi^k IF^k \tag{10.14}$$

Parameter ξ^k in [¢/kWh] converts the nature of customer inconvenience into a monetary value.

ICF well illustrates the inconvenience cost of waiting time for charging the battery. Thus, the extended PCC is designated to minimize the summation of two objective functions, household's total electricity payment and inconvenience cost of waiting for charging connected PEVs. The objective function of the extended PCC is presented as:

$$\min\,(CCF + ICF) \tag{10.15}$$

According to (10.14) and (10.15), ξ^k acts as a knob to control the trade-off between the two objective functions. The vehicle owner should decide how his vehicle charges (faster with probable higher cost, or slower with probable lower cost). Therefore, it is assumed that ξ^k is set by the vehicle owner of the *k*th PEV when he plugs in his vehicle to the outlet. Obviously, factor ξ^k affects the speed of charging directly. In practice, different choices of parameter ξ^k can be defined. $\xi^k = 0$ makes IF^k to be excluded from the objective function. As ξ^k is increased, the importance of *IF* for the *k*th PEV rises. This may lead to a higher *CCF* in return for providing costumer's comfort. ξ^k guarantees the convenience of the customers. As an instance, a customer who wants to charge his/her PEV sooner and does not care about the payment cost, sets factor ξ^k a high value. In the other side, another customer for whom the payment cost is more important, sets factor ξ^k zero or a negligible value.

In conclusion, the charge scheduling results are clearly obtained from a compromise between *CCF* and *ICF*.

10.5 Numerical Studies

This section analyses the basic and extended PCC formulations through different cases to demonstrate the effectiveness of the proposed method. Since, the optimization model is linear, linear programming techniques such as the interior point method [25] can effectively solve such problem. We use CPLEX solver of GAMS software to solve the problem.

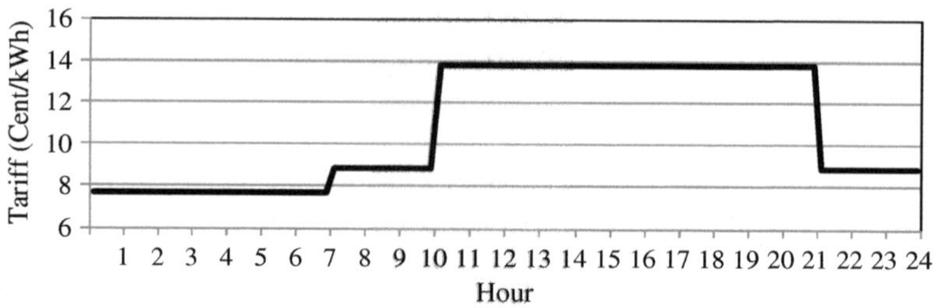

Fig. 10.3 Three-level tariffs for a day

The three-level TOU tariff utilized in the Baltimore gas and electric (BGE) company's program is taken from [26] and is used in the study results presented in this chapter. On-peak tariff, between 10 a.m. and 9 p.m. is 13.757 ¢/kWh. Mid-peak tariff between 7 a.m. and 10 a.m., and between 9 p.m. and 12 p.m. is equal to 8.866 ¢/kWh. In the remaining periods, the off-peak tariff is set to 7.67 ¢/kWh. This tariff is pictorially shown in Fig. 10.3.

As defined in (10.3), the IBR pricing can be applied to the TOU tariff, to provide a more proper tariff as the input of PCC problem. The lower level of the IBR tariff at hour t, α_t, is assumed to be 0.9 of TOU tariff at that hour. The price of the higher level of IBR, β_t, is considered to be 1.4 of α_t [18]. The threshold ∂ is assumed to be 2 kWh.

The following case studies are conducted to probe the PCC results in a smart home and its consequences on the distribution system characteristics:

Case I: Applying basic PCC to a home with one PEV;
Case II: Applying basic PCC to a home with two PEVs;
Case III: Applying extended PCC to a home with one PEV;
Case IV: Applying extended PCC to a home with two PEVs;
Case V: A residential test feeder with distributed PEVs at homes.

Note that, basic PCC only minimizes the payment cost, and extended PCC includes the payment and inconvenience costs.

10.5.1 Case I

This case investigates the proposed charge scheduling method for a PEV at home. This is a very simple case for which the correctness of obtained results can be readily verified. The PEV characteristics are assumed to be $C^1 = 15\,\text{kWh}$, $t_p^1 = 6\,\text{p.m.}$, $t_d^1 = 8\,\text{a.m.}$, $SOC^1(t_p^1) = 0.2$, and $\eta^1 = 0.88$. Also, as proposed in [30], the maximum charging level of the PEV is assumed to be 0.2 of its battery capacity; thus, $p^{1,\max} = 3$ kW. According to the SOC, the energy remained in the battery is 3 kWh and thus, the required energy to fully charge the battery is $E_c^1 = \frac{(15-3)}{0.88} = 13.64$ kWh.

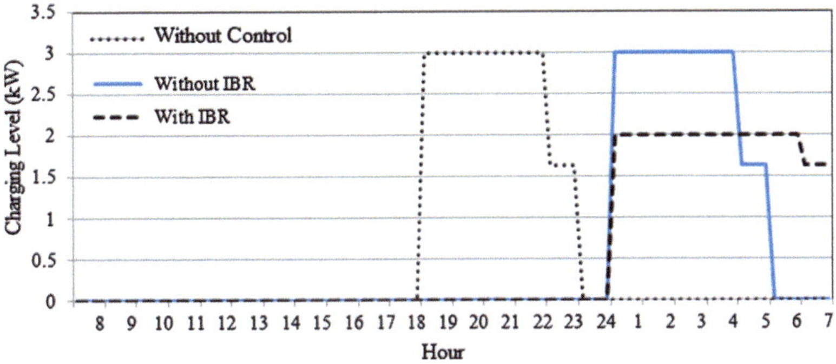

Fig. 10.4 PEV battery charging profile in the cases of uncontrolled charging and controlled charging with and without applying IBR to TOU tariff

The charge scheduling of one PEV at home with the goal of minimizing the charging cost is obtained by solving basic PCC problem. Figure 10.4 depicts the charging schedule of the battery with and without charging control, and in the case of charging control, with and without applying IBR pricing to the existing TOU tariff.

As can be seen in Fig. 10.4, in the case of uncontrolled charging, the battery starts charging immediately after plugged into the outlet (6 p.m.) with maximum charging level (3 kW). On the other hand, in the case of controlled charging, the PCS determines charging schedule based on the payment cost. It postpones the PEV charging to low tariff periods (hour 24), as shown in Fig. 10.4.

Moreover, Fig. 10.4 demonstrates that by applying IBR to TOU pricing, PCS limits the charging level of the battery below the threshold as much as possible in order to avoid paying 40 % additional cost of exceeding the threshold. However, without IBR application, PCS does not consider the charging level of the battery and charges the battery with the maximum possible level.

The payment cost due to charging the battery for the cases of uncontrolled, controlled without IBR, and controlled with IBR, is respectively ¢165, ¢104.6, and ¢94.1. The payment cost is high without any charging control compare to controlled charging. Since, in the case of applying PCC, the battery is charged during off peak hours. The cost would decrease further with applying IBR to TOU pricing due to capped charging level below the threshold and paying 10 % less (α_t has been assumed to be 0.9 of TOU tariff at that hour).

10.5.2 Case II

In this case, a house with two PEVs is considered. It is assumed that in addition to the PEV of the previous case, another PEV with 12 kWh battery size and 30 % SOC arrives home at 9 p.m. Its owner sets 6 a.m. as the departure time in the next day. So, $E_c^2 = \frac{(1-0.3)\times 12}{0.88} = 9.55$ kWh. Its maximum charging level ($p^{2,\max}$) is assumed

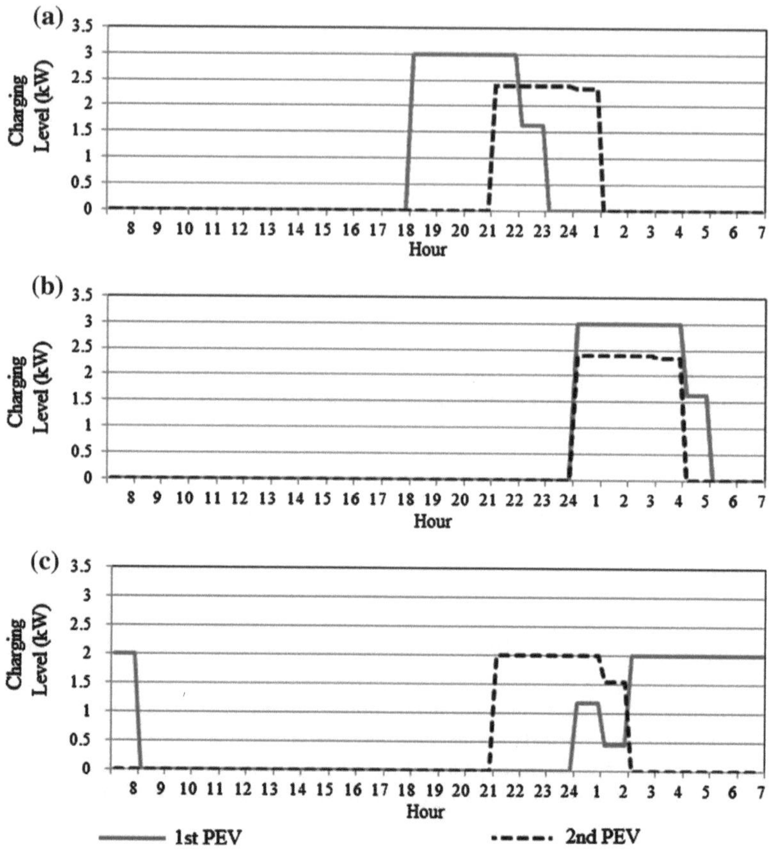

Fig. 10.5 Battery charging profile of two PEVs in the house, in the case of **a** uncontrolled charging, **b** controlled charging without applying IBR, and **c** controlled charging with applying IBR tariff

Table 10.1 Charging cost of two PEVs in the house

		Charging cost (¢)
Without control	1st PEV	165
	2nd PEV	131.3
Control without IBR	1st PEV	104.6
	2nd PEV	73.2
Control with IBR	1st PEV	99.5
	2nd PEV	74.2

to be 0.2 of the capacity, i.e. 2.4 KW. Figure 10.5 shows the results of uncontrolled charging as well as charge scheduling with and without applying IBR pricing. Moreover, Table 10.1 presents the charge payment cost of the 1st and 2nd PEVs in this case.

As depicted in above figure, without charging control, the batteries begin to charge immediately after plugged into the outlet (6 p.m. and 9 p.m. for the 1st and 2nd PEV, respectively) with maximum charging level (3 and 2.4 kW, respectively). This lead to significant increment in household peak load in addition to imposing high payment cost to the customer, as presented in Table 10.1. On the other hand, with controlled charging, not only the charging periods shift to medium and low peak periods which highly desirable from distribution operator point of view, but also the charging cost decreases significantly, which is a proper incentives for the customers to participate in this program.

Furthermore, Fig. 10.5b illustrates that without applying IBR, PCS postpones charging of both PEVs to the low tariff periods and charges them, simultaneously, from hour 24 with maximum charging level. However, in the case of applying IBR, PCS distributes the PEVs charging and limits the charging level of the batteries to 2 kW in order to avoid exceeding the threshold ∂. As a result, the second PEV is charged in the medium tariff periods to avoid charging more than the threshold. This would lead to a smoother household load demand. As shown in Table 10.1, the total charge payment cost of PEVs decreases as IBR is applied.

10.5.3 Case III

In this case, the proposed extended PCC method, which incorporates customer's inconvenience cost, is probed. Although this is a simple case, it could easily verify the correctness of obtained results and convey many interesting features of the proposed PCC formulation. The PEV characteristics are the same as that of case I. Different levels of customers' comfort are considered for the PEV and the extended OCC problem is solved for each to acquire the charge scheduling of the PEV at home. Figure 10.6 shows the charge scheduling of the battery with different value if ξ^1, i.e. 0, 1, 2, with and without applying IBR pricing to the existing TOU tariff. Table 10.2 also shows the payment and inconvenience cost of charging the PEV for the mentioned values of ξ^1.

Based on the presented results, the following remarks can be concluded:

- With $\xi^1 = 0$, the inconvenience cost of waiting time is not important for the vehicle owner and the problem is turned to the basic PCC problem. So, the PCS determines charging schedule just based on the charge payment cost and postpones the PEV charging to low tariff periods, as shown in Fig. 10.6.
- With $\xi^1 = 1$ and $\xi^1 = 2$, *inconvenience cost function* becomes important for the vehicle owner. Thus, as ξ^1 is increased, the PCS charges the battery sooner even in higher tariff periods in return for providing customer convenience. This results in an increase in the charge payment cost as shown in Table 10.2. Also, by a deeper analysis of the results, it can be deduced that when $\xi^1 = 2$, PEV starts charging with the maximum rate at the plug-in time to reduce inconvenience cost. This leads to a charge scheduling similar to uncontrolled charging in which the

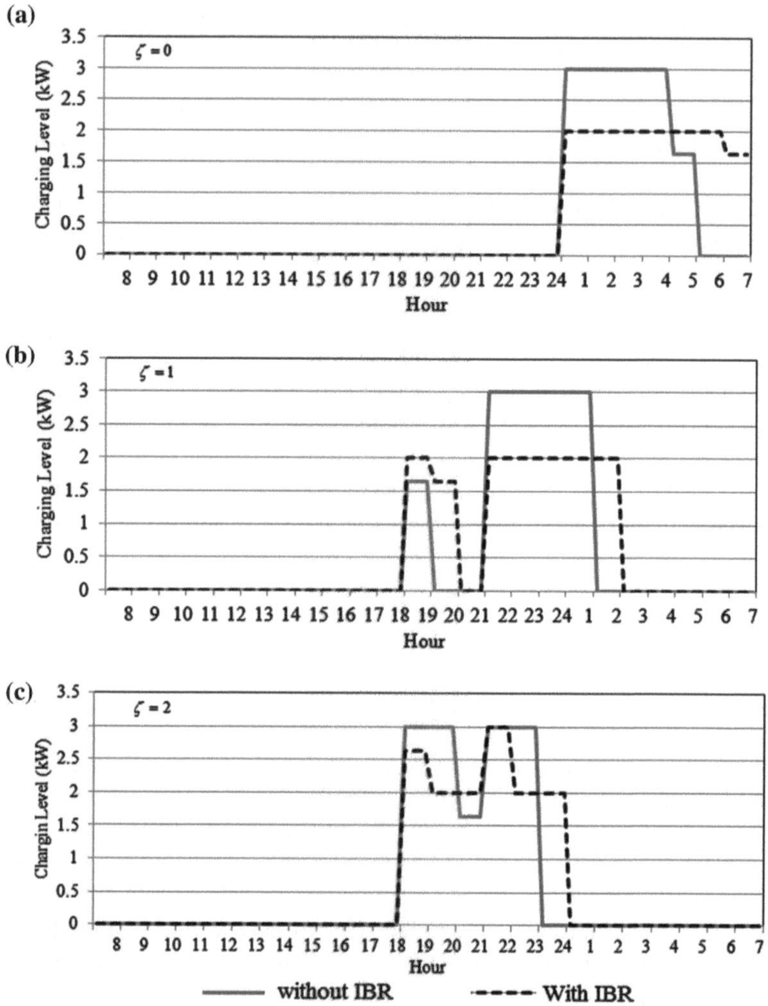

Fig. 10.6 PEV charge scheduling for **a** $\zeta = 0$, **b** $\zeta = 1$, and **c** $\zeta = 2$ with and without applying IBR

owner usually plugs in his vehicle at the time of arriving home and PEV starts charging with maximum rate from this time.

- Figure 10.6 demonstrates that with any value sets for ξ^1, by applying IBR to TOU pricing, PCS tries to limit the charging level of the battery below the threshold as much as possible in order to avoid paying 40 % additional cost of exceeding the threshold. But, without IBR application, PCS does not care about the charging level of the battery and since the waiting time to fully charge the battery is important along with payment cost, PCS charges the battery with the maximum possible level. In addition, with the presented combined tariff rates, the customer payment is decreased in comparison with the sole TOU application

Table 10.2 Charge payment and inconvenience cost of the PEV based on different ζ^1 with and without applying IBR [¢]

		ξ^1		
		0	1	2
Without IBR	Charge payment cost	104.6	125.3	158.3
	Inconvenience cost	0	47.5	48
With IBR	Charge payment cost	94.1	120.5	144.4
	Inconvenience cost	0	47.9	58.1

as presented in Table 10.2, since by restricting the charging level below the IBR threshold, the payment cost decreases by 10 % compare to sole TOU pricing.

- The inconvenience cost increases in the case of applying IBR; since, the PEV battery is charged later with IBR application compare to that of without IBR.

10.5.4 Case IV

In this case, similar to case II, the house with two PEVs is considered. The characteristics of both PEVs are the same as that of case II. Figures 10.7, 10.8, 10.9,

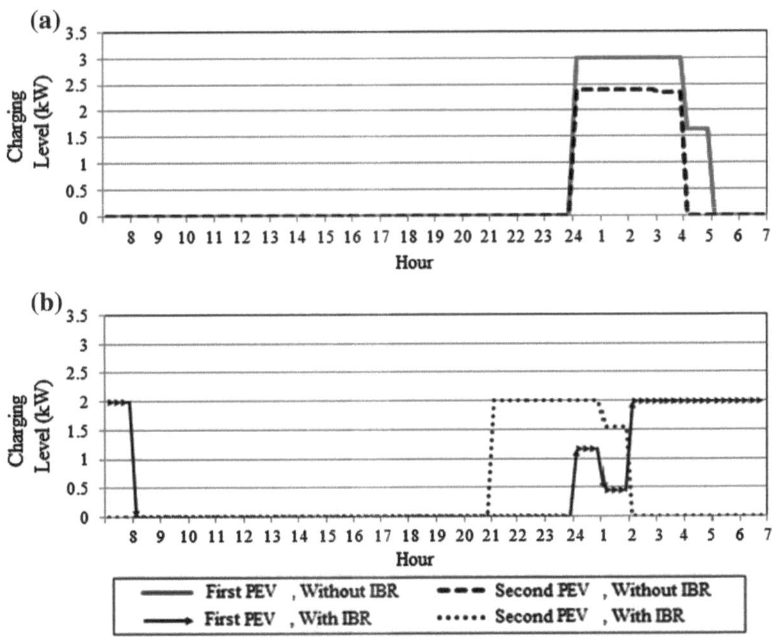

Fig. 10.7 Charge scheduling of two PEVs in the house with $\xi^1 = 0$, $\xi^2 = 0$ **a** without applying IBR and **b** with applying IBR tariff

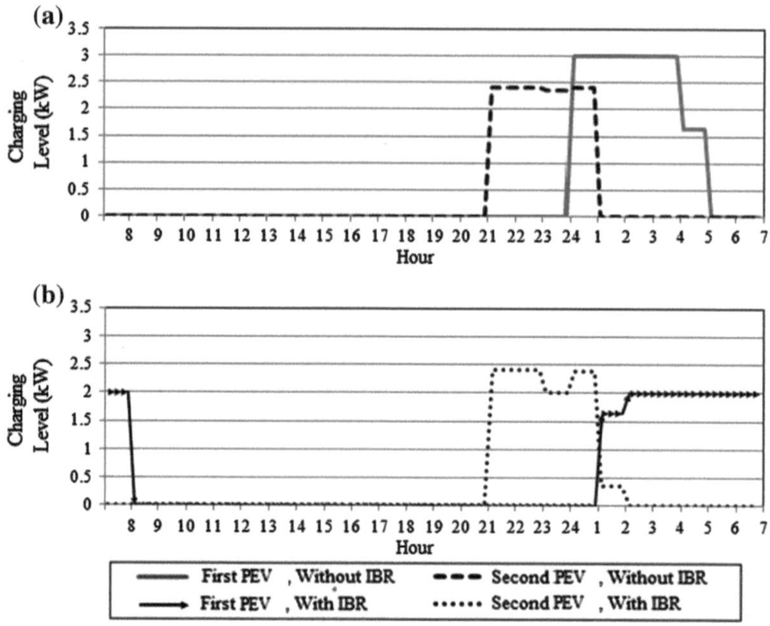

Fig. 10.8 Charge scheduling of two PEVs in the house with $\xi^1 = 0$, $\xi^2 = 1$ **a** without applying IBR and **b** with applying IBR tariff

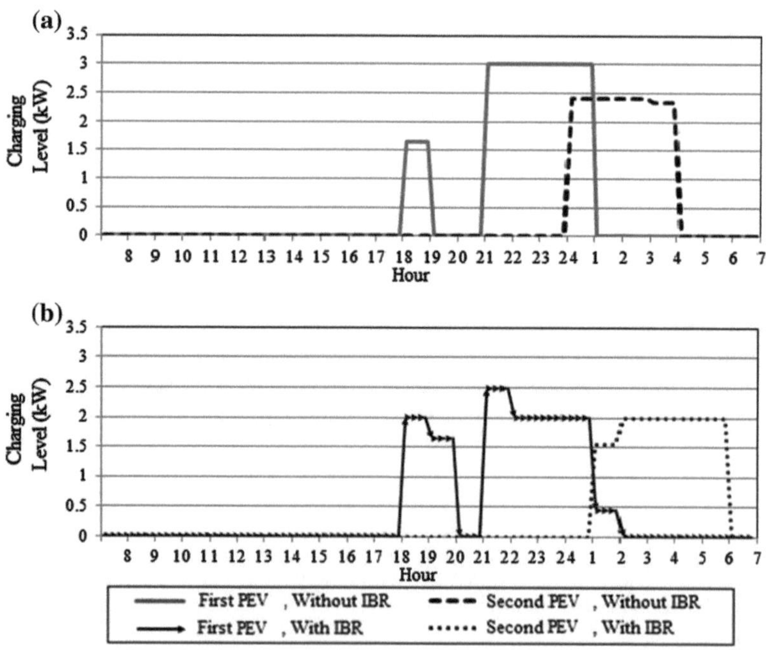

Fig. 10.9 Charge scheduling of two PEVs in the house with $\xi^1 = 1$, $\xi^2 = 0$ **a** without applying IBR and **b** with applying IBR tariff

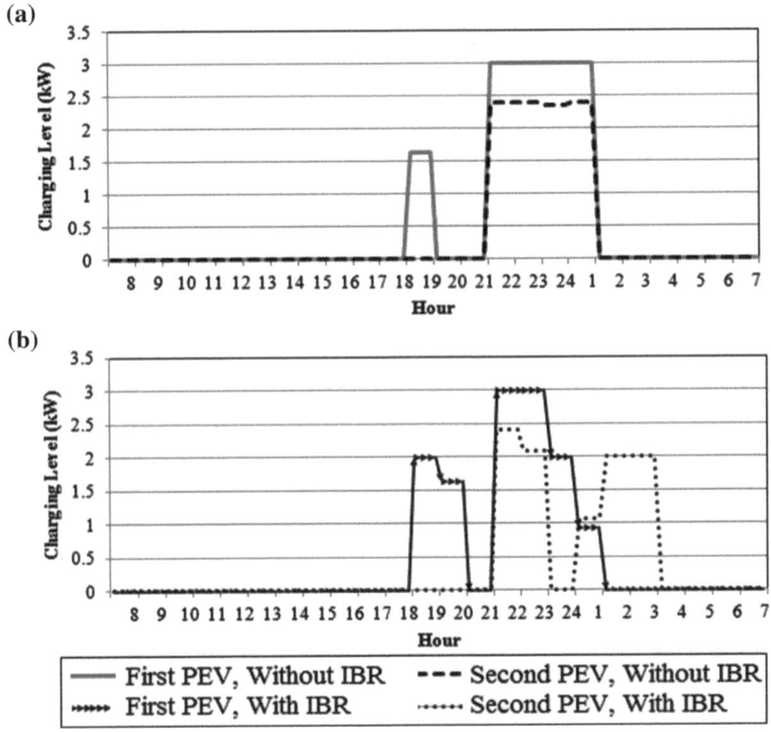

Fig. 10.10 Charge scheduling of two PEVs in the house with $\xi^1 = 1$, $\xi^2 = 1$ **a** without applying IBR and **b** with applying IBR tariff

Table 10.3 The charge payment and inconvenience cost of two PEVs in the house with and without applying IBR [¢]

ξ^1			0		1	
ξ^2			0	1	0	1
CCF (¢)	Without IBR	1st PEV	104.6	104.6	125.3	125.3
		2nd PEV	73.2	81.8	73.2	81.8
	With IBR	1st PEV	99.5	96.3	115.3	121.8
		2nd PEV	74.2	76.8	65.9	80
ICF	Without IBR	1st PEV	0	0	47.5	47.5
		2nd PEV	0	12.6	0	12.6
	With IBR	1st PEV	0	0	43.2	39.6
		2nd PEV	0	13.2	0	20.5

and 10.10 illustrate the results of charge scheduling of each battery with different values of ξ^k. Besides, Table 10.3 shows the charge payment cost and inconvenience cost of the 1st and 2nd PEVs.

According to above Figures, by applying IBR, the charge scheduling of the first PEV before the plug-in time of the second one (hour 21) is the same as that of Case III with different ξ s. But, after this hour, PCS reschedules the first PEV charging to minimize the cost of both PEVs charging.

It is of interest to deeply investigate the obtained results with different customer preferences in the following.

- $\xi^1 = 0,\ \xi^2 = 0$: In this case, actually the waiting time of charging the battery is important for neither of vehicle owners and as a result, the charge scheduling problem turns to the basic PCC problem in Case II. Thus, as shown in Fig. 10.7, without applying IBR, PCS postpones charging of both PEVs to the low tariff periods and charges them simultaneously from hour 24. However, in the case of applying IBR, PCS distributes the PEVs charging and limits the charging level of the batteries to 2 kW in order to avoid exceeding the threshold ∂ and paying 40 % more. As a result, the second PEV is charged in the medium tariff periods to avoid charging more than the threshold. This would lead to a smoother household load demand.
- $\xi^1 = 0,\ \xi^2 = 1$: According to Fig. 10.8, the second PEV begins to charge from its plug-in time, i.e. hour 21, due to importance of *inconvenience function*. Therefore, its charge payment cost increases compared to the case in which $\xi^1 = 0,\ \xi^2 = 0$, as presented in Table 10.3. Without applying IBR, the first PEV is charged with 3 kW charging level from hour 24; the same as $\xi^1 = 0,\ \xi^2 = 0$ case. With applying IBR, the first PEV waits until hour 1 when the second PEV is fully charged; then, it is charged with 2 kW charging level until its departure time, i.e. 8 a.m.
- $\xi^1 = 1,\ \xi^2 = 0$: The first PEV starts charging from its plug-in time, i.e. hour 18, since the *inconvenience function* is important for the first PEV in this case. Without applying IBR, the charge scheduling vector of the first PEV is the same as $\xi^1 = 1$ in Case III, depicted in Fig. 10.6. The second PEV is also charged similar to the case in which $\xi^1 = 0,\ \xi^2 = 0$. However, in the case of applying IBR, the PCS changes the charge scheduling vector of the first PEV at the plug-in time of the second PEV, i.e. hour 21, as illustrated in Figs. 10.6 and 10.9.
- $\xi^1 = 1,\ \xi^2 = 1$: In Fig. 10.10, without applying IBR, the charge scheduling vector of the first and second PEV is, respectively, like the cases of $\xi^1 = 1,\ \xi^2 = 0$ and $\xi^1 = 0,\ \xi^2 = 1$. With applying IBR, after hour 21, PCS simultaneously charges PEVs beyond the threshold for 2 h. This observation also certifies that *inconvenience function* is so important for PEVs' owner such that he is willing to pay more in return for providing comfort. According to Table 10.3, the charge payment cost of PEVs decreases compared to that of without IBR.

10.5.5 Case V

In this case, the IEEE 34-node test feeder [27] is selected as a residential distribution system to investigate the impacts of proposed PCC method on the distribution system load profile. This radial network is shown in Fig. 10.11. The network medium voltage is 24.9 kV and low voltage is 230 V. The network contains 33 load points, 8 of them including 810, 818, 820, 822, 826, 838, 856 and 864, are single phase and the others are three phases. Two houses are assigned to each phase of load points [10]. Thus, the total number of houses is equal to 166.

A typical household load profile is shown in Fig. 10.12 [28]. This load profile is assigned to each house.

The number of vehicles in the network and PEV characteristics including daily miles driven, last trip arrival time, types of vehicle, all electric ranges (AER), and charging levels are extracted from [2]. The departure time of vehicles, which has not been determined in that chapter, is assumed to be the first trip start time. The percentage of vehicles based on this characteristic is extracted from 2009 NHTS [31] based on the method and the obtained database explained in [2]. Figure 10.13 shows the percentage of vehicle versus their first trip start time.

In this chapter, 11.3, 35 and 45 % PEV penetration levels are considered [24]. Based on the PEV penetration levels, PEVs are selected randomly from the existing vehicles in the network. ξ should also be specified for each PEV. If the initial SOC of a PEV is more than 50 %, it can be assumed that the cost of charging would probably be more important for the vehicle owner than the speed of charging. Thus,

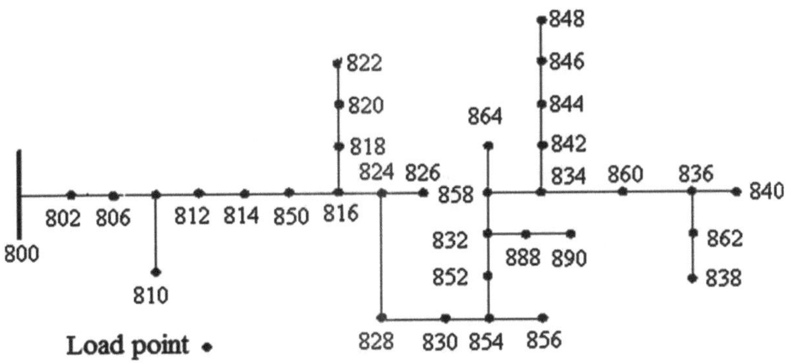

Fig. 10.11 IEEE 34-node test feeder [27]

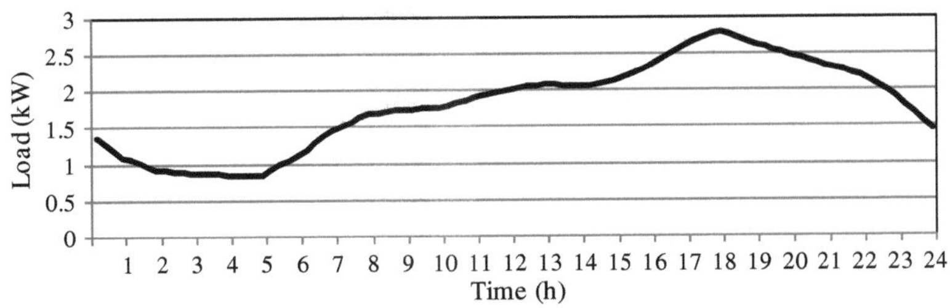

Fig. 10.12 An average household load curve in summer

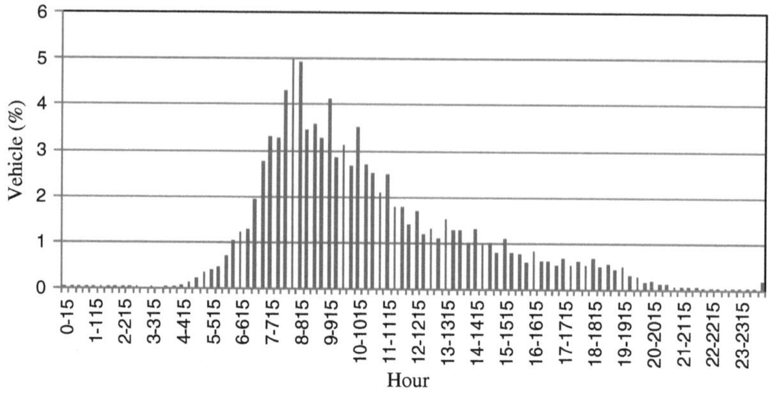

Fig. 10.13 Percentage of vehicles versus their first trip start time

ξ^k can be assumed to have a negligible value for such PEVs. In this study, for PEVs with more than 50 % initial SOC, ξ^k is selected randomly between 0 and 0.1. For the rest of PEVs, this factor is selected randomly between 0 and 2. Load flow analysis is performed to calculate the total load, feeder power loss, and voltage deviation in different load points of the network. In this section, open-source distribution system simulator (OpenDSS) is used to execute load flow analysis. OpenDSS is an electric power distribution system simulator for advanced analysis of distribution systems [29].

The proposed charge scheduling method is applied to each house. PCC problems are solved in each house and their impacts on the network load profile are explored as well. Figure 10.14 illustrates the network load profile with and without charging control. Table 10.4 concludes peak load, PAR of the system load demand, standard deviation of the system load profile, and the charge payment cost for all the PEVs, obtained from solving PCC in each house.

Accordingly, the following results can be presented:

- PCSs charge most of the PEVs in medium and low tarrif periods which coincides with the off-peak periods of demand. Therefore, charge scheduling of PEVs using the proposed method prevents significant peak load increment. Not only the peak load of the network is not increased, but also the PEVs are charged with lower cost. Table 10.4 clearly shows noticable decrement of charge payment cost by applying the proposed PCC algorithm.
- The proposed charge scheduling method makes charging periods shift to low load periods. This causes the average load to increase more than the peak load. Consequently, the PAR decreases significantly compared to uncontrolled charging. Moreover, Table 10.4 shows that by increasing the PEV penetration level, PAR decreases by applying the scheduling method. However, in the uncontrolled PEV charging cases, the PAR increases due to PEV penetration level increment.

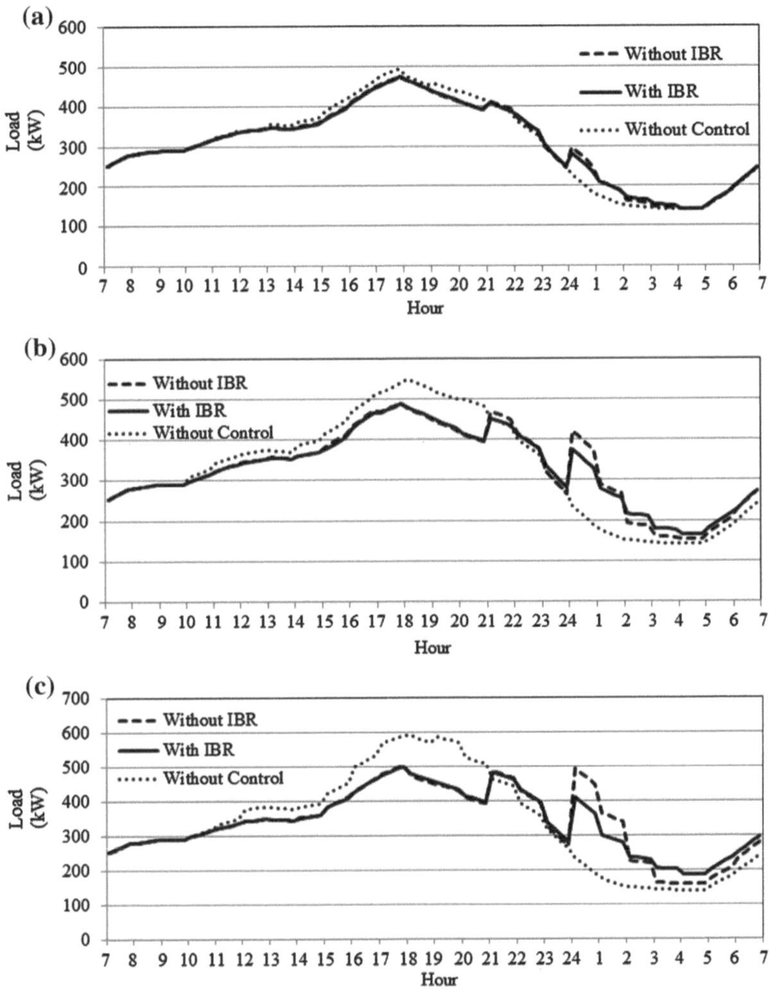

Fig. 10.14 Impacts of controlled and uncontrolled PEV charging on total load curve for **a** 11.3 %, **b** 35 %, and **c** 45 % PEV penetration levels

- The third row of Table 10.4 shows that the standard deviation of the system load profile decreases significantly by applying the proposed method compared to uncontrolled PEVs charging. It means that the system load profile becomes smoother. Also, by applying the IBR pricing, the standard deviation decreases compared to the cases without application of the IBR pricing. Since, the charging periods are distributed in each house to avoid exceeding the threshold ∂. This confirms the effectiveness of the combined TOU and IBR pricing.
- The fourth and fifth rows of Table 10.4 show that using the proposed method to schedule the charging time of PEVs in each household decreases the system

Table 10.4 Specifications of the distribution system with and without PEV charge scheduling with different PEV penetration levels

		PEV penetration level		
		11.3 %	35 %	45 %
Peak load (kW)	Without control	502	571	605
	Control without IBR	476	501	516
	Control with IBR	475	505	516
PAR	Without control	1.62	1.72	1.77
	Control without IBR	1.54	1.51	1.51
	Control with IBR	1.53	1.52	1.51
Standard deviation (% average load)	Without control	35.6	39.5	42.4
	Control without IBR	31.3	30.0	30.1
	Control with IBR	30.1	28.3	27.8
Total loss (%)	Without control	1.40	1.55	1.64
	Control without IBR	1.36	1.44	1.49
	Control with IBR	1.35	1.43	1.46
Max. voltage deviation (%)	Without control	3.42	3.96	4.24
	Control without IBR	3.22	3.41	3.52
	Control with IBR	3.21	3.44	3.53
Charge payment cost (Cent)	Without control	3,449	10,226	13,224
	Control without IBR	2,410	7,392	9,372
	Control with IBR	2,200	6,845	8,670

total loss and maximum voltage deviation of the load points, which is highly desirable from the DSO points of view. This result implies that although in the proposed method the social welfare and the distribution system specifications are not explicitly considered, applying the method implicitly improves these specifications.

10.6 Conclusion

This chapter proposes a novel method to schedule PEV charging in an online manner; the charging period and level of each PEV are scheduled from the time that it arrives home and then rescheduled later from the times that other PEVs arrive and are plugged into the outlet. The proposed optimization-based algorithm is designed to minimize payment and inconvenience costs. The proposed formulation gives the customer an opportunity to set his convenience importance. The proposed method is probed in different case studies. The implementation of optimization method in a smart home with one PEV represents that PEV charging periods shift to the off-peak tariff periods as much as possible. Also, as the importance of inconvenience function for the customer increases, the payment cost ascends and the charge scheduling vector of PEV becomes similar to uncontrolled charging. The results in a house with two PEVs show that the total charging payment cost increases due to increment of convenience importance for each PEV. In both cases, application of IBR to TOU pricing increases the positive impacts of the proposed method on the household load curve; since, IBR application causes charging profile to be distributed during the allowable interval of charging. Also, the presented results indicate that solving PCC problems in each house of a residential distribution system not only decreases the total charge payment cost compared to the uncontrolled cases, but also improves specifications of the distribution systems, i.e. decreases peak load, PAR, and standard deviation of load curve.

References

1. Ipakchi A, Albuyeh F (2009) Grid of the future. IEEE Power Energ Mag 7(4):52–62
2. Shafiee S, Fotuhi-Firuzabad M, Rastegar M (2013) Investigating the impacts of plug-in hybrid electric vehicles on power distribution systems. IEEE Trans Smart Grid 4(3):1351–1360
3. Clement-Nyns K, Haesen E, Driesen J (2010) The impact of charging plug-in hybrid electric vehicles on a residential distribution grid. IEEE Trans Power Syst 25(1):371–380
4. Darabi Z, Ferdowsi M (2011) Aggregated impact of plug-in hybrid electric vehicles on electricity demand profile. IEEE Trans Sustain Energ 2(4):501–508
5. Axsen J, Kurani K (2009) Anticipating PEV energy impacts in California. Institute of Transportation Studies, University of California, Davis. Working paper UCD-ITS-WP-09-03 (presented at EVS-24 in Stavanger, Norway)
6. Su W, Chow MY (2010) An intelligent energy management system for PEVs considering demand response. Paper presented at the FREEDM annual conference, Tallahassee, 2010
7. Palensky P, Dietrich D (2011) Demand side management: demand response, intelligent energy systems, and smart loads. IEEE Trans Ind Inf 7(2):1551–3203
8. Su W, Chow MY (2012) Performance evaluation of an EDA-based large-scale plug-in hybrid electric vehicle charging algorithm. IEEE Trans Smart Grid 3(1):308–315
9. Su W, Chow MY (2011) Performance evaluation of a PEV parking station using particle swarm optimization. Paper presented in IEEE power energy society. General meet Detroit, Michigan, USA, 24–29 July 2011
10. Sortomme E, Hindi MM, James MacPherson SD, Venkata SS (2011) Coordinated charging of plug-in hybrid electric vehicles to minimize distribution system losses. IEEE Trans Smart Grid 2(1):198–205
11. Masoum AS, Deilami S, Moses PS, Masoum MAS, Abu-Siada A (2011) Smart load management of plug-in electric vehicles in distribution and residential networks with charging stations for peak shaving and loss minimization considering voltage regulation. IEE Proc Gener Transm Distrib 5(8):877–888
12. Deilami S, Masoum AS, Moses PS, Masoum MAS (2011) Real-time coordination of plug-in electric vehicle charging in smart grids to minimize power losses and improve voltage profile. IEEE Trans Smart Grid 2(3):456–467
13. Clement-Nyns K, Haesen E, Driesen J (2011) The impact of vehicle-to-grid on the distribution grid. Electr Power Syst Res 81(1):185–192
14. Kristoffersen TK, Capion K, Meibom P (2011) Optimal charging of electric drive vehicles in a market environment. Appl Energ 88(5):1940–1948
15. Du P, Lu N (2011) Appliance commitment for household load scheduling. IEEE Trans Smart Grid 2(2):411–419

16. Fotuhi-Firuzabad M, Rastegar M, Safdarian A, Aminifar F (2014) Probabilistic home load controlling considering plug-in hybrid electric vehicle uncertainties. In: Karki R, Billinton R, Verma AK (eds) Reliable and sustainable electric power and energy systems management: reliability modeling and analysis of smart power systems, Springer, New york, pp 117–132
17. Rastegar M, Fotuhi-Firuzabad M, Aminifar F (2012) Load commitment in a smart home. Appl Energ 96:45–54
18. Mohsenian-Rad AH, Leon-Garcia A (2010) Optimal residential load control with price prediction in real-time electricity pricing environment. IEEE Trans Smart Grid 1(2):120–133
19. US Department of Energy (2006) Benefits of demand response in electricity markets and recommendations for achieving them. Report to the United States Congress, Feb 2006
20. Mohsenian-Rad AH, Wong V, Jatskevich J, Schober R (2010) Optimal and autonomous incentive-based energy consumption scheduling algorithm for smart grid. Paper presented in the IEEE PES conference in innovations in smart grid technologies, Gaithersburg, MD, Jan 2010
21. Reiss P, White M (2005) Household electricity demand, revisited. Rev Econom Stud 72 (3):853–883
22. Borenstein S (2008) Equity effects of increasing-block electricity pricing Centre for the Study of Energy Markets. Working paper 180, Nov 2008
23. Schneider K, Gerkensmeyer C, Kintner-Meyer M, Fletcher R (2008) Impact assessment of plug-in hybrid vehicles on Pacific northwest distribution systems. Paper presented in IEEE power and energy society general meeting—conversion and delivery of electrical energy in the 21st century, Pittsburgh, PA, July 2008
24. EPRI and NRDC Report (2007) Environmental assessment of plug-in hybrid electric vehicles-nationwide greenhouse gas emissions
25. Boyd S, Vandenbergher L (2004) Convex optimization. Cambridge University Press, Cambridge
26. Baltimore gas and electric three-level summer's tariffs (2012) http://www.bge.com/waystosave/residential/resprogramsresources/pages/time-of-use-pricing.aspx. Accessed March 2012
27. Radial Test Feeders—IEEE Distribution System Analysis Subcommittee (2014) http://www.ewh.ieee.org/soc/pes/dsacom/testfeeders/index.html. Accessed June 2014
28. Shao S, Pipattanasomporn M, Rahman S (2009) Challenges of PEV penetration to the residential distribution network. Paper presented in IEEE PES general meeting, Calgary, Alberta Canada, 26–30 July 2009
29. OpenDSS Software. http://sourceforge.net/projects/electricdss/. Accessed May 2014
30. Fernández LP, Gómez San Román T, Cossent R, Domingo CM, Frías P (2001) Assessment of the impact of plug-in electric vehicles on distribution networks. IEEE Trans Power Syst 26 (1):206–213
31. 2009 National Household Travel Survey User's Guide NHTS (2011) Accessed Feb 2011

MIX
Papier aus verantwortungsvollen Quellen
Paper from responsible sources
FSC® C105338

If you have any concerns about our products,
you can contact us on
ProductSafety@springernature.com

In case Publisher is established outside the EU,
the EU authorized representative is:
Springer Nature Customer Service Center GmbH
Europaplatz 3, 69115 Heidelberg, Germany

Printed by Libri Plureos GmbH
in Hamburg, Germany